BRCC

Adapting to Win

Adapting to Win

HOW INSURGENTS FIGHT
AND DEFEAT FOREIGN STATES IN WAR

Noriyuki Katagiri

PENN

UNIVERSITY OF PENNSYLVANIA PRESS

PHILADELPHIA

Copyright © 2015 University of Pennsylvania Press

All rights reserved. Except for brief quotations used
for purposes of review or scholarly citation, none of this
book may be reproduced in any form by any means
without written permission from the publisher.

Published by
University of Pennsylvania Press
Philadelphia, Pennsylvania 19104-4112
www.upenn.edu/pennpress

Printed in the United States of America
on acid-free paper
10 9 8 7 6 5 4 3 2 1

Library of Congress Cataloging-in-Publication Data
Katagiri, Noriyuki.
  Adapting to win : how insurgents fight and defeat foreign states in war /
Noriyuki Katagiri. — 1st ed.
    p. cm.
  Includes bibliographical references and index.
  ISBN 978-0-8122-4641-4 (hardcover : alk. paper)
  1. Insurgency. 2. Insurgency—Case studies. 3. Asymmetric warfare.
4. Asymmetric warfare—Case studies. 5. Guerrilla warfare. 6. Guerrilla
warfare—Case studies. 7. Non-state actors (International relations)
8. Non-state actors (International relations)—Case studies. 9. Strategy.
I. Title.
  JC328.5.K38 2015
  355.02'1801—dc23
                                                              2014012344

To Mariko

CONTENTS

1. How Do Insurgents Fight and Defeat Foreign States in War? 1
2. Origins and Proliferation of Sequencing 25
3. How Sequencing Theory Works 40
4. The Conventional Model: The Dahomean War (1890–1894) 63
5. The Primitive Model: Malayan Emergency (1948–1960) 79
6. The Degenerative Model: The Iraq War (2003–2011) 94
7. The Premature Model: The Anglo-Somali War (1900–1920) 115
8. The Maoist Model: The Guinean War of Independence (1963–1974) 131
9. The Progressive Model: The Indochina War (1946–1954) 150
Conclusion 169
Appendix A. List of Extrasystemic Wars (1816–2010) 191
Appendix B. Description of 148 Wars and Sequences 201
Notes 243
Bibliography 271
Index 293
Acknowledgments 299

CHAPTER 1

How Do Insurgents Fight and Defeat Foreign States in War?

How do insurgent forces fight and defeat foreign states in war? What can powerful states do to prevent policy disaster when they confront nonstate rebels in foreign lands? Recent conflicts in Iraq, Libya, and Syria—and Western experiences with them—have all underscored the importance of understanding how nonstate insurgent and guerrilla forces have dealt with enormous disadvantages in power to achieve their ends and what foreign governments and their powerful militaries can do to attain their own purposes.

These are not just policy questions. Until recently, few in academia believed in the power of rebel insurgents challenging powerful states in violent conflict. In 1967, Kenneth Waltz wrote that the "revolutionary guerrilla wins civil wars, not international ones" and that "the potency of irregular warfare (had) been grossly exaggerated."[1] At that time insurgency as a whole was such a small force in global politics that, even if some communist forces swept through parts of the Third World by guerrilla tactics, that would not pose a serious threat to American power. After all, guerrilla movements had little systemic effect on the bipolar stability between the United States and Soviet Union, at least until the Soviet withdrawal from Afghanistan in 1989.

Waltz's statement rings true to this day, except that it made a lot more sense for conflict through the early twentieth century. Since the mid-twentieth century, however, violent insurgent groups have done significantly better; they have made what were supposed to be "small wars" lengthy endeavors, raised the cost of war drastically, and won many of them quite impressively. Most recently, insurgent groups in Afghanistan and Iraq have managed to force the United States, arguably the champion of the post–Cold War international system, to suffer embarrassing if temporary setbacks. Forty years after Waltz made his argument, the *Washington Post* quoted Jason Lyall and Isaiah Wilson, close observers of this kind of war, as saying "although great powers are vastly more powerful today than in the 19th century... they have become far *less* likely to win asymmetrical wars. More surprising... the odds of a powerful nation winning an asymmetrical war decrease as that nation becomes more powerful."[2] In other words, things have changed dramatically in favor of insurgent underdogs in the international system. What explains this change? What does it mean to the future of great power politics?

In recent years, international relations scholars have made considerable progress in the understanding of many types of conflict: interstate, civil, and asymmetric. In a major study of asymmetric war, T. V. Paul explains how underdogs decide to go to war based on the perceived achievability of their political and military goals. More specifically, he argues that weak actors' choice for asymmetric war rests with their perceptions about the availability of external and internal support, short-term offensive capabilities, and their advantage in making the first strike.[3] Other scholars have followed, making arguments about weak resolve, strategic interaction, vulnerability of democracies to small wars, and mechanization of armed forces as major causes of upsets in asymmetric war.[4] These contributions, however, do not directly address a type of war that this book is concerned with: extrasystemic war. Drawn from the Correlates of War project, the terms "extrasystemic war" and "extrastate war" may sound confusing to some. What kind of war, one might ask, can be "extra" to the traditional state or international system? It is the specific type of the dyad between states and nonstate actors that makes extrasystemic war a distinct form of conflict. David Singer and Melvin Small define extrasystemic wars as those wars between a state member of the international system and a nonmember entity (nonstate actor) with a minimum of one thousand combat-related deaths per year.[5] In other words, it is a war between states and insurgent groups that commit "a violent, often protracted, struggle... to obtain political objectives such as independence, greater autonomy, or subversion of the existing political authority" that operate in a foreign territory.[6] While not all nonstate actors are insurgencies, I treat them synonymously in this book because most if not all insurgencies are belligerent nonstate entities seeking independence as the primary political ends and because the best analysis of violent insurgencies pitted against foreign governments comes from a collection of data on extrasystemic war.

Extrasystemic war shares some commonalities with interstate and civil wars, but as seen most recently in Afghanistan and Iraq, it poses a unique set of challenges that many have failed to appreciate. It is generally long in duration, involves conventional, guerrilla, and "hybrid" battles, and is highly political. Fairly common during the European imperialism of the eighteenth and nineteenth centuries, extrasystemic war continues to be a key security topic for many governments in the West. This is because most wars in the Third World involve Western governments with violent insurgencies, often pitted against local governments that are supported by powerful states intervening from outside. Furthermore, extrasystemic war is becoming more lethal, with the proliferation of small weapons among violent rebels and arms trade among insurgents, particularly for those who live in highly contested areas. In fact, Somalia, Sudan, Afghanistan, and Iraq are where we have seen extrasystemic wars recently, and these top the list of nations where minority groups are most exposed to dangers of genocide, mass killings, and violent repression.[7] Moreover, the United States left Iraq in 2011 and is scheduled to leave Afghanistan in 2014. Because violence continues to pervade these two states, foreign military intervention there is likely to stay on the table for major powers in the near future.

The central puzzle of this book is this: How do insurgent groups fight and defeat foreign states in war? What allows some nonstate insurgent groups to beat powerful states and others lose? The literature provides some insights into war between unequal powers, but none specifically for this type of conflict. In *Adapting to Win*, I answer these questions by exploring 148 cases of extrasystemic war and generating a set of distinctive patterns of how insurgent groups fight this kind of war. My answer is that successful insurgents tend to fight state adversaries in a sequence of actions that allows them to achieve their ends, whereas most unsuccessful groups end up adopting a sequence that does not. In other words, victory requires insurgents to evolve and do so in "right" sequences. I call this explanation the "sequencing theory," which posits that insurgent groups are likely to win extrasystemic war when their interactions with the states allow them to evolve into a powerful modern army capable of defending an emerging statehood. Growing powerful through iteration when confronting strong enemies is a challenging matter for any insurgent group. Because insurgents are generally weaker, most actually fail to evolve. However, quite a few have nevertheless succeeded through a set of sequencing patterns, and this book demonstrates how that happens.

In answering the main puzzle, *Adapting to Win* makes two contributions to the study of international security. First, it presents an alternative research project to the mainstream body of security studies that has until recently been fixated on great power interstate conflict and civil wars. Given the centrality of nation-states in the international system and given the growing relevance of internal war since the end of the Cold War, this fixation is natural. But it comes at the expense of analysis on extrasystemic war. To be certain, extrasystemic war does not make many headlines or affect the military balance of major countries. Held mostly in less attended areas of the globe, it is also closely associated with imperial and colonial conflict of the past. We must remember, however, that resources that states devote to small wars shape their balance of power with other states and affect the international system. The recent surge in the world's attention to violent insurgencies, in various parts of the Middle East, South Asia, and Africa, means that we must dispel the notion that it is peripheral to the interests of major powers. Indeed, as Lawrence Keeley argues, so-called primitive warfare has been "extremely frequent" in the history of mankind. Insurgents are highly aggressive: according to his analysis of fifty societies, 66 percent were at war *every year* and over 70 percent went to war at least once in every five years.[8] More broadly, because rebellions have a long history, extrasystemic war has been a recurring state of affairs since the birth of nation-states. Therefore, John Mueller is right that while great power war may be becoming obsolescent, civil war persists and so does "policing war," defined as militarized efforts by developed countries to bring order to civil conflicts in other parts of the world, which has a great deal of commonality with extrasystemic war.[9] More recently, extrasystemic war has been acknowledged for its relevance to other important strategic issues. Michael Horowitz, for example, shows that nonstate actors have actively evolved through organizational change, learning, and the building of linkages among themselves, as a way of adopting and carrying out new strategic innovations like suicide terrorism.[10]

The other contribution of this book is to enrich the policy-making community through the study of what lessons powerful states can learn to fight foreign insurgencies. The sequencing framework will inform statesmen and government officials about how to win through phases. I show what it takes for states to prevent policy disaster when they engage in asymmetric war. Therefore, although *Adapting to Win* largely takes on the perspective of the insurgent groups confronting states, it generates ideas for states in terms of how to execute extrasystemic war. Naturally, the book also considers implications about the kinds of government policy that are likely to prove effective and ineffective. For this reason, in the concluding chapter I examine the implications of America's conduct of the war in Afghanistan. Although it is too early to call a winner there, I argue that extremist insurgent groups such as al-Qaeda and the Taliban have largely failed in their attempt to generate effective sequences, while the United States/International Security Assistance Forces have arguably made some progress in making Afghanistan increasingly capable of self-governance and keeping the Afghan-Pakistani relations reasonably stable. Outside Afghanistan, however, insurgent groups have learned to fight through phases to inflict severe damage on those who intervene in their territory. From Somalia and Algeria to Pakistan, these groups have learned to become more persistent, adaptive, and innovative. This reality, along with tough economic problems at home and long-term security challenges from rising powers like China, means that the United States must find a way to fight small wars effectively. *Adapting to Win* proposes a sequential analysis as a new theoretical framework to generate key insights for U.S. security policy in the twenty-first century.

Conceptual Clarifications

Extrasystemic war is different from other types of war, such as interstate and civil. It differs from interstate war in that the latter is fought between nation-states while the former war involves states and insurgents. On the other hand, civil war is between government and nonstate groups in the same country. Civil war and extrasystemic war are interrelated, however, because they become inseparable when a foreign government intervenes in a civil war on either side. As Kristian Skrede Gleditsch, Idean Salehyan, and Kenneth Schultz argue, states experiencing a civil war are substantially likely to become involved in militarized disputes with other states, making war look extrasystemic. International disputes that coincide with civil wars are tied to the issues surrounding the civil war. Civil wars are likely to be internationalized when states seek to affect the outcome of the war through strategies of intervention and externalization.[11] In contrast, extrasystemic war has much to do with small wars and "hybrid" war. Small war is a "campaign other than those where both sides consist of regular troops," such as "operations of regular armies against . . . irregular forces," while hybrid war is a combination of traditional war with terrorism and insurgency.[12] Like small wars and hybrid wars, extrasystemic war involves the use of insurgency and guerrilla and terrorist tactics against states. Moreover, it is fair to say that extrasystemic war is a form of asymmetric war in which there are differences in the power of two sides. Of course, no war is fought between purely equal powers, so every war is asymmetric by definition. But because states generally have more resources at their disposal, reflected in the form of advanced military hardware and troops properly uniformed and professionally trained to fight capital-intensive wars, extrasystemic war distinctively favors them. One of the consequences of this gap is the power projection capability of nation-states, which often allows them to intervene in rebel territories and put insurgents on the defensive. This is why all extrasystemic wars have taken place on insurgents' territory.

Extrasystemic war takes both conventional and unconventional forms of violence.[13] Conventional extrasystemic war begins with both sides using standing armies in open terrain. The armies typically share the characteristics of massed lines, heavy fortifications, dependence on hardware, and physical force directed against combatants. In other words, force is used "directly" against the opponent.[14] The concept of conventional war derives broadly from a Western tradition that values weapons procurement, military education, and doctrinal development, based on what Rupert Smith calls the "paradigm of interstate industrial war: concepts founded on conflict between states, the maneuver of forces en masse, and the total support of the state's manpower and industrial base."[15] This tradition has more recently been passed on to U.S. forces and enshrined in the so-called American way of war.[16] Fred Weyand and Harry Summers argue that "we believe in using 'things'—artillery, bombs, massive firepower—in order to conserve our soldiers' lives. The enemy, on the other hand, made up for his lack of 'things' by expending men instead of machines, and he suffered enormous casualties."[17] Conventional forces generally include ground, naval, air, and marine components, but here they mean ground forces most of the time because insurgent groups tend not to have enough resources to field naval and air forces and because the army plays a decisive role in crushing the enemy's capacity to resist in decisive engagement. States typically prefer to fight conventional war because they train their forces to win it. The shift of American strategic focus from conventional force planning and nuclear exchange of the Cold War to less traditional missions like counterinsurgency (COIN) and counterterrorism is a relatively recent phenomenon. Seen from the long span of military history, most nation-states have consistently displayed a preference for conventional power. What is surprising, however, is that many insurgents have used conventional war. In fact, a number of extrasystemic wars in the nineteenth and early twentieth centuries were conventional.

The other form of extrasystemic war is unconventional, which is essentially guerrilla war. Samuel Huntington defined guerrilla war as "a form of warfare by which the strategically weaker side assumes the tactical offensive in selected forms, times, and places."[18] Guerrilla war has three forms of activity, according to Lawrence Keeley. The most common form is raids and ambushes in which a small number of men sneak into enemy territory to kill people, followed by large-scale battles and massacres and surprise attacks.[19] But generally, guerrilla strategy focuses on maneuver, speed, and stealth over formation and firepower. Rather than concentrating force, guerrillas disperse it to spread the enemy thin and target where the enemy is most vulnerable. Guerrilla strategy involves the use of soldiers as well as civilian populations, which provide supplies, information, sanctuary, training ground, manpower, and human buffers—all assets that can be used to neutralize the negative balance of power.[20] Of course, telling civilians from guerrillas in war is not easy, since all wars contain elements of both, which challenges the dichotomy between conventional and guerrilla wars, but for the sake of analytical parsimony I consider them to be alternatives.

What does it mean to win these wars? Scholars disagree over what defines victory and defeat, especially in insurgency environments. As William Martel argues, the term "victory" is used quite casually to express a generally successful outcome of a military contest. The literature does not have a language to describe victory in precise terms, so he examines victory in terms of achieving a set of political, military, territorial, and economic objectives on the tactical, operational, strategic, and grand strategic levels.[21] Dominic Johnson and Dominic Tierney define victory based on the achievement of political ends and material gains and losses made in the course of war, which are adjusted by the importance and difficulty of the missions. To measure progress in war, they use two methods: a "scorekeeping method" that focuses on actual material gains and losses and "match-fixing," in which evaluations become skewed by mind-sets, symbolic events, and media and elite spin.[22] Furthermore, former Secretary of State Henry Kissinger's famous dictum that "the enemy wins if he does not lose" reflects the inherent advantage given to the insurgents in our judgment about victory in asymmetric contexts. These considerations sound reasonable, but the problem is that they do not provide useful metrics to measure war outcomes in extrasystemic war.

However, all extrasystemic wars turned out to have clear winners, so it is enough to argue here that insurgent groups win extrasystemic war when they attain their utmost ends set at the beginning of the war while state sides fail to do so. Insurgents lose, in contrast, when they fail to achieve the objectives while states succeed. Of course, it is often the case that actors seek to achieve more than one objective in a single war and these objectives change before the war ends, which makes it difficult for us to assess mission accomplishment. But in most cases, territorial and political integrity is the main cause of extrasystemic war and actors regard sovereignty as the most important determinant of victory.

The Puzzle

Through the early twentieth century, governments defeated foreign insurgents lopsidedly in most encounters. A majority of these wars were acts of colonial conquest in which imperial Western powers traveled across the world and used brute force to subdue local subjects on a massive scale. At different times, Britain, France, Portugal, and other European countries waged wars of empire building almost nonstop in their scrambles for colonial possession. Once they settled the lands, they did everything they could to suppress revolts and exploit resources. The control was so systematic that whenever discontented elements rebelled against the settlers, they found themselves to be almost always on the losing side. Scholars stress the logic of power in explaining this pattern of insurgent defeat; insurgent groups lose because they are poor and powerless. Indeed, the international distribution of power favors nation-states at the expense of insurgents. Scholars have noted the general tendency for these disorganized forces to be weak on the battlefield. Mueller holds that a group of criminal thugs confronting a competent army is doomed to fail because they simply do not have the training, leadership, logistic support, weaponry, and morale.[23] To others, nonstate violence is not relevant anymore. According to Janice Thomson, nonstate violence, once highly marketized around the world, was even "delegitimized" and "eliminated" in early modern Europe when the sovereignty of nation-states became an established institution of the time.[24] Nonstate violence in its various forms, ranging from insurgency to terrorism, has long failed to serve the ultimate interest of those who execute it. More recently, Max Abrahms found that since 2001, terrorist groups have rarely achieved their political objec-

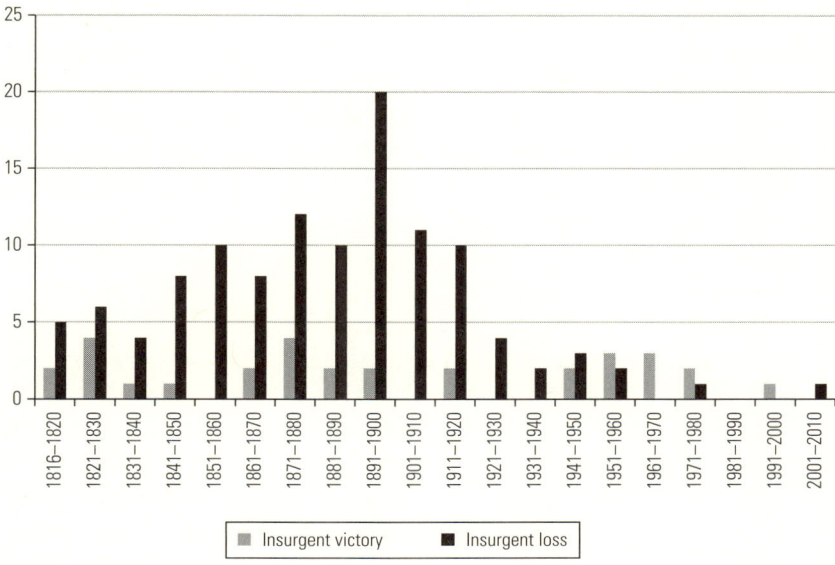

Figure 1. Extrasystemic war outcomes.

tives; they accomplished their policy objectives only 7 percent of the time.[25] Today, as great powers continue to build arms, train for new types of conflict, and become more powerful, this record of state victory appears embedded, in both theory and empirical evidence. Between 1816 and 1945, the victory rate for insurgents was only 15 percent, as they won only 20 of 130 wars.

Figure 1 indicates that since 1945, this trend has changed. For the past seventy years we have seen more insurgent groups overcome military inferiority to defeat external powers. Winning eleven of eighteen wars through 2010, the insurgents' victory rate has jumped from 15 percent to 61 percent. From Algeria and Indonesia to Guinea-Bissau, dedicated members of nationalist groups in colonial entities have engineered a dramatic shift in the strategic landscape by unseating leading colonial powers and taking charge of their newfound statehood. In a span of just a few years, European powers lost possessions in Africa, South Asia, Southeast Asia, and the Middle East. This trend was hardly limited to European experience; before the Soviet Union retreated from Afghanistan in 1989, the United States suffered through the Vietnam War, an experience that made the American public less willing to engage with tribal violence of the Third World and subsequently shaped the structure of the U.S. military for the remainder of the Cold War. America's victory in the Cold War did not mean that it could win small wars everywhere, as military missions in Somalia in 1993, the events of September 11, 2001, and the subsequent wars in Afghanistan and Iraq each proved extremely difficult and costly. Instead, these wars have opened a crack in U.S. forces and weakened the economy, reinforcing the general perception that even powerful countries like the United States find it hard to defeat a foreign insurgency.

This should not be surprising, because the growing success of weak actors in international politics has been recorded in recent research. Arreguin-Toft shows that in the past two centuries, materially weak actors have won only a third of all wars, yet since 1945 they have won more than half of them. The reason they used to lose these wars was because they would fight the same way as the other side did—whether using a conventional or guerrilla strategy. This has changed dramatically because weak actors have adopted military strategies opposite to those of powerful actors.[26] Lyall and Wilson find that before the First World War, strong nations used to beat irregular opponents at the rate of 80 percent, but the rate has declined to 40 percent since World War II. Non-great powers had similar experiences, defeating insurgents in 80 percent of pre–World War I cases but only 33 percent of post-1918 wars. This is mainly due to the increasing mechanization of government forces, which in turn weakened their ability to collect intelligence among local populations, differentiate combatants from noncombatants, and selectively apply rewards and punishments to the locals.[27]

The rising probability of insurgent victory coincided with the extension of war duration, which suggests that the longer the war becomes, the more likely insurgent groups are to win. This may be because weak actors are so determined that they are willing to endure the pain of war longer than strong actors. It may also be because the cost of a long war may be relatively lower for underdogs who have little to lose from defeat. Or it could be that people on the weaker side gain hope as war becomes longer while those on the stronger side tend to lose that hope. As David Galula argues, "The longer the insurgent movement lasts, the better will be its chances to survive its infantile diseases and to take root."[28] The relationship between war duration and outcome is important because in most extrasystemic wars, the state side is democratic and does not like long wars. In the short run, democracies may be more likely to win than their opponents because they choose the wars they fight and because they can mobilize domestic resources more effectively, but in the long run they are subject to electoral punishment for the conduct of overseas missions. As a result, democracies are more vulnerable to pressure to withdraw from unpopular wars.[29] As Scott Bennett and Allan Stam show, democracies begin to lose wartime advantage in capability and resolve roughly eighteen months into the war and at that point become far more likely to quit and more willing to settle for draws or losses.[30] So when democratic states win an extrasystemic war, it is generally a short war.

Figure 2 confirms this relationship. The average duration of extrasystemic war is 2.7 years, but in the postwar period when insurgent groups are most likely winners, it increases to 7 years. The wars became longest in the 1960s and 1970s when decolonization movements were the most intense. Other things being equal, the longer the war, the more likely insurgent groups are to

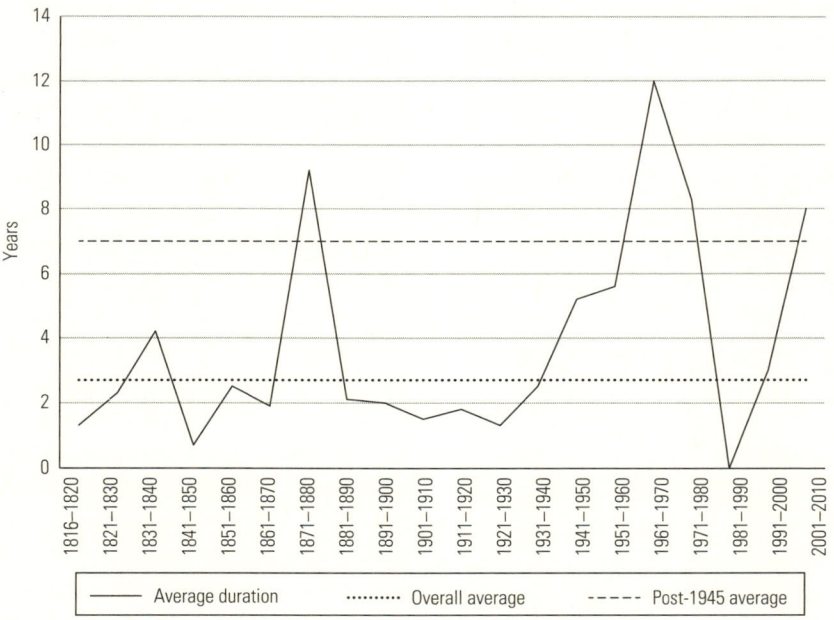

Figure 2. Average duration of extrasystemic war. Average duration is the average length of wars fought between each decade category. The decade category indicates the year when wars started. The overall average of 2.7 years is drawn from the average of all 148 cases, and the post-1945 average of 7 years is drawn from the average of 18 wars fought after 1945. Note the anomaly of the Aceh War (1873–1913), which dramatically increased the average duration of the 1870s.

succeed. The problem, however, is that war duration alone does not tell us anything about what happens in war. Consequently, the relationship is not causal, and the duration is not a cause of insurgent victory but only an *indicator* of their success. Yet duration is important because longer wars allow insurgent groups to evolve in a complex manner. The question, then, is what takes place in each of the wars that leaves some insurgent groups so much stronger that at the end of the war they find themselves to be victorious.

My answer is centered on sequencing theory. The theory operates under the assumption that extrasystemic war can unfold in multiple sequences. The very multiplicity of these trajectories assumed in the theory captures the multilinearity of extrasystemic war and explains the variation in the probability of insurgent victory.[31] Each sequence consists of up to three "phases," conventional war, guerrilla war, and state building, which represent a set of critical military, political, and economic factors shaping the strategic environment of extrasystemic conflict. The theory posits that insurgent forces may be able to boost their chance of victory when they evolve. They evolve by fighting their superior opponents in ways that transform them into a conventional army. They evolve as they use the gains they have made in an earlier phase relative to their foes and deploy them to their advantage as they fight on. Most of the time, they do not have enough capability to defeat their enemy in a single phase, so they are likely losers in most violent encounters. Through evolution, however, they can grow to be capable of moving on to a next phase where they are more likely to achieve their ends. In other words, the key variable for insurgent victory is whether the insurgents grow into an independent state with organized armed forces. Put simply, the main hypothesis of sequencing theory is that the more insurgents evolve, the more likely they are to win extrasystemic war.

The fact that there are several ways in which a "phase" combines with another phase means that there are as many sequences to consider. In this book I show six such sequences, or models, all of which evolve in dissimilar trajectories and result in different frequencies of occurrence. They are the (1) conventional, (2) primitive, (3) degenerative, (4) premature, (5) Maoist, and (6) progressive models. The conventional model describes a war in which states and insurgents fight by using organized forces. The primitive model depicts the execution of guerrilla war between the two sides from the beginning to the end. The degenerative model is a two-phase model, which begins with a conventional war (as in the conventional model) but turns into guerrilla war (primitive model) in the middle of the conflict. The premature model reflects a midwar transformation of guerrilla conflict (primitive model) into conventional war. The Maoist model shows that insurgents evolve from a small political party through the experience of fighting in guerrilla war, before the party becomes a conventional armed force. Finally, the progressive model describes a process of insurgent evolution from the period of guerrilla war through the years of state building into the establishment of modern armed forces. Because they are materially weak, few insurgent groups have managed to generate an ideal sequence, which explains why most groups have lost extrasystemic wars. The data show, however, that in the past several decades these insurgent forces have reversed the trend by evolving through the transformation of small guerrilla forces into a modern political and military system. They have built this system with support from local populations and have fought well on their home turf.

Sequencing theory presents but one of the several ways in which insurgent forces fight and defeat foreign states in war. I do not argue that the theory is the only way to explain extrasystemic war because existing explanations of asymmetric war may at times account for why some insurgents succeed and others fail in extrasystemic war. Sequencing theory, however, generates a useful analytical framework about how the order of sequences in conflict between states and nonstate insurgents can generate forces that empower the latter into victory. Furthermore, the theory posits that an insurgency's evolution or failure to evolve has a strong impact on its ability to achieve its goals. In other words, through evolution insurgents increase their chances to accomplish the utmost political ends. Yet sequencing theory is hardly focused only on insurgency. It argues that the evolution comes only with the state side co-opting the insurgent. Not every insurgency failing to appropriately evolve into a conventional force wins these wars because they fail to achieve the fundamental purpose of war. Thus, the causal logic of sequencing theory rests not with a tautological argument that insurgents' victory occurs when they evolve, but with whether the evolution they engineer goes through a successful sequence.

The Literature

Sequencing theory is a new theoretical framework of international security that fills a void in the existing theories of asymmetric war and counterinsurgency. The field of international security has generated a number of ideas about these issues, but sequencing theory presents an alternative means of analysis on irregular conflict and extrasystemic war. Hilde Ralvo, Nils Gleditsch, and Han Dorussen explore the relationship between democracy and colonial, imperial, and postcolonial wars, which concerns us here because many of these wars are extrasystemic at the same time, but they focus on the propensity of democratic countries to fight these wars rather than dealing with war *termination*.[32] Dan Reiter and Curtis Meek examine how factors like regime type, steel production, and national borders shape military strategies of states, but they do not investigate insurgencies.[33] Todd Sechser and Elizabeth Saunders examine conditions under which states seek to mechanize their militaries, but they do not discuss how nonstate actors do so.[34] To be sure, the literature on asymmetric war offers a set of ideas about how weak actors defeat strong ones, although it assumes no analytical boundary between interstate, civil, and extrasystemic types. Some of the major works in the literature address the cause of underdog victory in terms of (1) balance of resolve, (2) strategic interaction, (3) democratic weakness, (4) external support, (5) mechanization of armed forces, and (6) political objectives.

First, Andrew Mack writes that weak actors are likely to win when they are more resolved to withstand the cost of war than are powerful actors. The idea is that asymmetric war is a contest of will that favors the less powerful because they make greater efforts. Material power is considered indecisive because weak actors face the danger of annihilation and thus are determined to outlast the states that do not. This imbalance of resolve leads to a gap in the degree of vulnerability to domestic opposition. While weak actors maintain a high level of determination, stronger ones often see their resolve decline before they are forced out of war by their constituents.[35] In extrasystemic war, this means that insurgents win when their resolve is greater than that of nation-states. The theory does a good job of explaining a number of wars of decolonization where insurgent sides were presumably more determined to fight on to grab independence. The theory also works with sequencing theory in that it describes the fundamental asymmetry in power and resolve that drives some insurgents to seek military power and popular support as key ingredients in extrasystemic war.

Second, Arreguin-Toft posits that particular configurations of strategic interaction affect the likelihood of insurgent success. The idea is that weak actors are likely to win when they adopt a different military strategy than that of stronger actors. Specifically, they are likely to win when they adopt guerrilla strategy and the stronger actors adopt conventional strategy because they can out-will the stronger side. They are also likely to win when they adopt a conventional strategy and stronger actors adopt what he calls barbarism: targeting civilian supporters of the enemy. In contrast, they are likely to lose if both sides adopt similar military strategies because the stronger actors can simply outpower the weaker ones.[36] In short, strategic matching empowers the weak, whether states or nonstates, while strategic mismatch is likely to lead to the victory by the powerful. A key difference between this theory and sequencing theory is that while the former stresses the matching of different strategies as the determinant of war outcomes, the latter emphasizes the role of evolution as a key variable for insurgent victory against states. As I show in the case studies, the theory of strategic interactions does not account for the fact that actors change strategies in the middle of war, as many states and nonstate actors do in extrasystemic wars. The theory of strategic interaction is salient here, however, because many extrasystemic wars involve the use of both conventional and guerrilla strategies. As the wars have become more complex in recent decades, Arreguin-Toft's explanation does a good job of showing how different strategies between the two sides leave different implications for war outcomes.

Third, Gil Merom argues that weaker actors are likely to win when their opponents are democratic and susceptible to pressure from antiwar forces at home. Their chance of victory increases when middle-class constituents, who reject the high costs of lengthy war because the costs fall directly on them and because they oppose war on moral grounds, become strong antiwar forces to block the democratic government from raising the level of violence necessary to win. Soon enough, they begin to threaten to punish the government electorally. Merom's argument centers on an important dynamic surrounding the process of how democratic governments become weak through the mismanagement of war on the domestic front.[37] The theory speaks to a host of insurgencies in Africa, South and Southeast Asia, and Latin America where many Western democracies have underperformed because they succumbed to powerful pressure to withdraw from those wars. From a broader perspective, Merom's work advances an important research project on why democracies may fight poorly, especially in small wars and in comparison with nondemocracies. Research has shown that some democracies are prone to lose wars because they are more susceptible to electoral punishment than are nondemocracies. Voters and the mass media are sensitive to spectacular attacks that can shift voting patterns and constrain the ability of democratic leaders to wage war.³⁸ By looking specifically into democracies in small wars, Merom's work adds pessimism to the sense of vulnerability felt by many democracies constrained by the domestic effects of antiwar norms and institutions.

Fourth, Jeffrey Record argues that external aid is the key to underdog victory. Reviewing eleven insurgent wars from 1775 to 2007, he argues that external assistance correlates more consistently with insurgent success than any other explanation. Acknowledging some roles played by the resolve to fight, strategy, and regime type, he argues not that external aid is sufficient for insurgent victory but that it plays a crucial role in war outcomes between the strong and weak.³⁹ Similarly, Paul Staniland argues that resources and the structure of the preexisting social networks of insurgent groups are crucial parts of their mobilization and operation.⁴⁰ Material support may be especially significant if it comes from states that can offer firepower, funding, training, and intelligence far beyond the ability of these actors. Some states have incentives to support terrorist groups primarily for strategic reasons, such as to influence neighbors, topple regimes, counter U.S. hegemony, or advance ideologies.⁴¹ But state sponsorship is not always helpful to terrorist groups when they risk being the next target. Sponsors that provide safe haven can have incentives to provide information about the groups to the opponent in order to avoid getting into trouble with it.⁴² After all, as I show in the empirical chapters, the support-based theory functions as part of sequencing theory because they both show how resources enable insurgents to fight both guerrilla and conventional wars.

Fifth, Lyall and Wilson argue that states are likely to fail in COIN missions when they are obsessed with military mechanization. High levels of mechanization, along with external support for insurgents and government occupation of foreign territories, are associated with an increased probability of state defeat because they undermine the state's ability to collect intelligence from local populations and tell combatants from noncombatants, which increase the difficulty of selectively applying rewards and punishments to the populations. In other words, the modernization of armed forces, seen here in terms of the replacement of manpower with motorized vehicles, such as tanks, trucks, and aircraft, impairs the way state actors fight wars.⁴³ The theory can work in tandem with sequencing theory when the latter offers a chronological lens to deal with how extrasystemic wars have changed over time. Indeed, sequencing theory works where Lyall and Wilson's theory does not in terms of explaining why *prior to* the early twentieth century, military modernization did not always lead to state defeat. Rather than assuming that only in the *post*-twentieth-century period did major powers mechanize their forces, the sequencing perspective factors in the force structure of both sides as a key variable on the outcome of extrasystemic conditions. Mechanization not only has been a major part of extrasystemic war but also has always benefited states more than insurgents. Because the technological *gap* has remained largely constant for the examined period, if Lyall and Wilson are correct, states should have lost wars roughly at the same rate as today.

Last, Patricia Sullivan argues that the key to understanding victory lies in the nature of the political objectives states pursue through the use of military force. The effects of the factors above, such as strategic interaction and democratic weakness, are dependent on the nature of the states' political objectives. The way the states' political objectives align with the quality of their military strategy and resources determines war outcomes. Underdogs are likely to win when their political objectives are more closely aligned with military objectives and the balance of military capabilities between belligerents. In contrast, they are less likely to win when the objective relies on compliance from the target.[44]

These works have made important contributions to the expansion of our knowledge about asymmetric war. What these theories lack is an evolutionary perspective that insurgent growth has a strong bearing on insurgents' chances to achieve their purpose. In other words, none of the existing works takes into account the role of wartime evolution that often shapes the outcome of extrasystemic war. In this context, sequencing theory offers a theoretical challenge to the existing thoughts about asymmetric war by presenting a sequence as a means of analysis. By highlighting the way wars evolve through phases, this book shows how insurgents can increase the probability of victory in the exchange of violence.

Methodology

I take three steps to explain how insurgent forces fight and defeat foreign states in war. First, I examine 148 cases of extrasystemic war in the period between 1816 and 2010 and classify each of them into one of the six models. As a baseline I use existing datasets on extrasystemic war because, while they do not directly deal with insurgencies per se, they provide a list of nonstate actors fighting states in the closest way possible to extrasystemic wars.[45] However, they contain a number of inaccuracies about outcomes, duration, and categorization, so whenever necessary I have recoded and updated the datasets.[46] The result is Appendix A, which lists all 148 wars and presents information on duration, winners, and models used.[47] Second, I supplement the raw data with a set of descriptions of each war, drawn from primary and secondary sources in historical literature and compiled in Appendix B. The descriptive analysis is followed, third, by comparative historical case analysis designed to provide an extra layer of robustness and substantiate the six models of sequencing theory.[48] I investigate the six models, respectively, in the (1) Dahomean War, (2) Malayan Emergency, (3) Iraq War, (4) Anglo-Somali War, (5) Portuguese-Guinean War, and (6) Indochina War.

The rationale for choosing these 6 out of the 148 cases is as follows. Each is best suited to provide empirical support for the six models. The cases also clarify conditions for success and failure of insurgencies in extrasystemic war. Furthermore, they highlight the stark difference between the colonial and imperial era and the decolonization period. Specifically, the case studies contrast the key tendency of insurgent forces before the 1940s to fight mainly using the conventional model and the tendency of state forces to lose extrasystemic wars against those insurgencies in the decolonization era. In other words, the six cases help achieve my aim to maximize the temporal and geographic distribution of the survey while reflecting the number of models of sequencing theory. In data collection, I carried out archival research, examining sources including memoranda, meeting notes, transcripts of conversations, war and diplomatic communiqués, as well as memoirs and secondary historical literature. Taking these three steps has proven to be an effective way of exploring questions central to this book. How exactly do insurgent forces fight foreign states when they do? If and when they defeat their opponents, what strategy have they used? When they lost the war, what did they do wrong? And ultimately, what causes the variation in the sequences that insurgent forces end up adopting in extrasystemic war?

There are three caveats in regard to the data collection and analysis. First, the existing dataset does not fully capture differences between cases. For example, while some insurgencies turn out to be relatively short-term rebellions in South Asia, others are modern extensions of ancient African kingdoms defending what little autonomy they have left, such as the Dahomey, Ashantis, and Mahdists, and these do not appear to be insurgencies. Furthermore, there is more than one type of rebellion and insurgency; in postcolonial Africa alone, William Reno argues a variety of rebel groups can be categorized into anticolonial, majority rule, reform, warlord, and parochial types.[49] In this book I do not use Reno's typologies, but they reinforce my point that these differences make cross-case comparisons difficult. What I do to deal with this issue is to first eliminate cases that do not conform to the category of extrasystemic war and then ascertain that all the other cases share common features of extrasystemic wars, with insurgent groups being illegitimate members of the international system at the time of war and nation-states being such members. In other words, to carry out comparative case studies effectively, I treat these cases in a uniform manner under the category of extrasystemic war and by exploring the central questions consistently across the empirical chapters.[50] This method allows us to draw theoretically informed and substantially meaningful implications for the recent and ongoing wars in Iraq and Afghanistan.

The second issue concerns the conditions under which wars take place. In some cases insurgent groups opt not to go to war when conditions for it exist. Given the enormous challenge of extrasystemic war, there are incentives for them not to fight in general. Nonviolent resistance may in fact be preferred over violence because it presents fewer obstacles to moral and physical involvement, information and education, and participator commitment. High levels of participation in nonviolent protests also enhance group resilience, tactical innovation, and opportunities for civic disruption and reduce incentives for the regime to fight.[51] This issue raises the question of what constitutes necessary and sufficient conditions for war, which needs to be solved in order to avoid generating a no-variance research design. I solve this question by counting conditions for war when it takes place and by considering why some insurgents choose not to fight at all when conditions for war exist and why other insurgent groups do fight when conditions do exist. At the same time, because my data do not speak for all cases under which war should have taken place but did not, I do not make claims about whether they are sufficient conditions for war. But I assume that they demonstrate a set of conditions that enable us to create a series of distinct patterns of war. Doing so allows me to make a contextually restricted yet substantial contention about conditions for war.

The last issue is with counting "models" of sequencing theory. A single war may account for more than one model while the model overlaps with another. For instance, a war that I code as a conventional war model may actually count as a degenerative model if insurgent groups might have tried to go degenerative but failed in the conventional war before they could reach a second phase. Similarly, a war may account for both primitive and Maoist models if the insurgent group fights like guerrillas first and then fails to move on to a next phase. The overlapping issue may potentially complicate my counting rule for the models. I solve this problem first by ensuring that each case is completely different from the others and second by counting models in which we see hard evidence that insurgent groups made an effort to reach the next phase. Therefore, if evidence proves that the groups tried but failed to reach a phase, then the case counts toward a more complicated model, while if the groups did not try to attain another phase, then it counts as a single-phase model.

Plan of the Book

This book has ten chapters. In the second chapter, I trace the intellectual origins of sequencing theory in the application of evolutionary biology to political science, coupled with analysis of the works of Lenin and Mao, who made use of phases and sequences in revolutionary situations. I also examine how early ideas of sequence in war became consolidated and spread into limited parts of the postwar globe, which enabled a few insurgent leaders of Third World independence movements to achieve decolonization. The idea of fighting through sequences, however, has hardly been uniform; the sequencing method has generated variation in the way that these groups evolved. In fact, the very multiplicity of sequences explains the variation in war outcomes, with each "model" having different frequency levels and war outcomes. I also discuss factors that shape the probability of insurgent victory and show that although insurgent victory is associated with duration, geography, and norms of self-rule, none fully explains the outcome. Instead, insurgent victory rests most closely with the proliferation of the idea and practice of sequencing among successful groups. Thus, my discussion of the spread of the sequencing method centers on the ability and willingness of insurgent groups to "learn lessons" in the middle of war and afterward. In the third chapter, I explain exactly how sequencing theory works. I show three major components of the theory and how they combine to make sequences and subsequently categorize extrasystemic war into six sequencing models. The first four models (conventional, primitive, degenerative, and premature models) are popular among many insurgents but are likely to fail, while the last two (Maoist and progressive) models are successful but rare.

The fourth to ninth chapters examine the six models of sequencing theory. In each chapter, I test the three most relevant theories of asymmetric war against sequencing theory. In the fourth chapter, I explore the Dahomean War of 1890 to 1894 to show that insurgent groups generally fail when they use the conventional model. The central proposition of this chapter is that, as seen in the abortive Dahomean effort to challenge the French army, war against a stronger foreign power is impossible to win if insurgents fight like an army. This case study clearly shows that the Fon fielded a conventional army that was two times *larger* than that of the French, and this manpower advantage allowed them to survive a couple of years, but they failed to go beyond the conventional phase and win the war. In the fifth chapter, I examine the Malayan Emergency of 1948 to 1960 during which members of the Malayan Communist Party confronted British forces in a war for independence, only to be defeated. The main cause of this outcome was not the level of resolve or power or strategic choices but the fact that the insurgent group failed to build strong support bases and move out of guerrilla insurgency. This case shows that a guerrilla war campaign can be lengthy, but contrary to the conventional wisdom, a long guerrilla campaign does not instantly make states the loser. In fact, this case study underscores the dilemma of insurgent groups stuck in an attritional guerrilla war and unable to do anything to move beyond that phase.

In the sixth chapter, I explore the Iraq War of 2003 to 2011. Iraq illustrates precisely why the two-phase degenerative model does not work well for insurgents. In the first phase, Saddam's ground forces and Ba'ath Party fought the U.S.-led coalition forces in the desert of Iraq in conventional warfare before their incomplete destruction drove the war into a new phase. The second phase was guerrilla combat in which the coalition forces initially suffered from their failure to adapt to the changing environment. Yet U.S. forces reorganized around the population-centric COIN operations that marginalized the insurgents starting in 2007. While these factors played an important contextual role in the war, I show that it was the shift from conventional to guerrilla war that proved critical for the insurgents' performance. The coalition left Iraq in a shambles but in a state of affairs that allowed their departure—however temporary that may be. Of course, there are reasons to be concerned that the war may turn for the worse, as the current Iraqi prime minister, Nuri al-Maliki, may not be able to hold various sectarian interests together. The insurgents have also claimed victory on their own terms as they "succeeded" in achieving their mission in terms of expelling the coalition forces (although the departure was carried out on the coalition's timetable). But because the insurgents split along internal and tribal lines, because either they or Ba'athists no longer control much popular support or the government, and because they have yet to field an army capable of seriously challenging the Iraqi government, prospects for their resurrection to the degree that it might seriously impair the coalition's political capital in the future have dramatically shrunk.

In the seventh chapter, I discuss the Anglo-Somali War of 1900 to 1920 to explore why the premature model does not work. The Somalis fought through the transition of guerrilla war into conventional war without making visible efforts to build a state. The so-called Dervish initially fought well to survive four British expeditions by adopting an elusive guerrilla strategy. But the situation turned sour in the second phase when they found themselves exposed in open desert to British firepower and airpower. Without building institutions that would have consolidated support bases and increased their military capability, the Dervish fell quickly once Britain became serious in Somalia. This case study is useful in showing that, again, what appears to be a very long (twenty years) guerrilla military campaign overseas can actually be understood as a simple two-phase conflict in which the state side emerged victorious in the end.

The eighth and ninth chapters provide a set of contrasting portraits in which insurgents adopted models that worked. In the eighth chapter, the Portuguese Guinean War of 1963 to 1974 provides evidence for the Maoist model since the war lasted long enough for us to examine the case in three phases, and because the Guinean side was least expected to generate the outcome it did. I discuss how a guerrilla force called the African Party for the Independence of Guinea and Cape Verde (PAIGC) managed to overcome its military disadvantage to defeat colonial Portuguese forces over three phases. The first phase was a period characterized by Guinean success in building institutions, especially the party itself, which allowed it to push the war into a guerrilla war phase, where the PAIGC gained strong popular support. The PAIGC's success in "winning hearts and minds" brought the war to the final phase of conventional war where its modernized army forced Lisbon to withdraw from the territory. I argue that while each of these phases—state building, guerrilla war, and conventional war—had an important influence on the war, it was the very evolution of insurgency across the three phases that shaped the outcome.

In the ninth chapter, I study the progressive model using the Indochina War, during which Vietminh forces fought French troops between 1946 and 1954. I deal with this war as a case study of the progressive model because it provides ample evidence that the Vietminh experience was a clear departure from the Maoist model and because, as above, the war lasted long enough for us to see the effect of changes across three phases. The Vietminh won the war because they took three steps in the right order; they began as a guerrilla movement, did everything they could to gain popular support, and built the Indochinese Communist Party (ICP) during the state-building phase, before acquiring a modern military force to defeat the French at Dien Bien Phu. This case study is important because sequencing theory shows that even though the war lasted only eight years, it passed through three different phases of important strategic changes that became a key factor for the Vietminh victory.

I conclude the book by outlining a set of implications for the broader strategic visions of the United States. Here I draw a set of security policy implications for the United States, using the war in Afghanistan/Pakistan as a strategic context. Although the war is more complicated than simply extrasystemic, it has a strong commonality with extrasystemic wars in that they are fought between state and insurgents and both conventional and guerrilla strategies have been used. The Afghanistan and Pakistan theater has been improving for the U.S.-led International Security Assistance Force (ISAF) since the 2011 death of Osama bin Laden, although it remains complex and fluid. At this time the war has manifested itself in a primitive pattern, with the US/ISAF coalition using a mix of regular and irregular forces to fight the Taliban starting in late 2001 in mostly guerrilla-counterguerrilla interactions. Afghan President Hamid Karzai's regime remains corrupt, fragmented, and weak, whereas neither al-Qaeda nor the Taliban nor the Haqqani network is able to produce an effective and centralized counterweight to govern the territory. The primitive model draws a *theoretical* implication for an eventual victory for the coalition after a series of complicated negotiations over the status of the insurgents and Afghan governance.

I show, however, that it is a challenge to analyze this war along with other cases of extrasystemic war because it represents one of the increasing numbers of insurgencies that have little interest in governance but instead seek an absence of governance as a means of achieving their ends. By the logic of Henry Kissinger's dictum that "the enemy wins if he does not lose," the Taliban and al-Qaeda may actually have won the war in Afghanistan because they have not been defeated and because the U.S. and coalition forces are leaving, which is one of their objectives of the war. The insurgency could also win insofar as it achieves its principal goal of removing, regardless of on whose terms, U.S. and coalition forces from Afghanistan. Furthermore, these groups have little interest in replacing the central government with another such structure. Thus, they have proclaimed victory on their own terms on numerous occasions as the coalition forces prepare to depart from the country in 2014.

In the meantime, the United States remains mired in economic problems at home and faces the daunting prospect of major cuts in defense spending and a resultant reduction to its capability to project power overseas. Of course, these extrasystemic wars in Iraq and Afghanistan alone are unlikely to have significant systemic implications on the international system. Yet the way the United States fights wars does shape the military balance between major powers. American performance in war will help determine the nature of future force planning, structure, and military strategy outside of the theaters of war. Given the rise of China as a great power, continuing financial troubles in the United States and Europe, and growing military volatility in parts of East Asia, South Asia, and the Middle East, we must explore the meaning of the shift of strategic resources for American national security.

CHAPTER 2

Origins and Proliferation of Sequencing

> All guerrilla units start from nothing and grow.
> —Mao Zedong[1]

Evolutionary Origins of Sequencing Theory

As this book's title suggests, the intellectual roots of sequencing theory lie in the application of evolutionary thought to the field of international security. Sequencing theory draws from the combination of two propositions in the field—Darwinian adaptation by natural selection and Lamarckian inheritance of acquired characteristics—that are often considered to be intellectual opposites. First, Darwinian selection operates on the logic of competition in which species, in this case insurgent groups, cope with a hostile environment by making a series of adjustments to survive. Competition and adjustment are no strangers to scholars of international relations; Waltz discusses the need of the state to adapt to the structure of international politics by developing capability and balancing power, or face the fate to "fall by the wayside."[2] In terrorism studies, Bruce Hoffman writes that "an almost Darwinian principle of natural selection . . . seems to affect terrorist organizations, whereby every new terrorist generation learns from its predecessors. Terrorists often analyze the mistakes made by former comrades who have been killed or apprehended."[3]

Darwinism is an appropriate tool with which to explore extrasystemic war because it shows how insurgent groups fight powerful adversaries through three mechanisms—variation, selection, and replication—all of which will be seen as key operational assumptions in the empirical chapters that follow.

Variation means that insurgents consist of a diversity of combatants seeking to master mutation and innovation for the sake of survival. In selection, states impose pressure on insurgents because the former is stronger than the latter, resulting in faster adaptation by the insurgents who survive. In replication, insurgents are exposed to combat for a long time because they fight on the home territory for years at a time, which helps promote their experience, learning, and innovation, while state forces rotate soldiers on short combat tours to different regions. These conditions are the key to the success of insurgents. Dominic Johnson writes that after all, "selection effects favor weaker sides, such as insurgents and terrorists, because they are more varied, are under stronger selection pressure, and replicate successful strategies faster than the larger forces trying to defeat them, such as the US army in Iraq. To put it simply, large 'predatory' forces cause their 'prey' to adapt faster than they do themselves."[4] In extrasystemic war, therefore, insurgents have the natural advantage over states, although that does not mean that they will always win when they fight. Instead, they are more likely to win when they adapt and evolve. Only successful insurgents survive selection pressure and end up generating models of strategic behavior that can be replicated elsewhere. Darwinism alone, however, is not sufficient to explain variation in extrasystemic war because if it were, the logic would show that insurgents would always be the winner. We need Lamarckism to account for why some insurgents succeed and most fail.

Lamarckism posits that individual efforts are the main driver of species adaptation. Today considered an obsolete theory, especially in comparison to Darwinism, it nevertheless provides a key impetus to sequencing theory. While Darwinism posits that external environment imposes the rule of competition that shapes the chance of species to survive, Lamarckism argues that species acquire the skills of survival and pass the traits to their offspring in the process of evolution. In other words, only a handful of insurgent groups acquire traits and pass them on over time. In this book, I treat the inheritance of acquired characteristics as part of the temporal framework; that is, insurgent groups acquire and develop traits that are then passed from time A to time B in ways that shape their performance in war. Evolution is an inherently progressive process.[5] The literature of evolutionary biology treats Darwinism and Lamarckism as being mutually competitive, but treating them in isolation from each other overlooks essential characteristics of war. Investigating three decades of neurobiological research, Peter Hatemi and Rose McDermott write that neither alone holds the key to understanding human behavior like war. Rather, it is necessary to understand the underlying characteristics of both.[6] Therefore, I see Darwinism and Lamarckism to be complementary with respect of extrasystemic war.

Sequencing theory operates on the combination of Lamarckism and Darwinian selection in a theory called neo-Lamarckism, which maintains that genetic changes are influenced by environmental factors and exogenous forces. In so-called gene-environment interplay, genes provide the platform for the synthesis of proteins, which triggers a series of chemical processes and informs actor behavior in interaction with environmental stimuli, which generate various neurological, cognitive, and emotive implications. The behavior of individual insurgents, the environments they are exposed to, and interaction with others end up shaping the gene expression.[7] Therefore, Darwinism and Lamarckism play crucial roles at every step of the development of sequence. Insurgent efforts at adaptation and innovation are constantly tested by selection, which occurs in an environment where the actors fight in ways that shape the environment itself and generate opportunities for them to win the war. In this competitive environment, successful insurgents are those that adjust well to changing demands of war and generate a self-sustaining capacity for a later period, while unsuccessful insurgents are those that fail to do so. As Rafael Sagarin writes, "A fundamental tenet of evolutionary biology is that organisms must constantly adapt just to stay in the same strategic position relative to their enemies—who are constantly changing as well."[8]

Of course, sequencing theory does not capture the full developmental paths of all species. Efforts to theorize war come at the cost of sacrificing a large amount of information. The theory ameliorates this, however, by treating conflict as a sequence. Doing so will allow us to focus on a set of key factors in the order of phases as the determinant of war outcomes. At the same time, the theory does not claim that all insurgents evolve like biological mechanisms. What it does is use the growth of insurgent forces relative to states as a causal explanation for how they fight and defeat states in war. The way insurgents evolve through a series of interactions with state foes shapes the war outcome. Rather than modeling war as a single-shot lottery, I assume that a series of changes occurs in the duration of conflict that strongly shapes the outcome. This view is consistent with that of Scott Gartner, who argues that actors conduct wartime strategic assessment and form beliefs about their likelihood of success from what they observe during war. This assessment, based on what he calls the "dominant indicator approach" because decision makers use numeric indicators of success and failure to make such an assessment, enables decision makers to assess if strategy is working, use various metrics to evaluate the effectiveness and the pace of acceleration, and, if necessary, make changes in strategy to influence war.[9] Insurgents conduct wartime strategic assessment in order to move to the next phase, a collection of which will allow us to see war as a sequence. As a theory of social science, sequencing theory helps us simplify the complex reality by focusing on a host of events in a given period.

Sequencing in Revolution and War

The courage and recklessness of the weak fighting the strong are well documented in a number of historical texts, ranging from the story of David and Goliath in the Bible's book of First Samuel to the late nineteenth-century revolutionary works of Marx to the modern literature on decolonization. Yet pioneer works on the use of sequence did not emerge until the turn of the twentieth century when Vladimir Lenin showed how members of the suppressed class would revolt.[10] Of course, the revolutionary leader made little reference to evolutionary thought at the time, although the fundamental ideas had a great deal to do with basic evolutionary concepts. Even then, the use of sequence was not the central thesis of Lenin, who was keener on social revolution and the impact of international capitalism on weak economies. Nor did he intend to help boost backward societies' chances in war in the first place. As a result, through the twentieth century there was little literature on how to win a war that revolutionary leaders had access to. Insurgents of all sorts had no intellectual text to resort to when devising a strategy for the weak, which is partly why, combined with material weakness, many of them lost extrasystemic wars throughout the nineteenth century.[11]

Lenin's notion of carrying out revolution through phases derived from what he called the "revolutionary situation." By this concept he defined revolution in terms of significant changes in mass mobilization and group struggle. These changes resulted from strategic innovations that produced cycles of development from primitive insurrectionary movements into professional rebellions. In what looked like an orderly progression of events, Lenin formulated a strategy for the suppressed class of social discontent to launch a series of revolutionary movements in the colonial periphery of imperialism through the creation of the revolutionary situation, which was made of the endangered ruling class, dissatisfied people, and politically active workers.[12] By then revolutionary movements had concentrated among a small number of underground circles based more on personal relationships than rule-based structures. These movements gradually became specialized into the "vanguard," who took up challenging roles in the planning and execution of revolutions. One of the most notable innovations of the time was the concept of party-state. Insurgents began to realize that the development of a state was key to winning a revolution, although they still considered the state to be an objective rather than a means to victory. As it turned out, state building became one of the most critical factors for underdogs facing powerful foes. In this way, revolutionary groups evolved into a set of increasingly competitive uprisings of mass violence and became the basis of political and military organizations ready to fight modern war.[13]

Slowly, Lenin's thesis spread across the world to shape the way successive leaders devised strategies. One such leader was Mao Zedong, who developed revolutionary ideas into a strategy of people's war, guerrilla war, and protracted war through his experience with Japan and the Guomindang in the 1930s. Mao's works covered many other issues than just fighting against imperial forces. His teachings over the following decades spanned issues like leadership, organizations, culture, and political change.[14] Yet on guerrilla war, Mao argued that programs of national liberation and anti-imperialism set the stage for the creation of communist parties before they adopted a semidictatorial form of internal domination justified in the name of progress toward socialism.[15] In a people's war, the vanguard party would mobilize the masses and prompt them to plant the seeds of resistance. They would do so in order to overthrow the urban centers of capitalism by encircling these centers with strategic footholds in the periphery. In short, Mao put forth a framework that can be considered a classic sequencing approach; in addition to his statement that opened this chapter, he argued that there "must be a gradual change from guerrilla formations to orthodox regimental organization."[16]

Naturally, sequencing theory draws from Mao's concept of "stages" in people's war, which proceeded in three stages. The first stage was characterized by what Mao called the "strategic defensive," in which guerrilla forces would withdraw from the front lines to secure strategic locations, establish bases, and carry out operations on exterior lines behind the enemy. It was followed by the "strategic equilibrium" phase, in which the guerrillas would bring the war to parity by extending the combat area into a stalemate of attrition before they would begin to match the enemy in battles. This phase would leave the enemy frustrated with endless maneuvers of evasion and make it difficult for the enemy to operate effectively, a stage best characterized by efforts to protract the war and gradually reverse the balance of power between the two sides. The last stage was "strategic counteroffensive," in which guerrillas would form a regular army to overrun enemy forces in conventional battle. At this stage, guerrilla warfare would be rendered a part of people's war with a reduced role and would be followed by conventional war. Thus the essence of people's war was to develop a chronologically ordered, symbiotic relationship between the three phases.

Sequencing Theory in the Postwar Era

The end of World War II unleashed a wave of decolonization movements in the 1950s and 1960s, which prompted scholarly research about how these movements evolved and generated a number of approaches in the Western scholarship to deal with the movements. One of the most prominent works of the time that used sequences in insurgency environments was that of Roger Trinquier, a French army officer who witnessed debacles in Indochina and Algeria and saw insurgency movements develop in three phases. First, insurgent groups would conduct isolated raids to attract attention from the population. Second, they would carry out terrorism selectively against enemy personnel. Finally, they would install a small armed band through guerrilla warfare.[17] Another major figure was David Galula, whose *Counterinsurgency Warfare* has been read widely; he saw communist insurgents in China develop through five phases: (1) creation of a communist party, (2) buildup of a united front, (3) execution of guerrilla warfare, (4) movement warfare, and (5) annihilation campaign. For counterinsurgents, he provided a strategy in as many as eight steps: (1) concentrate enough armed forces to destroy or expel the main body of insurgents, (2) detach for the area sufficient troops to oppose an insurgent's comeback in strength and install these troops in the hamlets, villages, and towns where the population lives, (3) build contact with the population and control its movements in order to cut off its links with the guerrillas, (4) destroy the local insurgent political organizations, (5) set up new provisional local authorities through elections, (6) test these authorities by assigning them various concrete tasks, replace the incompetents, support active leaders, and organize self-defense forces, (7) group and educate the leaders in a national political movement, and (8) win over or suppress the last insurgent remnants.[18] Other experts used different sequences than these, adding alternative paths to the growing body of scholarly research. Multiple methods of insurgency generated a set of resultant diverse responses by the government, which in turn allowed insurgents to evade uniform government counterattack.

Sequential strategies worked in tandem with changes in postwar international politics. Calling for independence, nationalist leaders in respective colonies around the world seized the emerging consensus about the immorality of colonialism, Wilsonian self-determination, human rights, and justice. European powers faced the destruction of colonial justification and the emergence of institutions like the United Nations that demanded the secession of power to the suppressed colonies in Africa, Asia, and the Middle East. Independence movements in the colonies now had strong support from the combination of decolonization processes, norms of self-determination, and international organizations encouraging the transfer of power. During the 1950s and 1960s, norms of sovereignty and institutions emboldened these forces and shaped the international atmosphere for mass decolonization from Kenya to Indonesia to India. One of the key elements in this movement was the power of narrative and argument. Ethical arguments regarding slavery and colonialism fostered changes in long-standing practices, arguably the greatest changes in world politics to occur over the past five hundred years.[19]

Yet because the influence of norms and institutions was hardly uniform across various colonial societies, it was not a decisive factor for insurgent movements to succeed in extrasystemic conflicts. There was remarkable variation in the normative power of decolonization between, for instance, Latin America and Africa. Although in Latin America, the Dominican Republic aligned with the interest of colonialist nations, Guatemala, Mexico, and Haiti proved to be strong anticolonialists at the UN. African countries were even more emotional and vocal than those in Latin America because they embraced more ethnic and cultural ties with the dependents of the territories.[20] As a result, wars did not always break out because colonial control in many places ended rather peacefully. After World War II, more countries became independent peacefully, including French colonies like Guinea and the former Soviet republics, while just sixteen anticolonial movements turned violent before they became independent. Therefore, the postwar spread of norms and institutions for independence notwithstanding, most insurgencies seeking decolonization ended up opting against violent recourse. This may have had to do with the recognition that civil resistance would be more effective than violence in achieving decolonization aims and because civil resistance presented fewer obstacles to moral and physical involvement, information and education, and participator commitment. Indeed, political participation contributed to enhanced resilience, tactical innovation, and opportunity for civic disruption, and provided less incentive for the regime to maintain the status quo.[21] Even where movements did achieve independence, there was a limit to the strategic effect of these norms and institutions. Table 1 shows that since 1945, insurgencies have won eleven of eighteen extrasystemic wars, but they have also lost seven: in Madagascar, Malaya, Hyderabad, Kenya, Cameroon, East Timor, and Iraq. Furthermore, norms and institutions of independence were not always the same thing as insurgent victory. While victory in extrasystemic war did lead to independence most of the time, insurgents' *defeat*, too, sometimes gave them freedom, as seen in places like Cameroon in the 1960s. The norms and institutions did not empower the anticolonial insurgency in Angola and Mozambique, either, because the outcome hinged more strongly on the Carnation Revolution in Portugal, which was caused primarily by a few select junior officers in the Portuguese military dissatisfied with the way they were fighting in Guinea Bissau. For these reasons, while the normative and institutional explanation may be relevant to wars of independence, it is not sufficient to explain variation in extrasystemic war outcomes.

Sequencing theory presents itself as an extension of academic discourse about analyzing conflict through stages. In recent years, there has been an increasing number of such works that use phases as a means of analysis. In addition to Gartner, Marc Sageman investigates a three-phase evolution of terrorist groups in order to follow the development of al-Qaeda.[22] Not surprisingly, the need to confront these untraditional security challenges has spawned a sequence-based response from military analysts. David Kilcullen discusses al-Qaeda's four-phase evolution in terms of what he calls the "accidental guerrilla syndrome."[23] Military practitioners and policy makers have similarly highlighted the importance of strategic adaptation in more contemporary experiences. John Nagl's work has called for the U.S. army to change and adapt, arguing that organizational culture is key to the ability to learn from events, which is why the British army successfully conducted COIN operations in Malaya and why the American army failed to do so in Vietnam. The British army, because of its role as a colonial police force and the organizational characteristics created by its history and culture, was

Table 1. Extrasystemic Wars After 1945

#	Name	Warring Sides	Duration	Winner	Independence After War?
1	Indonesia	Netherlands vs. Indonesia	1945–1950	Insurgent	Yes
2	Indochina War	France vs. Vietnam	1946–1954	Insurgent	Yes
3	Third Madagascan	France vs. Madagascar	1947–1948	State	Yes
4	Malayan Emergency	Britain vs. Malay Communists	1948–1960	State	Yes
5	Indo-Hyderabad	India vs. Hyderabad	1948	State	No
6	Tunisia	France vs. Tunisia	1952–1956	Insurgent	Yes
7	Mau Mau Rebellion	Britain vs. Kenya	1952–1960	State	Yes
8	Morocco	France vs. Morocco	1953–1956	Insurgent	Yes
9	Algeria	France vs. Algeria	1954–1962	Insurgent	Yes
10	Cameroon	France vs. Cameroon	1955–1960	State	Yes
11	Angola	Portugal vs. Angola	1961–1975	Insurgent	Yes
12	Guinea-Bissau	Portugal vs. Portuguese Guinea	1963–1974	Insurgent	Yes
13	Mozambique	Portugal vs. Mozambique	1964–1975	Insurgent	Yes
14	East Timor	Indonesia vs. East Timor	1975–1978	State	No
15	Namibia	South Africa vs. Namibia	1975–1988	Insurgent	Yes
16	Soviet-Afghan	Soviet Union vs. Afghanistan	1979–1989	Insurgent	n/a
17	Somalia	United States vs. Somalia	1992–1995	Insurgent	n/a
18	Iraq	United States vs. Iraq	2003–2011	State	n/a

better able to quickly learn and apply the lessons of COIN.[24] Evidence shows that the U.S. military has relearned lessons from recent COIN experiences, improved its ability to conduct stability operations, changed the institutional bias against counterinsurgency, and figured out how to account for successes gained from the learning process.[25] The U.S. military has also successfully adapted to untraditional security missions in Afghanistan and Iraq. Learning from two centuries of small wars and nation building, the United States increasingly has emerged successful from insurgency wars, carrying out the surge in Iraq most recently.[26] These research findings buttress the importance of organizational adaptation in COIN settings, especially the U.S. military's adaptation to the changing environment of the Iraq War, in which "successive and sequential historical decisions shaped how the army would and could function in Iraq, particularly as its missions changed during the course of the war."[27]

Learning and Proliferation of Sequencing Practice

What explains variation in the outcomes of extrasystemic war is the way both sides of war fight in sequence. But how do insurgent forces know about sequencing? The answer rests with their ability and willingness to learn to put the war in sequence. Insurgent forces need to learn from wars of the past and adapt to their specific environment and spread a demonstration effect from one battlefield to another, from one country to another, in order to generate a cascade of insurgency victory across the world. But because there are constraints in the process of learning, receiving and accepting new ideas, and ultimately spreading them, few can consciously adopt effective models. These constraints include the availability of communications technology and the possibility that actors interpret things differently. Actors also learn selectively; research shows that humans learn much more from their own experiences than from those of others.[28] Furthermore, learning is not the same thing as adopting a new policy; one can learn something but decide not to use it. Of course, these problems do not hamper only insurgents; states, too, have had trouble figuring out how to innovate. That is why, again in Algeria, French generals did not necessarily apply the lessons they learned from their experiences in Indochina. Yet one of the major reasons why many insurgent groups have lost wars over the years had to do with the fact that they never learned from their mistakes. Indeed, evidence shows that many of them have *repeatedly* used the wrong strategy against the same enemies, even though they had been losing such wars for decades. Therefore, many groups actually fail to evolve across generations, which explains why some groups "fight the last war" again and keep losing. Table 2 indicates that throughout the history of extrasystemic war before 1945, thirteen of sixteen groups that lost once went *back* to fight the same foe using the same strategy.

There are at least two reasons why these groups used the same strategy repeatedly. The first is the availability of military technologies that favored a conventional strategy and force structure. As noted earlier, global attraction to modern military technology in part explains why many insurgent forces fought extrasystemic wars in mostly conventional fashion in the nineteenth century. The other reason is the limited access the insurgent forces had to information about foreign wars. Shortages of communications devices, networks, and printing presses through which to learn from other groups gave insurgents little choice but to fight wars the way they knew how. As information technology, overseas travel, and printing presses became available over time, some insurgent leaders shared ideas through publications, meetings, interviews, and speeches. These leaders received an early education in Europe, where they exposed themselves to a diverse set of ideas including Marxism, learned Western languages and customs, and embraced nationalist thought before they returned to their motherlands. While many of them embraced class struggle and the philosophy of Marx and Lenin, the proliferation of learning devices coincided with the time of Maoist people's war. While Mao's ideas have influenced many revolutionary leaders generation after generation, they were made available only in the early twentieth century and came too late for many insurgent leaders who had little access to technology and information. Furthermore, these ideas proved to be so hard to replicate outside China that they impacted other revolutionaries only variably. Variation in insurgents' access to technology, press, and information naturally had dissimilar effects on insurgent movements around the world. Among a diverse set of anticolonial forces in Southeast Asia to Latin America to Africa, Mao proved to be a key shaper of decolonization movements, but only in limited ways. As a result, revolutionaries across the globe embraced the sequential ideas with local characteristics to fit the latter, creating variation in areas like Southeast Asia, Africa, and Latin America.

In Southeast Asia, major extrasystemic wars of recent decades have included the Malayan Emergency of 1948 to 1960 and the Indochina War of 1946 to 1954, both of which demonstrated that learning and adaptation of

Table 2. Repeated Extrasystemic Wars with the Same Strategies

#	Name	Years	Model	Winner
1	First Bolivar Expedition	1817–1819	Conventional	State (Spain)
	Second Bolivar Expedition	1821–1822	Conventional	State (Spain)
2	First Burmese	1823–1826	Conventional	State (Britain)
	Second Burmese	1852	Conventional	State (Britain)
	Third Burmese	1885–1889	Conventional	State (Britain)
3	First Ashanti	1824–1831	Conventional	State (Britain)
	Second Ashanti	1873–1874	Conventional	State (Britain)
	Third Ashanti	1895–1896	Conventional	State (Britain)
	Last Ashanti	1900	Primitive	State (Britain)
4	First Zulu	1838–1839	Conventional	State (Britain)
	Second Zulu	1879	Conventional	State (Britain)
	Third Zulu	1906–1907	Conventional	State (Britain)
5	First Afghan	1838–1842	Conventional	Insurgents (Afghan)
	Second Afghan	1878–1880	Conventional	State (Britain)
	Third Afghan	1919	Conventional	Insurgents (Afghan)
6	Seventh Kaffir	1846–1847	Conventional	State (Britain)
	Eighth Kaffir	1850–1853	Conventional	State (Britain)
	Ninth Kaffir	1877–1878	Conventional	State (Britain)
7	First Sikh	1845–1846	Conventional	State (Britain)
	Second Sikh	1848–1849	Conventional	State (Britain)
8	First Senegalese	1854–1865	Conventional	State (Britain)
	Second Senegalese	1887–1890	Degenerative	State (Britain)
9	Ten Years	1868–1878	Conventional	State (Spain)
	Little	1879–1880	Primitive	State (Spain)
	Spanish-Cuban	1895–1898	Primitive	Insurgent (Cuban)
10	First Egypt-Ethiopia	1875–1876	Conventional	Insurgents (Ethiopia)
	Second Egypt-Ethiopia	1885	Conventional	State (Egypt)
11	First Boer	1880–1881	Conventional	Insurgent (Boer)
	Second Boer	1899–1902	Degenerative	State (Britain)
12	First Mahdist	1882–1885	Conventional	Insurgents (Mahdists)
	Second Mahdist	1896–1899	Conventional	State (Britain)
13	First Madagascan	1883–1885	Primitive	State (France)
	Second Madagascan	1894–1895	Conventional	State (France)
	Third Madagascan	1947–1948	Primitive	State (France)
14	First Ethiopia-Italy	1887	Conventional	State (Italy)
	Second Ethiopia-Italy	1895–1896	Conventional	Insurgents (Ethiopia)
15	First Matabele	1893	Conventional	State (Britain)
	Second Matabele	1896–1897	Conventional	State (Britain)
16	First Sino-Tibetan	1912	Conventional	Insurgent (Tibet)
	Second Sino-Tibetan	1918	Conventional	State (China)

sequencing, or lack thereof, was a critical determinant of war outcomes. In Malaya, Chin Peng led a communist rebellion against British colonial forces under the banner of the Malaya Communist Party (MCP). For much of the emergency period, Chin maintained limited contact with Mao, received little support, and ignored the need to "win hearts and minds" of the population. A few years into the war, the MCP realized that its guerrilla strategy was not working, so it issued a policy directive in 1951 to urge its members to refrain from coercive measures on the population. The MCP also received advice from China and Russia to modify the struggle in line with the growing emphasis on "peaceful coexistence" with the population, but the MCP minimally adopted these measures. By the late 1950s the independence movement was practically over. In Indochina, Ho Chi Minh and Vo Nguyen Giap learned a great deal from Mao, copied some ideas but generated their own versions, and successfully led a phased war against the French in their quest for independence. One of Ho's first official statements after the outbreak of war was that Vietminh leaders would follow the people's war. Another Vietminh leader, Truong Chinh, authored a pamphlet called *The Resistance Will Win*, which drew extensively from Mao's writings on guerrilla war. In contrast to the MCP, the Indochina Communist Party (ICP) worked closely with the Chinese Communist Party (CCP). Vietnam had assisted Mao during the Chinese civil war, and China reciprocated during the Indochina War. In fact, the CCP's victory in 1949 became a catalyst for the Vietminh to resurrect the ICP in 1951. The Vietminh closely studied the CCP's wars against Japan and Chiang Kai-shek while translating documents and training materials into Vietnamese and distributing among troops.

In Latin America, Mao's ideas influenced the thoughts of revolutionaries such as Che Guevara and Regis Debray about how to carry out guerrilla wars through stages. Guevara made clear the origin of his ideas on guerrilla war when he said that "we have always looked up to Comrade Mao Tse-tung. When we were engaged in guerrilla warfare we studied Comrade Mao Tse-tung's theory on guerrilla warfare. Mimeographed copies published at the front lines circulated widely among our cadres; they were called 'food from China.' We studied this little book carefully and learned many things."[29] A number of chapters in his selected works resonated closely with the ideas of Marx, Lenin, and Mao, including the concept of vanguard, people's war, and people's army. Guevara praised Giap in his prologue to Giap's book when it was brought into Cuba and published in *La Guerra del Pueblo: Ejercito del Pueblo*.[30] Regis Debray, too, displayed a considerably detailed knowledge of the associated readings, citing authors like Marx, Engels, Lenin, and Mao, as well as Cuban national hero Jose Marti.[31] In fact, Mao's ideas became so popular that local leaders applied them in many instances against their own governments. Gilberto Vieira, the general secretary of the Communist Party of Colombia, outlined five phases of civil war: (1) preparation and organization, (2) large-scale program of psychological action against the government, (3) isolation of armed forces, (4) division of armed forces, and (5) economic, political, and social reconstruction of the zones of operations, using American aid.[32] These leaders communicated closely with their revolutionary counterparts elsewhere as they commanded their wars of liberation.

In Latin America, however, local characteristics and geography intervened to generate a diverse effect. In particular, Guevara's famed "foco theory," which argued that revolutions would begin with a small core of discontented elements and blossom over time into an organized group, stood out as a clear deviation from people's war. Accordingly, Guevara anticipated that a revolution in Cuba along with Fidel Castro would proceed through three stages, with the first phase including small-sized guerrilla units that would mix with the physical and human conditions of the battlefield. Unlike the Maoist concept of "base areas," the guerrilla units would carry out limited attacks at this phase. In another important contrast to Mao's theory, revolutionaries would not wait until all conditions were met but instead would allow a small guerrilla nucleus—*foco*—to operate liberally. In the process of guerrilla growth, the battle would reach a point where commanders would move around to spread violence. This spread of violence would establish a rough parity of power where a compact group would emerge and seek to dominate the war. The final phase would consist of the rebels successfully capturing large cities and overrunning the army. In the other contrast to Mao's theory, mobile operations introduced in this phase would *not* replace guerrilla fighting; instead, regular forces would be a supplement to guerrilla forces.[33] It was clear that Guevara's theory had a sequential element, but it apparently developed on a different path.

In Africa, there were numerous extrasystemic wars in recent history, including the Ashantis, Zulus, Boers, and Senegalese. Here again, we see the impact of learning and proliferation of sequencing ideas on the way local insurgent forces fought extrasystemic wars. Many of these tribal groups had little luck throughout the European colonial period and until the 1960s when decolonization movements became widespread. Among the prominent leaders were Amilcar Cabral, Kwame Nkrumah, and Frantz Fanon. Leading an insurgency in Guinea-Bissau against Portugal, Cabral read Mao's writings when he was in China in 1960 and studied them further before he opened his war in 1963.[34] Fighting a long but increasingly successful struggle for independence, he praised Mao, Guevara, and Nkrumah as champions of Third World revolution.[35] In Ghana, Nkrumah shared the idea of fighting over multiple phases as the key to liberate the territory when he followed Lenin's masterpiece—*Imperialism: The Highest Stage of Capitalism*—to criticize the residual colonialism in his aptly titled book *Neo-colonialism: The Last Stage of Imperialism*.[36] In Algeria, Fanon developed what was essentially a sequencing strategy, a little-known fact for a psychiatrist best known for the treatment of peasant rage and emotions in suppressed societies. During the Algerian war for independence, Fanon was a key ideological leader behind the operations of the FLN and National Liberation Army (ALN), which also advocated a multiphase approach that differed from Mao's. He anticipated that in the first phase, a small group of frustrated individuals would make a spontaneous attack on colonial forces in their local areas. This period would be characterized by small-scale movement; "the aim and the program of each locally constituted group was local liberation." In the second phase, colonial forces would respond with a series of large-scale offensives. Through these offensives the forces would turn into guerrilla operatives, similar to those seen in a peasant revolt. In the final phase, the revolt would transform itself into a revolutionary war. The insurrectionists would evolve into warriors of liberation strong enough to decimate the colonial leaders and regain sovereignty. Through three phases, according to Fanon, a war of liberation would be complete.[37] Of course the Algerian war per se did not proceed like he anticipated; it was mostly a series of insurgent operations using local cells and individual networks in hit-and-run operations in avoidance of frontal attacks on French troops. Although in later phases Algerians made efforts to modernize forces, the center of gravity lay in the underground terrorist cells. Therefore, the war was highly divergent from the people's war concept, although it had apparent links with sequencing ideas. These revolutionary leaders inherited the intellectual impetus from Lenin and Mao and localized the practice in ways that fit their strategic environments. Sequencing strategies gradually disseminated in small pieces from the mainstream approach while retaining distinctive revolutionary characteristics.

CHAPTER 3

How Sequencing Theory Works

What Makes Sequencing Theory?

Recent proliferation of sequential strategies, both in practice and through academia, attests to the growing recognition of the importance of using sequences in conflict. The existing ideas and works, however, need an overarching framework, which requires us to explore precisely how sequencing theory works. To answer this question I disaggregate its components in the context of extrasystemic war. In so doing, I reveal the presence of three phases—guerrilla war, conventional war, and state building—that compose the theory. I also show that sequencing strategies are likely to emerge when insurgent leaders put together a set of these phases in order to evolve during the war, creating a host of sequential patterns distinct from other strategies designed to fight stronger enemies. Indeed, a number of factors encourage insurgent leaders to choose a sequencing strategy over others, including external support, internal popular support, leadership attributes, and internal weaknesses of the states they fight. At the center of insurgent leaders' decision to adopt sequencing strategies is the general acceptance of these strategies as a means of overthrowing powerful rivals. At the same time, however, there are different requirements for each of the six sequencing strategies. In this section, I clarify a set of different requirements for those phases to constitute a sequence.

Guerrilla War

The first of the three phases of sequencing theory is guerrilla warfare, where we see both sides engage in hit-and-run operations for a sustained period. In this phase states and insurgents exchange low-intensity violence while mobilizing the population as human buffers and sources of supplies and striving to build informal networks as support bases. Guerrilla war is, according to Samuel Huntington, "a form of warfare by which the strategically weaker side assumes the tactical offensive in selected forms, times, and places."[1] The core of guerrilla warfare rests with the competition for popular support because it is the population that provides the sine qua non of guerrilla war—food, lodging, sanctuary, intelligence, recruitment sources, and legitimacy. Thus Chalmers Johnson states that "guerrilla warfare is civilian warfare..., conflict between a professional army... and an irregular force, less well trained, less well equipped, but actively supported by the population of the area occupied by the army."[2] Of course, in practice the distinction between combatants and the population is difficult to draw because the former effectively integrates the latter in combat, and one can hardly distinguish the innocent from civilian-clothed militias. Thus the separation between guerrillas and the population is analytical. Needless to say, the primacy of popular support has been widely acknowledged in the literature.[3] For Mao, "weapons are an important factor in war, but not the decisive factor; it is people, not things, that are decisive. The contest of strength is not only a contest of military and economic power, but also a contest of human power and morale."[4] David Galula notes that "military action is secondary to the political one, its primary purpose being to afford the political power enough freedom to work safely with the population."[5] In "colonial struggles against imperial powers," writes Huntington, the war "begins with the mobilization of new groups into politics and the creation of new political institutions."[6] In short, the key to success in the phase of guerrilla warfare is popular support. The side that seizes popular support and builds strong bases is likely to dominate the guerrilla war phase.

There are many factors that shape a successful guerrilla insurgency, but one of the key factors is external support. State support is critical for insurgents to be successful because outside resources increase their internal sustainability.[7] Insurgents also benefit from other groups that might coalesce to generate a united front. Some groups ally with weak states in the neighborhood. Otherwise unable to challenge enemies on their own, insurgents exploit these states that are too weak to prevent their access, states that seek to foster instability in their rivals, and large refugee diasporas that allow the groups to establish bases. States that host them tend to intervene in negotiations between governments and these groups and block progress toward peace when they pursue their own agendas and because these sanctuaries complicate intelligence gathering, COIN operations, and peacemaking.[8] External aid comes in both wartime and peacetime, but it is critical for insurgents to receive aid during wartime when material shortages are prevalent.

Yet state sponsorship may not always be helpful. Having a sponsor that provides safe haven actually increases the risk of being eliminated by the target because sponsors may be tempted to provide information to the target to avoid potential costs from military operations within their territory.[9] External support can backfire, furthermore, under moral hazards when an expectation of such support encourages groups to fight recklessly in the hope of getting it. Material aid gives them the expectation that a war will be easier to win, so they may become unnecessarily aggressive in the assumption of false insurance.[10] Finally, getting such support is not easy. Winning outside support is highly competitive and uncertain. Even when provided, such support is not a philanthropic gesture but an exchange based on the relative power of each party to the transaction, and its effect is more ambiguous than is often acknowledged. Competition for foreign intervention occurs in a context of economic, political, and organizational inequality that systematically advantages some challengers over others.[11] External support is not always the decisive factor in extrasystemic war.

Conventional War

In guerrilla warfare, it is the population that plays a central role while the army provides a supportive function. In conventional war, it is the opposite. The army takes the main role in crushing enemy capacity in decisive engagement and gaining control of the population, territory, or vital industrial and communications centers while protecting the population behind the front lines. But conventional war does not serve states and insurgents equally. Military history makes it clear that it is states, rather than insurgents, that enjoy the benefit of using regular armies in war because they have the inherent advantage in resources and organization.[12] Western powers taxed their citizens and monopolized the production and dissemination of advanced weapons. In contrast, nonstate insurgents did not have any other means but to pick up arms after battle, use aid from allies, and secretly purchase weapons from foreign agents. Naturally, the quality and quantity of weapons imports and advisers did not match those of states.

Conventional war has evolved to become what it is today. Risking oversimplification, we know from history that by the fourteenth century, not just European powers but also colonized peoples in non-European territory had developed armies of heavy cavalry and infantry, but Europe led the way to add the power of the siege cannon in the mid-fifteenth century. Later in that century came the creation of standing armies that dramatically increased the size of force structure along with the expansion of government expenditures on military affairs.[13] The evolution of land and sea transport after 1815 had a strong association with the development of the steam engine, which affected the balance of power between European powers and their colonial subjects. The railway in Europe then eliminated prolonged marches and allowed huge armies to be moved quickly.[14] Growth in military technology and European imperial motives worked hand in hand to produce an overwhelming firepower advantage that brought down a host of political entities in Africa, South and Southeast Asia, Latin America, and Middle East in the following centuries. At the center of this European expansionism was the conventional military power that drew on the dramatic innovation of weapons.[15] Weapons evolved through stages of gunpowder, cannons, crossbows, and atomic weapons, which reflected the notion that development in military technologies changes the nature of military organizations and modern war.[16] Until World War I armies consisted mostly of infantry and were organized around the principle of "foraging," in which the bulk of their supplies was obtained from local populations. Foraging involved the use of monetary payments, forced requisition, and simple looting to acquire provisions from populations near the conflict zone.[17] The end of World War I saw the replacement of manpower with motorized vehicles like tanks, which increased the mobility and survivability of military forces on industrial-age battlefields.

State Building

Military power is a determinant of victory, but it is not the only one. As Lenin's party-state concept attests, we must add state-building efforts to the equation. Both sides engage in building government structures and seek to monopolize the means of violence over a given territory. Central to the process of building a state is the control of institutions and the growth of organizational capacity to govern the territory. Generally understood as a set of formal and informal rules that constrain and reward human behavior,[18] institutions here are a generic group of formal organizations such as political parties and civil service, police, financial, legal, and educational systems that facilitate government functions. Institutions are quite visible in major centers of activity, such as the capital. Yet the process of state building is more than just the establishment of institutions. Insurgent groups must build national unity with popular consensus, win international recognition, yield economic productivity, collect revenue, and pay for the war. State building may take place simultaneously with the phases of conventional and guerrilla war because they are not mutually exclusive.

The role of state building has been overlooked in the study of irregular war because it does not center on the destruction of enemy forces. Yet state building is a quite distinctive phase, as it belongs to a separate analytical category from fighting war.[19] Michael Doyle singles out this category as a key to the success of independence movements in their endeavor against empires as they manage to institutionalize the participation of their newly mobilized citizenry.[20] Organization is the key to rebellion in terms of control, finance, recruitment, violence, and resources.[21] While state building is a more subtle phase than guerrilla war and conventional war, we see the phase of state building when these institutions perform a central role during the war.[22]

States have resources to build a government in colonial territories, but they face challenges in the process. Governments face two types of "dilemmas" when they deploy forces overseas and intervene in foreign nations, a common scenario in extrasystemic wars. The "duration dilemma" means that states that occupy foreign territories provide security for the locals, but the welcome is likely to decrease over time as populations seek to reassert their control and press the state to leave. The "footprint dilemma" shows that while foreign states' presence is needed for security in local societies, it may increase the danger of stimulating nationalist resistance, as we saw vividly in the 2003 Iraq War, for instance.[23] These dilemmas challenge foreign governments' efforts to be adept at building a state in conflict zones. Thus occasionally insurgents outperform foreign governments in building institutions, resulting in a more equal balance of political power and a growing ability to fight war better. This scenario may be likely when states have not built a counterbalancing structure. In what Tilly calls "multiple sovereignty," insurgents obtain recognition and receive support from the population. The war becomes a polarized venue for the contention of political control between states and insurgents who compete for the public mandate and negotiate autonomy.[24]

How Do Insurgents "Adapt to Win"?

In a highly contested series of violence against government opponents, insurgents boost the chances of achieving their goals by evolving into an organized army with serious work for statehood. The process appears as war moves from one phase to another in the temporal sense. The decision to move is mutual, because it rests with the insurgents' ability and willingness to do so, coupled with the ability and willingness of the states to block the evolution. In other words, evolution is not complete unless the state side follows the insurgent's move to a next phase, and vice versa. The state side often uses its enormous resources to dictate the term of violence in order to veto the insurgent move. Otherwise, the war develops through combinations of the following three phases—guerrilla war, conventional war, and state building. Insurgent forces grow more organized and militarily capable, and consequently more likely to make a transition from one phase to another and achieve their ends. There are, however, different sets of requirements to meet, depending on between which phases the transition takes place.

First, a transition from guerrilla war to conventional war occurs under two conditions. First, insurgent forces must gain a significant advantage in the level of popular support over their opponents. Popular support is the key ingredient here for the transition to take place because the population generates manpower for the army, raises money for the insurgency, and provides legitimacy for the military operations it conducts in the conventional war phase. Without it, the army would have to fight the war as its social foundations decay quickly. The other condition is that, once the war moves beyond the conventional phase, the insurgents' armed branch must have sufficient resources to continue to protect the population and territory. In other words, insurgent forces need both popular support and military power to evolve from a guerrilla group to a modern army. We see a good example of this transition in segments of many extrasystemic wars; for example, the Somali-British War of the early twentieth century demonstrates how difficult it is to complete the transition if the insurgents do not have adequate resources even if they initially managed to win local support. The Somali insurgency lasted nearly two decades but eventually collapsed because they became vulnerable to growing British firepower and airpower despite the fact that the transition from the guerrilla war phase to conventional war was successful.

Similarly, a transition from the guerrilla war phase to state building takes place under two conditions. On the one hand, insurgents must win a popular mandate to build a state so that the population will support their emerging institutions. Popular support is the key ingredient here because, without it, government institutions would lack legitimacy and resources to become an independent state, a necessary ingredient to win extrasystemic war. The other condition is that the institutions they create in the second phase have enough resources to reinforce the support base. Stable institutions are the key here because, without them, the insurgents would soon lose a key popular support base to continue their state-building efforts. In other words, insurgent forces need to outperform their state foes in terms of popular support and institutions in order to evolve. A good example of this transition is the Indochina War, in which the insurgent Vietminh forces initially succeeded in winning hearts and minds of the people. This victory at the local level allowed them to launch a nationwide political effort to reestablish the political party to drive the war effort. The party leadership generated a powerful effect to centralize the war effort because the population supported efforts to build an independent state.

A shift from state building to conventional war has two requirements. On the one hand, the political institutions built in the first phase must generate resources, policy, strategic direction, and administrative support for military operations to empower the insurgency. Institutional stability and resourcefulness are critical ingredients here because, without them, armed forces will likely collapse. On the other hand, insurgent forces must develop the capability to protect these institutions as they fight for independence. Short in military power, insurgent groups would lose policy direction and fall into aimless mass killing. In other words, for the transition from state building to conventional war to work, insurgents need both military power and institutions.

Fourth, a transition from conventional war to guerrilla war occurs when insurgents already have a fairly developed and organized army that can protect the population and maintain order and territorial integrity against state aggression in a volatile environment. Military power is the key factor that enables insurgents to consolidate gains made in the conventional war phase into the guerrilla war phase because insurgents would otherwise be too powerless to protect noncombatants and leave themselves vulnerable to enemy attack. At the same time, the army must have a significant level of popular support because the population generates manpower for the army, raises money for the war, and increases legitimacy for military operations it conducts in the conventional war phase.

Finally, insurgent groups may take the war from the state-building phase to guerrilla war when the institutions are capable of governing the people and winning popular support for the insurgency. Stable institutions are the key here because without them the insurgents would soon lose a key support base to continue their state-building efforts. In other words, they need both popular support and institutions in order to evolve. On the other hand, the insurgents must win popular support to build a state and make the institutions legitimate. Popular support is the key ingredient here because without it institutions lack legitimacy and resources to function as an independent state. Thus the insurgents must generate sufficient social support for the state-building process.

Adaptation and evolution are key causal factors of insurgent victory because the transition from one phase to another empowers the insurgents in two critical ways. First, groups elicit greater expectations of success from the population, who in turn invest more in their efforts. The stakes become higher, and the insurgents become more determined to win and more aggressive on the battlefield. This has negative consequences for states. Because insurgents fight harder now, they are harder to defeat and ready to fight longer. They begin to win more battles, making it difficult for states to score victories. As a result, the probability of state victory diminishes, increasing the length of time necessary for states' potential victory. Second, now that the insurgents have greater chances of winning, the population realizes that their chances of victory are greater, so they support the insurgents more. After all, more insurgent victories mean that states have to do more to win, while it becomes harder for them to do so, discouraging them from continuing the war. Thus, as the war moves from one phase to another, insurgents gain a greater likelihood to achieve their ends. As they evolve, they grow stronger than they were in the first phase, and more capable and willing to take advantage of the gains they made and more likely to move on to the third phase.[25] Leaders of successful insurgencies, from Indochina's Ho Chi Min to Amilcar Cabral of Portuguese Guinea, have recognized these dynamics and instituted them. Cabral, for one, stated that "the successes won during the past year and the objective factors we have already established and consolidated enable us to look to the future with confidence."[26]

Models of Sequencing Theory

Sequencing theory consists of six models so as to reflect the number of ways that insurgents evolve. Not all extrasystemic wars proceed in linear fashion, but there are six major paths that they can take as they progress. However, not all models are evenly used. Over the period of 1816 to 2010, the relative weight of the models shifted. In much of the nineteenth century, most extrasystemic wars could be described in terms of army-to-army combat that I call the conventional model. During this period, as seen in Figure 3, a majority of insurgent groups built regular armies and used them against foreign states in open terrain. The population was protected behind the front lines and played a marginal role in combat. Most colonial European states were better armed, organized, and stronger, but many insurgents were bold enough to challenge them.

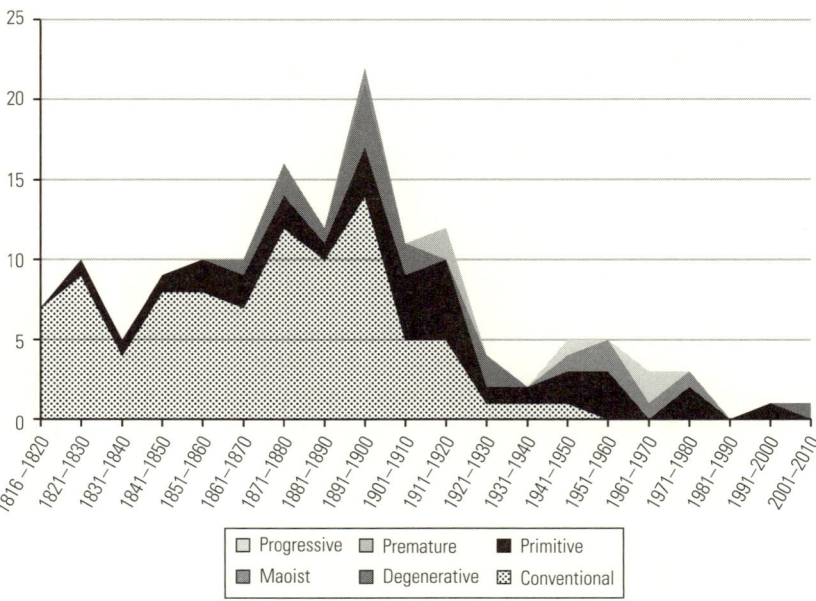

Figure 3. How extrasystemic wars change over time.

Table 3. Six Models of Extrasystemic War

Model	First Phase	Second Phase	Third Phase
Conventional	Conventional war		
Primitive	Guerrilla war		
Degenerative	Conventional war	Guerrilla war	
Premature	Guerrilla war	Conventional war	
Maoist	State building	Guerrilla war	Conventional war
Progressive	Guerrilla war	State building	Conventional war

From the middle of the nineteenth century, however, we see a slowly increasing number of guerrilla wars, as denoted in the primitive model, which challenged the primacy of the conventional model in the early twentieth century. This made the conventional and primitive models the two most dominant types of extrasystemic war in much of the nineteenth and early twentieth centuries. In other words, having no more than just a single phase, these models depicted where insurgent groups did not evolve. In recent years, however, extrasystemic wars have become complex. A small number of what I call the degenerative and premature cases emerged in the late nineteenth century to add a layer of complexity to the overall picture of extrasystemic war. By the middle of the twentieth century, furthermore, wars became even more complex. The primitive model had become stable, but a small number of so-called Maoist and progressive wars emerged to replace the degenerative and premature wars. Table 3 summarizes the six models. In the following sections I describe how each of the models works.

Conventional Model

In the conventional model, states and insurgents use armies to fight in open terrain from the beginning to the end. The key to victory in this classic battle of position is the destruction of enemy forces; war ends when one side's army annihilates another. Yet conventional war is hardly a product of one side's military prowess. This model reflects the commitment of *both sides* to fight in similar fashion and assumes that they end up taking similar actions to make a conventional war. When, for example, insurgent groups consider adopting a guerrilla strategy while state forces adopt a conventional strategy, the model posits that both sides become inclined to fight in an orthodox fashion. This is because they have enough resources to build an army and history of using armies in past wars. The Dahomean War, which I examine in the next chapter, is a good example. Conversely, even when states consider adopting a guerrilla strategy and insurgents adopt a conventional strategy, the war is likely to be conventional because states have inherent advantage in resources and organization and know that they will fight better in a conventional war. In this model, neither side changes the conventional military strategy in the middle of the war because neither is sure if changing the strategy will improve their situation. Under these conditions, conventional military strategy becomes the dominant strategy for both sides.

Insurgent groups all around the world have adopted a conventional strategy even though the guerrilla strategy has always been available. Globalization of conventional war dates as far back as when nonstate groups began to embrace military technologies. The attraction of orthodox war was so strong that some of these groups were even quicker than states to appreciate new weapons once they seized them.[27] Insurgent preference for conventional war is by no means accidental.[28] Insurgent groups of different sizes have displayed a remarkable similarity in the structure of military organizations across the globe, following the patterns of capital-intensive militarization and proliferation of standing armies.[29] Insurgents' propensity toward conventional war is based on the inherent inclination of "the poor and weak and peripheral copy[ing] the rich and strong and central."[30] The converging power of conventional strategy also represents the fact that military technologies and organizational features of developed and developing states have been similar. This similarity may in part be due, among other things, to the attraction of conventional weapons; as Michael Adas argues, "No invention elicited as much astonishment and respect from Africans as European firearms."[31]

Used in 63 percent of the entire samples of extrasystemic war, the conventional model is the most popular model of extrasystemic war for several reasons. For one, it brings stability to the military organization. Because it requires a high degree of training and discipline, it forces leaders to seek firm control of their army when the war becomes difficult. It is favored also because, as organization theory posits, militaries have an institutional interest in the autonomy, survival, and expansion of their organization by way of conventional strategy. Successful execution of conventional war rests with fast battlefield decisions, so the organization must have adequate decision-making autonomy. It requires investments in advanced military equipment, which justifies the need for an increased size and budget for the organization. This concept is also consistent with important elements of most military cultures, as they seek effective ways to minimize casualties, facilitate the seizure of initiative, and deliver quick and decisive victory. Indeed, some insurgents are found to be equal or superior to their civilized counterparts in four respects: (1) devotion to offensive strategy, (2) use of surprise, scouting, and intelligence, (3) use of terrain, and (4) tactical mobility.[32] Finally, conventional war is popular because it raises a symbolic value as a civilized nation. Only an army organized, trained, and uniformed along European lines firmly under the command of the standing officer corps is considered to be able to mount a challenge against a Western power. A concentrated field army implies refinement and attracts the world's attention. From this standpoint, conventional war is seen as an extension of military modernization. It is also a signal to external audiences that the rebel group can fight a stronger foe on the same level.

Many insurgent groups resorted to the conventional model in the nineteenth century as part of a global trend in favor of using traditional military resources to fight wars. The conventional model takes place under the condition of sufficient military resources, past experiences with conventional war, and the assumption that the insurgent groups have little knowledge about the utility of different war strategies including guerrilla strategy. The groups must have access to enough resources to build up an army of soldiers, train and arm them with weapons, and organize and discipline the forces to fight modern enemies. Insurgent forces also use the conventional model when they have a history of having used it to win a war. Most extrasystemic wars took the form of the conventional model until Mao introduced and proliferated guerrilla war as a popular military strategy. Conventional extrasystemic wars therefore were fought repeatedly by many insurgent groups during the nineteenth century even though they found themselves almost always on the losing side.

While many insurgents prefer conventional war, they are generally not successful at it. While they have fought conventionally in 63 percent of their wars, they have won only 15 percent of them. Historians and ethnographers provide a set of reasons for the poor performance of insurgents. Among them are Quincy Wright and Harry Turney-High, who list nine reasons for their failure: (1) poor mobilization of manpower and reliance on voluntary force, (2) inadequate supply and logistics, (3) inability to conduct protracted campaigns and lack of strategic planning beyond the first battle, (4) no organized training of units, (5) poor command and control, (6) lack of discipline and low morale, (7) shortage in weapons and neglect of fortification, (8) lack of professional warriors and specialization, and (9) ineffective tactics.[33] Turney-High, an anthropologist and a principal architect of the concept of "primitive war" and "submilitary combat," further accuses insurgents of viewing war as a social institution and a diversion for a variety of nonmilitary functions.[34] His view is familiar in extrasystemic wars. Insurgents generally lack unit discipline, which means they are short on training, structure, and physical compulsion, whereas states spend enormous resources on these assets. Insurgents are unable to plan for long wars. Weakness in command and control mean that they have doubtful grasp of individual soldiers, who are prone to acting badly toward civilian populations and generating a great deal of anxieties among the civilians. Soldier misbehavior leads to the decline of popular support, which in turn undermines nonstate operations by reducing manpower and supplies. Although this model has been observed most frequently in the history of extrasystemic war, it is largely a model of nineteenth-century warfare.

Primitive Model

The second model is the primitive model, represented by the execution of guerrilla war from the beginning to the end. It takes place when the guerrilla strategy is the dominant strategy of both states and insurgents. On the one hand, insurgents fight guerrilla war when they seek to take advantage of their access to people, support bases, and topographic features like mountains and urban areas. They may favor the guerrilla strategy even if their support base is not strong because the alternative—conventional war—would only weaken such bases and because there is still a chance to regain people's trust if they fight well. After all, the conventional model rarely works for insurgents, so many of them may prefer guerrilla war. On the other hand, state forces employ this strategy as well if they have enough support or think they can gain it over time, or if the terrain favors guerrilla operations. They make this choice in response to the insurgents' adopting of guerrilla strategy and based on the realization that conventional military strategy would not be effective against guerrillas. Therefore, even when insurgents consider using guerrilla strategy, if states adopt a conventional strategy, the war tends to be primitive. Similarly, even if states use a guerrilla strategy and insurgents adopt a conventional strategy, the war may become primitive because insurgents would make a switch in hopes of improved chances. In this model, neither side will change its strategy in the middle of war because, on the one hand, insurgents are comfortable with it or do not have enough resources to do so and, on the other hand, state forces believe that they are approaching victory with it. Under these conditions, guerrilla war becomes the dominant strategy of both sides.

In addition, the primitive model takes place under several conditions. Guerrilla war is likely when geographical conditions in the battlefield support operations in jungle areas, mountains, villages, and urban areas. Guerrilla operations in extrasystemic war became prominent also once Mao began to spread the idea of fighting in asymmetric conditions and in small wars against big powers. Thus, the primitive model assumes that insurgent leaders choose the option of fighting in more clandestine and asymmetric conditions over the option of facing the enemy face-to-face in open-terrain conflict. Furthermore, insurgent leaders are likely to adopt the primitive model when they do not have enough military resources to build conventional armies they can feel comfortable with, to train soldiers in satisfactory manner, and when they have developed mistrust in using the conventional method in fighting more powerful adversaries.

The primitive model does not work for insurgents most of the time. Of course, through guerrilla activities insurgents can prolong the war, undermine enemy willpower, and sometimes win outright. Yet guerrilla war is a challenge for insurgents for three reasons. First, unlike in conventional war where warriors can capture portions of territory to defend, insurgents must trade territory for performance in order to maintain the strategic parity. This repositioning may prove to be highly unpopular and cost the insurgents their support bases. Keeping the masses on one's side requires a large amount of energy and resources. J. Boyer Bell cautions that the masses in guerrilla war are "mere mouths to feed, not assets but responsibilities" as they "have a reluctance to sacrifice for a distant grail, a distaste for a duty seldom properly understood, and they rarely live a life so intolerable that death is preferable."[35] Second, people may rescind support for insurgents when they fear government retaliation or appreciate government actions for them. On the one hand, because revenge is a powerful deterrent, most people naturally prefer to carry on their daily lives without threats on them. On the other hand, states may offer civilians a better alternative in the form of increased spending, infrastructure building, and job security to undermine sources of grievance. Popular support declines in long campaigns where people become tired and demand a quick end to war. Even if guerrilla war becomes longer, it does not always make them more likely to win. For these reasons, strategists have warned against the careless use of guerrilla war. As far back as in 1906, Lenin argued that "the party of the proletariat can never regard guerrilla warfare as the only, or even as the chief, method of struggle; it means that this method must be subordinated to other methods, that this method must be commensurate with the chief methods of warfare."[36] Mao found guerrillas in general to be so vulnerable that he did not endorse reliance on the guerrilla strategy alone and instead insisted on combining it with the use of base areas and regular armies.

Degenerative Model

The degenerative model is a sequence of actions that combine the conventional and primitive models. In the first phase, both sides engage in conventional war where they use regular armies to fight pitched battle. Most states have the advantage in this environment, so insurgents are likely to collapse at this point, as in the conventional model. The degenerative model departs from the conventional model, however, when either side determines that it can no longer sustain its operations but has just enough resources to shift to guerrilla war. Therefore, when nearing defeat in the conventional phase, insurgents use their remaining resources to disperse their forces into the jungle, mountains, and urban areas. State forces respond similarly if they expect to have sufficient popular support for COIN operations. As stated above, the choice of guerrilla strategy is conditional on the fact that insurgents have advantage in deciding the nature of combat and that states in turn believe that the conventional strategy will not be effective for guerrillas. Even though they survive the first phase, however, they are considerably weakened by the time they reach the second phase. After all, this model is referred to as degenerative not only because the war moves "backward," but also because conventional war in the first phase has the effect of undermining support bases and destroying resources that could be used in the second phase.

There are several conditions that must be met for insurgent groups to fight in the degenerative model, which include the knowledge of group leaders to fight using both conventional and unconventional methods, availability of weapons to fight in capital-intensive combat and willingness to fight protracted war, and the leaders' realization that there is a limit to fighting conventionally. Geographic features must also allow them to fight both in open terrain at one time and in jungles, urban areas, high mountains, and villages at another. The other key condition is that insurgent leaders believe in the superiority of conventional war in order to choose it as the initial method of fighting, only to change their mind later that they need to replace it with an inferior military strategy. Thus, the degenerative model operates on the assumption that the insurgent and state leaders have a degree of strategic and operational flexibility in the course of war.

The degenerative model has never worked for insurgents, for three reasons. First, neither the conventional war nor the guerrilla war phase is self-sustainable. Most insurgents are highly likely to be decimated in the conventional phase for the above-mentioned reasons. Even if they somehow reach the second phase, they are likely to lose, as in the primitive model. Second, the two phases are not mutually supportive when they are put into *this* sequence. The initial phase generates high-intensity combat that puts innocent civilians at risk, destroys villages, and kills the livestock. The burden of destruction in the first phase falls on the people who are forced to pay for war. Thus local support declines and guerrilla operations decrease. Even if the war moves to the second phase, insurgents suffer from the drainage of popular support, followed by the reduction of military capability. Finally, the second phase is unlikely to go well because the population draws negative impressions from the process. Collapse of conventional strategy in the first phase sends them a signal that the war is not going well. They learn that an initial military effort has failed despite their effort and sacrifice, so now they are less likely to support guerrilla causes. States welcome this change of popular sentiment, which enhances their resolve to win and, naturally, their chances to dominate this phase. Therefore, the degenerative model does not work.

Premature Model

The premature model is made of a sequence of actions that expand the primitive model into the conventional model. It reflects a form of strategic evolution in the direction opposite to the degenerative model. In this model, both sides start a guerrilla war, but insurgents realize that the war is not decisive enough, so they crawl out of the jungle and mountains and use their remaining resources to organize and arm soldiers. By the time they reach the second phase, however, any army they organize is likely to be overwhelmed because the preceding guerrilla activities had drained insurgents' resources and resolve, boosted public frustration, and diminished popular support. The state side responds to the shift because it knows that, having the advantage in resources, training, and organization on the battlefield, doing so will increase their chance of victory. The war will remain irregular, however, if the state sees no value in deploying regular armies and is able to destroy the insurgents at that point. In short, the model characterizes the insurgent prematurity in the level of preparation for conventional war.

The premature model takes place when the transition occurs "prematurely." That is, insurgent groups end up fighting like this when they do not have resources or willingness to build a state even if they do have capability and willingness to fight both like an army and guerrilla forces in different time periods. Alternately, the premature model may occur when the state side succeeds in sabotaging insurgents' early efforts to carry out a counterrevolution, stopping their evolution short of becoming a mature political organization. Because of these potential dangers associated with the model, the insurgent groups need to be flexible and resourceful enough to fight in both conventional and unconventional manners. They also need to initially act in the belief that the guerrilla strategy suits the particular terrain they fight on and is militarily more effective than the other options, only to realize later that switching to a conventional war strategy makes more sense and only if the state side reciprocates in creating such a transition.

There are two extrasystemic cases where insurgents fought along this model, but none has been successful. By the time a war reaches its second phase, most insurgents normally have used up their resources to build a powerful army after protracted guerrilla combat, while the state side keeps its material advantage. There are few resources left in the insurgents' governing organizations to collect taxes and sustain the activities. Another reason is that the two component phases of the premature model are not compatible with each other when they are put into *this* sequence. The initial phase depicts the execution of low-intensity combat for a sustained period of time when the population gets involved in cross fire and makes sacrifices in the food and shelter. Even if insurgents survive the first phase, the war takes heavy tolls on them and destroys resources needed for the second phase. Thus insurgent bases become precipitously underresourced and weakened along with the reduction of operational ability and popular support.

Maoist Model

Unlike the previous four models, the Maoist model proceeds in three phases. In the first phase of state building, insurgents retreat into a safe area and build a united front. This phase entails efforts on both sides of war to "build the state" by establishing and strengthening political parties and puppet regimes, bureaucracy, education, and constitutional systems and unify the movement across these institutions. Successful countermeasures by the state side, such as securing the occupied areas and boosting puppet regimes, may make the insurgents too weakened at this stage to proceed to the next. In other words, insurgents attempt to build the institutions in order to establish a state entity with sovereign recognition in mind. In extrasystemic war, the base area functions as an important foundation for mobilization, empowerment, and deployment. In Guinea-Bissau, for instance, rebels built up the African Party for the Independence of Guinea and Cape Verde as well as various political systems as the organizing body to lead the insurgency against the Portuguese.

The institutions give insurgent groups adequate resources to engage in low-intensity conflict. These resources also allow them to move their battles from rural areas to towns, give access to the urban population, gain publicity, and enhance political appeal. Johnson writes, "The establishment of rural revolutionary base areas and the encirclement of the cities from the countryside is of outstanding and universal practical importance for the present revolutionary struggles of all the oppressed nations and peoples."[37] This phase has an important function of reinforcing the state-building effort. Success in building a state structure in turn allows insurgents to make a transition to the third phase of conventional war where they have gained a great deal of manpower and firepower. Military modernization allows them to carry out counteroffensive campaigns involving both mobile and positional forms of combat. Battle areas expand in size. As such, this sequence of actions has a semblance to people's war, in which insurgents "promote the mobilization and organization of peasants in lands subject to imperialist interference, leading to guerrilla warfare and finally to regular warfare against the forces of imperialism and their local allies."[38] It starts with the institutions playing a key role, followed by guerrilla war and conventional war to complete a sequence.

A rare sequence, the Maoist model works through the mechanism of mutual reinforcement between phases. The first phase of state building is sustained by the second phase of guerrilla war in that the institutions allow insurgents to wage guerrilla resistance. We see the consolidation of political power in insurgency, as institutions provide resources to help the population so as to elicit their support. In turn, popular support strengthens *back* the institutions as the institutions become embedded in the society through public discourse. This mutual reinforcement mechanism works between the second and third phases, too, in that insurgents are more likely to win conventional war if they gain advantage in military power over states. In turn, armed forces supplement the insurgency by offering firepower and protecting the vulnerable. This socialization allows the armed forces to become the principal instrument of destruction. As Mao said, "Without a people's army, the people have nothing."[39]

Specifically, the Maoist model generally occurs when both states and insurgent groups have access to adequate resources and willingness to fight in all three phases. There must be sufficient realization on the part of both state and insurgent leaders that they must focus their resources on building a state in the first phase of nation building. The political effort must correspond with insurgents' efforts to fight the war on the military front, starting with guerrilla war. Geographic conditions should support guerrilla movements, but simultaneously there must be geographic conditions that support armies to transform heavy weapons, organize, and fight in conventional manner in the final phase. To make it even more difficult, the insurgent leaders at least must have the recognition that they have to proceed in sequence and must put their war efforts into proper sequence.

Progressive Model

The progressive model proceeds in three phases. The first phase of guerrilla war sees both sides fighting primarily in search for popular support. Achievement of adequate support propels them into a second phase, where they strive to consolidate their political party, build bureaucracies, generate resources, distribute wealth across the society, and appeal for foreign recognition. The combination of guerrilla war and state building is the cornerstone of many independence movements in the postwar era. The two phases alone, however, are not sufficient for insurgencies to win; they must be reinforced by military power. Thus at this stage, they will gain control of arms, which will be a key determinant of battles as well as a key consolidator of the political process. By way of using foreign aid and material resources, insurgents become better armed, trained, and organized, while popular consensus emerges across the territory in the direction of independence. Furthermore, their army develops more efficient units, with chain of command and control improving between the field and central authority. State attempts to arrest this development by the use of brute force often fail as most of them find their forces stretched thin in enemy territory. The third phase characterizes the shift of military balance from the state to insurgents.

Specifically, the progressive model is likely to become the strategy of choice over other options when state and insurgent leaders possess both capability and willingness to fight in all three phases. Ideally, the terrain they fight on supports guerrilla operations as well as large movements, eventually allowing insurgent groups to develop capability, transport heavy weapons, organize, and fight in conventional manner. Yet there must also be sustained efforts by at least the insurgent side to build a range of political and economic institutions to support their war effort in such a way as to transform guerrilla operations into more conventional military missions. To make it even more difficult, the insurgent leaders at least must have the recognition that they have to proceed in sequence and must put the war efforts into proper sequences.

There are several challenges associated with insurgent efforts to adopt this sequence. To start with, it is difficult to mobilize the population because people have often been socialized into colonial control and tend not to have the courage to rise. That makes the population pool too small to draw on for mobilization and recruitment. On the other hand, state forces seek to arrest the growth of insurgency at this early stage, but it may be the hardest stage to detect it because the latter may deliberately take a low profile or simply fail to garner much attention. Even if they do get noticed, it is difficult to know which insurgents will develop into a significant threat that will justify a quick and definitive reaction by states. At the same time, however, states find it relatively easy to attack the insurgency because the group is small, inexperienced, geographically dispersed, and therefore vulnerable. Along the same lines, insurgents must undergo a dangerous path of extinction unless they have a growing population. This makes it ideal for the states to strike fast, hard, and early.[40] Again, failing that, states will find that the longer the war gets, the more difficult it becomes.

In the two-hundred-year history of extrasystemic war, there have been three instances of insurgents fighting along this model, and they have all completed the transition to win them. The progressive model works through the mechanism of mutual reinforcement. The insurgents must do well in the first phase to do as well in the second. For this transition to succeed, insurgents need an organization strong enough to help the population. On the other hand, strong popular support will help the insurgents to consolidate the political process, as the use of violence against enemies often has a unifying effect.[41] Thus, primary emphasis is placed on the government's collaboration with the population. Similarly, in the third phase of conventional war, the insurgents must use government and party resources to support the warfighting apparatus. History is replete with botched efforts toward independence because the leaders were powerless.

Strategic Evolution and the Likelihood of Victory

The main argument of this book is that the likelihood of insurgent victory is a function of insurgent ability and willingness to adapt and fight state adversaries in an evolutionary method. The relationship between the way insurgents evolve (or do not evolve) and victory is causal because the choice of sequencing models has a powerful impact on the insurgents' chances of achieving their political ends. The relationship between evolution and war outcomes is also empirical because the popularity of the models is closely associated with the likelihood of insurgent victory. Figure 4 and Table 4

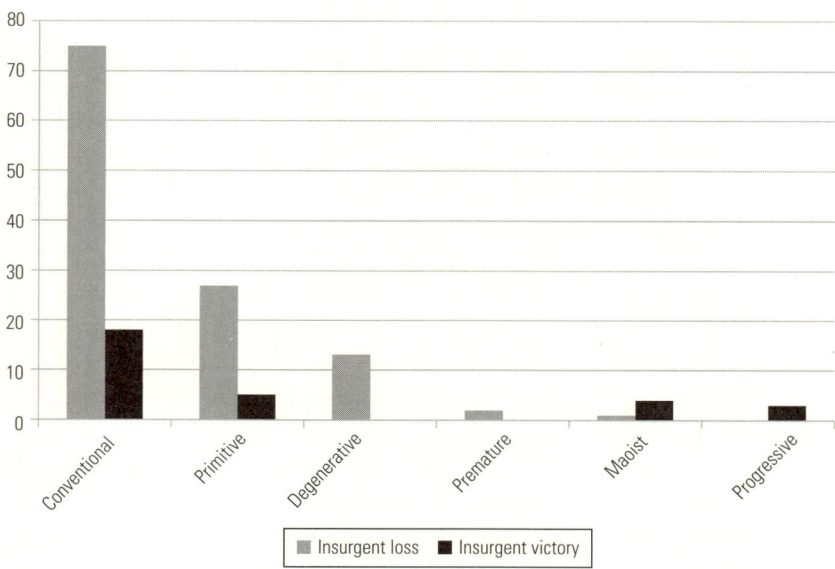

Figure 4. The likelihood of insurgent success.

Table 4. Chance of Insurgent Victory

Model	Conventional	Primitive	Degenerative	Premature	Maoist	Progressive
Total use	63% (93/148)	22% (32/148)	9% (13/148)	1% (2/148)	3% (5/148)	2% (3/148)
Frequency	High	Medium	Medium/low	Low	Low	Low
Insurgent victory	19% (18/93)	16% (5/32)	0% (0/13)	0% (0/2)	80% (4/5)	100% (3/3)
Likelihood of success	Unlikely	Unlikely	Unlikely	Unlikely	Likely	Very Likely
Examples	Dahomey, Zulu, Kaffir, First Afghan, Tunisian	Algeria, Hut Tax, Caco, Malayan Emergency	Iraq, Maori, Aceh, Saya Sen	Somalia, Congolese, Syrian, Italo-Libyan	Guinea-Bissau, Indonesian, Mau Mau	Indochina, Angola, Mozambique

indicate that while the conventional model is most popular, at 63 percent, followed by the primitive model at 22 percent, their probabilities of success are 19 percent and 16 percent, respectively, essentially closely associated with failure. The degenerative and premature models are less common, but they have not been successful even once. In contrast, the Maoist and progressive models are rare, with only 3 percent and 2 percent of all the samples, but they bolster insurgent performance the most. In other words, the conventional model is the most frequent but unlikely to work for insurgents, while the primitive and degenerative models are moderately frequent but unlikely to lead to success. It is also clear that the premature model is rare but unlikely to work, whereas the Maoist and progressive models are rare but very likely to work. Thus the relationship between frequency and success is inverse, which suggests that these last two models are associated with the recent increase in the probability of insurgent victory. In wars through the 1940s, insurgent groups tended to fight wars quite simplistically. When wars were simple, the insurgents were the losers. Since the 1940s, however, insurgent groups have evolved by adopting successful models.

These findings generate a set of key insights into our understanding of guerrilla war, conventional war, and state building. In most extrasystemic wars, guerrilla war alone does not work for insurgents. Insurgents have learned this only in recent decades. Instead, a guerrilla strategy works well as a *supplement* to conventional war and state-building phases. On the other hand, state building has become a necessary factor for insurgent victory. A successful state-building program will boost insurgents' ability to fight guerrilla war and conventional war, but it needs to be augmented by good performance in the phases as well. The dynamic relationship among popular support, military power, state building, and the likelihood of victory indicates that the more support insurgents have, the more institutions they build, and the more power they have, the more successful they are.[42] As insurgents evolve from a group of guerrilla insurgents into a modern state with an army, they have a greater chance to win.

CHAPTER 4

The Conventional Model: The Dahomean War (1890–1894)

In this chapter, I deploy the conventional model as a theoretical framework to examine the Dahomean War. The central proposition of this chapter is that war against a stronger foreign power is an impossible task if the insurgent side adopts a conventional model. According to this model, insurgent forces engage in open-terrain violence with their counterpart until one side is defeated. The conventional model is characterized by the direct use of force where military power plays a central role in the outcome. As such, rebel forces in Dahomey, located in what is now Benin in West Africa, mounted a brave challenge to the French army before they were ultimately crushed. The Dahomeans fielded an organized army that was twice as large as that of the French but still lost the war through a series of ground battles in two major campaigns. Dahomeans' initial advantage in manpower size allowed them to prolong the war by a couple of years and bring a draw to the first campaign, but their failure to evolve throughout the duration cost them the war. In this context, the Dahomean War serves as a good case study for this book. It illustrates some of the key problems associated with the conventional model, such as insurgent failure to adopt guerrilla strategy, build a state as a means of war, and increase military capability. This case study shows that the Dahomeans, relatively understudied in the field of international security, confronted a powerful French adversary, used modern weapons and strategy in orthodox combat, and were defeated. The case study proves my argument that war without adaptation is a suicidal endeavor even though the insurgent side has a numerical advantage in manpower.

This chapter proceeds in four sections. First, I trace the war from 1890 to 1894 to present a brief background. Next, I use the conventional model to explain the war outcome. In the third section, I connect Dahomean defeat to the lack of strategic evolution on the part of the insurgents, namely the absence of state-building efforts and guerrilla war. In the final section, I examine the existing theories of asymmetric war. While this case study alone does not fully refute these theories, it demonstrates why none of them adequately explains the cause of Dahomean defeat.

Background

The Dahomey war originated in the broad historical context of Western imperialism across nineteenth-century Africa. In the 1880s, rivalry between European powers grew at a time commonly known as the Scramble for Africa. Africa as a whole generated an image of abundant resources to be exploited, and the kingdom of Dahomey became an attractive site of its own for the lucrative slave business and palm oil trade, inducing Britain, France, and Germany to seek to conquer the territory. More important, it stood as an extension of French economic and strategic interest in colonizing West Africa. Among these imperial powers, France was particularly motivated to conquer Dahomey for both financial and strategic reasons. Famine in the 1880s had hurt French agriculture and generated a mercantilist movement known as the *pacte colonial*, which asserted that colonies would provide markets and raw materials and could become part of a greater French civilization.[1] France was strategically motivated, too, because resources that Dahomey offered could be used for war, because colonies provided bases of operation, and because they enhanced national reputation.[2] Penetration of the Dahomey hinterland would also allow France to alter the strategic landscape in West Africa in its favor. Dahomey would present a takeoff point for a move into the Niger Bend and an opportunity to navigate portions of the Niger River. Possession of Dahomey was also key to France's success in imperial competition, while loss of it would mean a relative decline of French power. The strategic benefit of colonialism, coupled with prestige and pride, was therefore an important part of French national interest. The operational drive centered on the so-called Chad Plan, which centered the idea of unifying all forces under French "assimilation" all the way from the Mediterranean to Chad.[3] France's adventure was led by the premiership of Jules Ferry, who argued that "for the time being, forget revenge and concentrate on the expansion of the empire."[4] As France began to prepare for invasion, agreements with Britain in the 1880s gave it a virtual free hand over Dahomey. French invasion was vehemently opposed by Dahomean King Glégle, who asserted the right to collect customs at Cotonou, a naval kingdom neighboring Dahomey, threatened a massive retaliation, and preemptively raided Porto Novo.

Glégle's death in 1890 led to the outbreak of the Dahomean War. Unlike most other wars, in this war it was the *weaker* Dahomean side that had numerical advantage in manpower. The Dahomean army in 1890 had around 8,000 troops, before the number doubled in two years, against the initial French contingency of 3,450 men,[5] composed mostly of Africans led by French officers. Dahomey had about 2,500 female soldiers who had physiques superior to those of men, fought ferociously, and therefore were well cherished. Dahomey's numerical advantage continued until the end of the war, but their overall military deficit led to their defeat. As this chapter demonstrates, much combat took place in the form of battles in two major campaigns between Dahomean and French armies rather than in small-scale guerrilla operations because the war was more conventional than unconventional. The possession of an army did not mean that Dahomey was a state. In fact, it would never develop a mature state structure beyond a loosely controlled kingdom and instead functioned more like a group of nonstate warriors confronting the powerful intervention of French troops invading their territory. Therefore, the Dahomean kingdom was never recognized as a member of the international system.

Conventional Model and the Dahomean War

The war began in 1890 when negotiations for peaceful resolution to French demands for territorial control over Dahomey broke down and several thousand Dahomeans charged into Cotonou to confront French forces in a firefight. With the French launching the so-called first campaign, the Battle of Cotonou left several hundred Dahomeans dead and forced survivors to withdraw, while the French sustained few casualties. Soon Dahomey regrouped and sent a detachment south for a rematch at Atchoupa. After receiving reinforcements from Porto Novo, the French ordered four hundred men to march north and intercept the Dahomeans. At the Battle of Atchoupa, the Dahomeans destroyed five hundred warriors, but French troops held their ground and formed a defensive position there. They launched more charges and pushed the French all the way back to Porto Novo before breaking off the attack and retreating without taking the city. Refraining from launching further attacks, Béhanzin signed a treaty recognizing Porto Novo as a French protectorate and ceded Cotonou in exchange for a large indemnity. Béhanzin kept his kingdom intact, managed to avoid colonization, and prepared for another campaign, believing that the treaty would not last long. The year of 1891 was peaceful, as Dahomey used the brief recess to revive the slave trade in an effort to buy weapons as part of its rearmament program. The temporary "draw" left the French forces so embarrassed, however, that they changed command and assigned a higher-ranking officer, General Alfred-Amedee Dodds, to the area, although they never resolved their numerical disadvantage. When the second campaign started in 1892, Dodds arrived with a force of over 2,000 legionnaires, marines, engineers, and Senegalese soldiers, while Porto Novo added some 2,600 porters. The Dahomean army was still a few times larger, totaling around 12,000 men and armed with modern carbines. Despite French reinforcements, therefore, it was the Dahomeans who kept the numerical advantage and had perhaps grown confident because they had forced a draw with a more powerful enemy.

Yet the Dahomeans proved to be no enemy of France. Their inferiority in military power forced them to break off quickly at Dogba. After the defeat at Dogba, Béhanzin himself took up arms and attacked French forces moving upriver to Poguessa, although the Dahomean charges were fruitless in the face of French bayonets. French victory of Poguessa was followed by another French victory at Adégon. At this point Dodds decided that his troops must make a decisive march on Abomey, the capital, to end the war.[6] Not surprisingly, the French overran the Dahomeans, now numbering just fifteen hundred men, and marched on to force Béhanzin to burn the capital and flee. Upon capturing Béhanzin in 1894, Dodds proclaimed victory. In less than seven weeks of fighting in the second campaign, the Dahomean army had lost between two and four thousand dead and between three and eight thousand wounded. French casualties were far fewer, with a few dozen dead and wounded.

The conventional model best captures the Dahomean War, which assumes that both sides' strategies converged in fighting face-to-face, open-terrain combat in a period of consistently conventional conflict. But the two sides had different motives to fight that way. On the one hand, the French settled on conventional war because it allowed them to use their power advantage. Doctrine, training, operations, and weapons procurement all suited a conventional strategy and were deeply embedded in the French military. Having inherited the Napoleonic tradition that relied on artillery, square formation, and rigid doctrine, national leaders resisted changes and only slowly integrated new battle methods. Instructors had taught this approach at military schools for decades, and this ideological conservatism lasted through the 1890s. Anywhere France fought, it embraced a high expectation for a decisive victory on the battlefield. Thus the French military institutions kept up a conventional doctrine, which built on movements inland from coastal areas and small maneuvers by mostly indigenous infantrymen and levies. The pattern was followed by the destruction of African polities by these columns often assisted by people who had been abused by those polities. Furthermore, Dodds emulated the strategy of British General Garnet Wolseley, who twenty years earlier had led a successful expedition into nearby Ashantiland. In British strategy, a small force with plenty of firepower would move quickly to destroy enemy forces and dictate terms of control and avoid the impact of weather, disease, and ambushes. To that end, Dodds concluded, the infantry must travel light and fast.[7]

On the Dahomean side it was a different story because the notion of fighting the more powerful French forces face-to-face was suicidal. As an underdog the Dahomeans had every reason to *avoid* direct confrontation and use the "weapon of the weak" by fighting like guerrillas. Fortune-tellers had advised Béhanzin against waging pitched battles in favor of stealthy ambushes and night raids.[8] The Dahomeans, however, remained committed to conventional battle because it had been institutionalized through their history of conflict with neighboring kingdoms and because of the general appeal of military modernization. Conventional doctrine and armies provided symbolic meaning as a mature polity and reasonable justification for the existence of Dahomey as an aspiring modern nation that could challenge an equivalent. By way of modernity, conventional war offered an illusion that Dahomey could fight on the same level as the strong. Furthermore, the plain topography of Dahomey favored infantry and cavalry operations rather than jungle and urban wars suited for guerrilla missions. Thus in 1891 Dahomey purchased from the Germans and Portuguese a variety of modern weapons, including flintlock muskets, blunderbusses, and other weapons like crossbows, arrows, spears, and machetes. At least one German soldier was constantly in Dahomey to train Dahomeans in the use of the new weapons, siege tactics, and physical fitness. Indeed, the Dahomean army had been known for its military potency. Archibald Dalzel, British governor of Benin in the 1760s, wrote that Dahomey maintained a considerable standing army where the king could gather his forces quickly, with officers armed like regular troops.[9] Dahomey was "the strongest indigenous military power on the west coast of Africa."[10] Béhanzin organized his forces according to the principle of "levee en masse" to recruit all combat-ready adults. He put a commander called Gau in charge of planning military strategy, logistics, deployment, and command and control.[11] Béhanzin's campaigns thus became standardized. Between one and three campaigns normally took place in the dry season. Mobilization calls would come by drum, forcing all villages to respond or face collective punishment. Force sizes ranged between twenty and two hundred men; all the rest were reserves who were nevertheless well trained and armed with their own weapons.[12]

Conventional war generally leads to insurgent failure after a few battles, and Dahomey was no exception. When the war started, it quickly revealed five major flaws with the Dahomean army, starting with the shortage of resources. The first campaign broke out during the planting season in Dahomey, interfering with the normal production of crops because the military commandeered the needed laborers. The series of French attacks not only destroyed the soil but also prevented farmers from cultivating it. The war also instigated a revolt among the neighboring Yoruba slaves, making it difficult for the Dahomeans to collect food near their areas. Therefore Dahomey's wartime harvest declined below the average and forced soldiers to prepare their own food and raid neighbors for slaves and capital. Soldiers seized provisions while leaving little behind for the farmers.[13] The slave trade, while bringing in cash, took away laborers available for mobilization and undermined sources of recruitment for war. Furthermore, because previous wars had been shorter in duration, preparations for this war were more pervasive. Food importation from neighbors became difficult due to the wars, the slave trade, and Béhanzin's hostility toward them. To bring more land into cultivation in the hope of increasing productivity necessitated more labor. Finally, harsh taxation and mobilization calls proved to be such heavy burdens that villagers gradually resisted calls to contribute to the kingdom. As a result, Dahomey had to turn to women soldiers and the slave trade that gave preference to and therefore reduced the number of able-bodied men.

The second problem was that French firepower effectively offset Dahomey's numerical superiority. French weapons, such as the Maxim gun, which fired much faster and at longer range than Dahomey's blunderbusses and muzzle-loading flintlocks, decimated repeated Dahomean charges before the warriors could get within musket range. The French mixed this technological advantage with their maneuvers to generate maximum combined effects, overcome the burden of carrying heavy weapons, and move quickly against Dahomean efforts to cut them off. There were reports of some successful Dahomean ground combat,[14] but these instances were strategically isolated. In the Battle of Poguessa, French bayonets proved to be highly effective. French rifles with fixed bayonets outreached Dahomeans' machetes.[15] These technological advances allowed a smaller number of soldiers to transport a high volume of combat power during expeditions. All this enabled a small number of French troops to defeat a quantitatively larger army.[16]

The third problem with Dahomeans fighting conventionally rested with the fact that they viewed war generally as a social, rather than military, enterprise. In peacetime, they devoted themselves more to court ceremony than to military training. While Dahomey's military organization kept the guise of a modern army, the division of labor served more ceremonial purposes. The army was divided into the right and left wings because on ceremonial occasions it formed two symmetrical sections placed to the right and left of the king. Such formation was justified to reflect one of the important dualities of the kingship. Since war had a major social value in itself, the army was made congruent with the organization of the kingdom. Thus while the French army was disciplined to fight, the Dahomean army was used for social purposes. This widespread social ideology had conceptualized war rather as a form of literal manhunt aimed more at capturing individuals than killing them and occupying territory. It also encouraged the Dahomeans to use prisoners of war for the slave trade and to buy guns to ensure a supply of human sacrifices. Not surprisingly, the Dahomean practice of surprise and night raids included surrounding a town in the darkness and forcing entry, not necessarily to kill but to capture as many people as possible. As a result, if their attacks did not succeed, or if they themselves were taken by surprise, the Dahomean army fell into confusion quickly.[17]

Fourth, differences in French and Dahomean battle formations posed problems for Dahomey. On the one hand, the French army used a square formation, which allowed them to protect vulnerable sections and prevent other units from being cut off. This formation worked quite well when moving through unknown country with unreliable maps. The army organized sentries so well that it fended off the Dahomean army's persistent surprise night attacks. It also concentrated firepower and directed it so well that few Dahomeans got anywhere near the front lines. In addition, the army made certain to protect French supply routes by building defensive staging posts, which neutralized Dahomean advantages in maneuver. In contrast, Dahomey's problems with the French were exacerbated by their adoption of an arc formation with the most important chief on the right and lesser chiefs on the left. These two divisions were composed of several units led by village chiefs. Under this structure, the war chiefs of local groups brought their kinsmen and slaves into the field. Yet this recruitment method did not work well because responsibility to the army for war rested with the individuals, and as a result they were often poorly armed.[18] As Ross shows, French bayonet charges drove away the Dahomeans from any defensive position and their stands at natural and man-made hazards hardly worked.[19] Last, Dahomey's weapons were not immediately effective; the Dahomeans needed time to get accustomed to the weapons they bought after the first campaign.[20] While France did not suffer this problem, Dahomey did as an importer of weapons. For a war of just four years, an interval between arms purchase and adoption proved to be too long. The military "generation" of the first campaign accustomed to the old weapons had to spend time and receive training to make proper adjustments before they could use new ones. If a considerably different weapon was adopted, it was inherently incongruous with existing weapons and doctrine. The "assimilation" problem made the Dahomean rebels hesitant to coordinate the new weapons with earlier ones and difficult to use them effectively. The result of this problem was widespread confusion within the armed force that was difficult to solve in a short period. Thus, when the second campaign opened, the Dahomean army found itself quickly on retreat.

There was no surprise that the war went awry for Dahomey, but French victory did not come easily, for three reasons. The first problem was friction between the French civil service and colonial forces who disagreed over the use of armed forces as a means to achieve national objectives. Commanders on the ground refused to obey orders from civil officers, such as Eugene Etienne, the powerful Undersecretary of State for Colonies, who wanted a strong imperialist approach. The colonial lobby had forced parliament in 1892 to establish the Ministry of Commerce, Industry, and Colonies, but it could not force colonial troops to follow its orders as they were responsible only to the navy. Etienne nevertheless demanded civilian control of the military to assert commercial interests.[21] The colonial forces also believed that the regime was skeptical of them and inclined to disapprove ambitious overseas operations. As Douglas Porch shows, these operations were often carried out in haste, without adequate support or intelligence, and largely for political reasons. The civil-military friction was exacerbated by the lack of colonial doctrine. Fighting for almost a century outside France, the French confronted opponents with different levels of organization and capability. Some were well organized and armed, like the Dahomean army, but the French also fought guerrillas elsewhere. The lack of uniform doctrine generated incentives for the army to seek solutions that spoke more to their interest in Europe. Indeed, the army remained keen to keep European military practices out of the colonies, rather than letting colonial experiences influence the European way of warfare. The persistent use of heavy columns and reliance on machines meant the European doctrinal encroachment of African wars, which proved to be ineffective. The system was also inefficient because "too much precious manpower was expended in convoy duty and in guarding posts on the line of march," making the so-called tail-to-teeth ratio "excessive."[22] Thus, the French way of fighting did not conform to the economical use of limited resources.

The second problem was domestic politics in France. Much disagreement among parliamentarians with regard to the legitimacy and conduct of war severely undermined French operations. At the time of the war, imperial expansion remained controversial, and French politicians reacted according to their political motives. If a colonial mission turned out badly, parliament heavily punished the leaders who promoted it. For this reason, French officials continued to be sensitive to the political climate in Paris. Views of one colonial minister differed radically with those of his predecessor. Largely for this reason French colonial policy remained inconsistent during this period. Coupled with civil-military discord, this meant that France had no long-term colonial policy to dictate through the chain of command. The final problem was conscription in France, which proved to be necessary for these wars but apparently unpopular in the periphery. Unpopularity meant low combat morale among the officers. Hubert Lyautey referred to officers "who know their horses better than their men," and a general staff looking for opportunities to avoid service.[23] By the time France began to accept volunteers in 1893, their numbers were far below the requirement for colonial service.

This experience prompted the French authorities to reconfigure each colonial unit and increase the use of native forces. These native forces were useful because they brought in local knowledge, enhanced French "divide and rule" policy, presented a scapegoat for French brutality, and reduced costs.[24] They also meant that France could provide a concrete solution to the recruitment problem, let alone meeting actual combat demands, only when a large number of pacified locals supported them. As Raymond Betts shows, however, it was apparent that this solution was insufficient to keep many other colonial possessions safe for French interest, which therefore required the country to overhaul the mobilization system, starting with the need to address the recruitment shortage.[25] The colonial army's efficacy was thus undermined.

In sum, the Dahomean army was able to challenge the French in the first campaign and sign a cease-fire treaty on an almost equal basis, arresting the French expectation of decisive victory. Dahomean military power, however, was not sufficient to defeat the French. A year of cease-fire was followed by the outbreak of the second campaign, where military parity was restored briefly but quickly overturned by the French. What exactly was missing for Dahomey? In the next section, I deploy the conventional model of sequencing theory and discuss two necessary factors—state building and guerrilla war—that turned out to be missing for a successful sequence. The key problem for Dahomey, of course, was the adoption of the model itself as it confronted the French forces.

Necessary Conditions for Sequencing Theory

The Dahomean War is a good example of the conventional model because it clearly shows that neither side made serious efforts in state building or fighting guerrilla warfare. All visible state-building efforts failed utterly in Dahomey, generally identified as one of the major problems facing a number of weak states. Executed properly, this process would have involved the establishment of institutions necessary for the Dahomey kingdom to function like a state, including law, administration, public health, education, and utilities. Throughout much of history, African nations have "lacked the capability to formally control large amounts of land beyond the center of the polity because they could not project power in other ways."[26] Governments did not worry much about what local territories did as long as they received tributes, and there were not always imminent threats to the survival of kingdoms. What would later follow as nation-states in Africa were characterized not just by borders and citizens with national identities but also by bureaucracies and representative systems. Once in place, these systems allowed national leaders to use resources to provide support for the population and maintain territorial integrity. However, because of the mix of underpopulation, vast territory, varied environmental and geographical conditions, it was always difficult for political administrations in Africa to be built and then to exert control over a number of people.[27] Dahomey was no exception. Neither side actively invested in building institutions in Dahomey as part of war efforts, although it turned out to be the French who were more or less successful in undermining the Dahomean regime. The French were in Dahomey primarily to exploit resources and fend off competitors, so they made sure that the Dahomean regime would remain weak relative to their control. The aim of French operations, after all, was not to build a strong colonial government but to place Dahomey under rule. Remaining instability in Dahomey, in particular northeastern districts, meant the difficulty of controlling some of the outlying regions, forcing French officers to resort to harsh military rule. Dahomey's military problems stemmed from the lack of centralized institutions. A centralized regime would have provided public goods in society.

Dahomey was a kingdom, but it was not recognized as an international nation-state. Dahomey's institutions included a loosely structured hierarchical system, with the *Migan* at the top commanding the army's right wing and assuming key roles as prime minister and supreme judge and in charge of all Dahomeans outside the royal family. Under the *Migan* was the *Meu*, who designed budgets and supervised ceremonies. Both *Migan* and *Meu* had veto power over foreign treaties.[28] The royal court was a function largely of the administration of the kingdom rather than acting as an independent entity running state affairs. Thus there was little in the way of an independent judiciary system to exercise power to check and balance the actions of the royal family, which had grown in number to twelve thousand, as well as its war conduct.[29] A number of chiefs called the *Togan*, which literally meant "chief of the land," collected taxes, recruited men for the military, supervised agricultural projects, among other things.[30] Institutions inherited from the past were never replaced by any new administrative structure for the purpose of war. Beyond the existing institutions, however, it remained a decentralized regime. Repeated wars in neighboring areas and ransacking for the slave trade had partly alienated the regions. Existing research on political development in Africa points to the population density as a causal factor for the absence of institutions in Africa.[31] Like those of most of its neighbors in precolonial Africa, Dahomey's political structure was kept extremely loose relative to its regions.[32] Thus as seen in the secession of Porto Novo and Cotonou before the war broke out, breakaway movements were common in what John Hargreaves calls the "African partition of Africa."[33] The remaining institutions in the kingdom, such as the system of warlordism, were a structural hindrance to central control. Exercising little control over provinces and military operations, no centralized institution was built anew for the purpose of war. Although Béhanzin remained the most important figure in the kingdom, he was also *among* a number of leaders, and he was not as powerful as his predecessors had been.[34] With as many as 120 different tribes in Dahomey, provincial leaders proved to be too strong for the center to hold together diffuse interests.[35] Provincial institutions were hardly united at the national level and generated little collective force to defeat France. There was no centrally controlled political hierarchy. While Dahomey was tyrannical, it was the least limited and regulated monarchy.[36] The central administration was largely powerless in strategic and operational decision making. Moreover, Dahomey failed to wage the war with close support from the citizenry through the execution of guerrilla warfare. Resources accumulated at the center of kingdom should have been distributed among the populace to guarantee the steady supply of recruitment, as Robert Bates argues "those who held positions of privilege had to insure that the benefits created by the states were widely shared" because otherwise they would be left without popular support.[37] In sum, inattention to political and social dimensions was a critical part of Dahomean failure.

The other missing factor was guerrilla warfare. Neither France nor Dahomey bothered to train forces for untraditional missions; a brief guerrilla skirmish reportedly initiated by Dahomey in 1892 invoked no French reciprocation and disappeared before long.[38] Thus there was little effort made to address popular support on the part of Dahomey. Small-scale mobilization did take place, but it was meant not for war but to meet the need of recruitment or ceremonial services. Dahomey remained a conventional force for three reasons. First, it did not have the human resources necessary for guerrilla war. Slave trade in coastal areas precluded efforts to win over local hearts and minds. Conventional war also necessitated a concentration of manpower in the army. Second, guerrilla war may have been seen as insecure because it meant voluntarily ceding territory to the enemy. Loss of territory would then weaken Dahomey conventional army operations. Béhanzin preferred to have his army stay conventional in order to protect his own land rather than letting the French exploit it. Finally, Dahomey was a decentralized society, and Béhanzin was probably unsure if he would be able to draw on strong popular support during guerrilla warfare without inadvertently increasing the power of regional leaders relative to his and endangering the chance of keeping the civilians on his side.

To be fair, one should not expect Béhanzin to have known much about guerrilla warfare. There is no evidence of his knowledge about it. Although France had fought guerrilla wars in Spain earlier, Dahomey had no recorded experiments with guerrilla war to date. There was no widespread use of communication devices in Dahomey that would tell the insurgents how to fight like guerrillas. No individual leader stood up to adopt a guerrilla strategy, either. So Dahomey had no prior history of or exposure to irregular conflict. There is little evidence that, despite the fortune-tellers' advice to the contrary, Dahomey recognized the importance of winning civilian support. Few visible efforts were made to accommodate different interests of 120 tribes residing in the territory. The Dahomean army captured slaves and sold them so as to buy arms. Thus the army's reserve in manpower was depleted to meet immediate financial goals. Furthermore, the climate made conditions for local mobilization difficult in the savanna. The double rainy season left two dry seasons that were too long and too dry to support a dense forest with a high canopy. Thus weather and vegetation severely constrained the Dahomean army's effort to secure stable bases of mobilization.[39] Because Dahomey's population in 1890 was estimated at 150,000, from which an army of 12,000 men was drawn, only about 15 percent of the population was in the army.[40] The rest, about 125,000 people, were at best partially and temporarily mobilized.

Other Explanations

The literature of asymmetric war provides a host of explanations for how the weak beat the strong. While other works do not address the Dahomean War per se, they offer basic ideas we can draw for the war. I select here the theories that pose the greatest challenge to sequencing theory. First, the balance of resolve theory posits that in war between unequal powers, the weaker side is likely to win when it is more determined to withstand the cost of war.[41] From this perspective, the Dahomeans lost the war because they were not as determined as the French. In other words, they would have *won* the war if they had had greater determination to fight than the French had. This theory is plausible because French victory was partly due to Paris's motivation, both financial and strategic, to conquer Dahomey and the powerful colonial lobby at home. However, there is good reason to believe that Dahomey had stronger resolve at several points in the conflict. Dahomean soldiers were highly motivated because they were trained by the chiefs bringing in their family and followers to serve in the army. Most soldiers in combat units were trained and disciplined by dignitaries and selected to fight for the survival of the kingdom.[42] The draw after the first campaign also bolstered Dahomean resolve because it gave them confidence for the second campaign. This coincided with the possibility that resolve among the French troops fluctuated over time. Two issues challenged their imperial drive. One was the difficulty to govern the native population in a faraway place in ways that would boost French colonial interest, especially in terms of cost and benefit of keeping other parts of its empire, such as Algeria, Tunisia, Sahara, and Morocco. Beyond North Africa, France would soon extend control over Western and Central Africa by the turn of the century, but growing military expenses called for strategic restraint. Over two dozen newspapers came to note the ephemeral nature of the conflict. The Colonial Department in France became risk-averse and began to recommend that troops avoid involvement in deep conflict. Risk aversion was closely associated with the costly adventure in Dahomey. The financial crisis of 1890 had shaken European economies and frightened most businessmen. France's colonial party was dominated by left-to-center republicans who represented industries with vested interests in an open slave trade and against the economic disruption brought on by military conquest.[43] Imperial skepticism deepened when Alexandre Ribot, minister of foreign affairs, disapproved of the war and called for a quick end to it.[44] Because cautionary attitudes began to spread among lawmakers, when they transferred the war command to the navy, they did so with a condition to obtain a negotiated settlement with Béhanzin, rather than a decisive victory.[45] In 1893, the parliament rejected a merchant marine law designed to promote government subventions to merchant ships trading along West African coasts.[46] Growing public dissatisfaction with conscription also forced France to adopt a volunteer law the same year. French resolve deteriorated further when doctors presented the public a series of negative assessment about the hygiene conditions in Dahomey.[47] In addition, public dissatisfaction with the war grew in direct proportion to the vigor and despair of African resistance.[48] Despite this, the Dahomey did not defeat the French in the early phase of the war. Of course, the fact that the theory is weak in Dahomey's case does not refute the theory completely. Indeed, I argue that the theory provides a useful analytical framework. It explains that the French forces suffered a draw in the first campaign in part because of low resolve and domestic problems.

Second, the theory of strategic interaction posits that the weaker side is likely to win the war if it adopts military strategy that is different from that of the strong. According to this theory, the Dahomeans lost the war because they used the same conventional military strategy as the French. To put it another way, the Dahomeans would have *won* the war had it adopted conventional strategy *and* the French used guerrilla strategy, or had they adopted the guerrilla strategy *and* had the French used conventional strategy.[49] On the surface, this theory appears valid because the Dahomean War was conventional from the beginning to the end. While the theory posits that the stronger side wins in a conventional matchup because it simply overwhelms the weaker side, it does not explain why the first campaign ended in a draw. In reality, French military power was so indecisive that Paris had to sign a cease-fire, keep Dahomey outside the French dominion, and fight again. Yet it is also true that the theory does explain the outcome, which is also consistent with the expectation of the sequencing theory. Indeed, I argue that the theory of strategic interaction fits well as part of the framework of sequencing theory, in that the former provides a clear explanation for how and why underdogs in conventional settings tend to lose the war in a manner consistent with the conventional model of the sequencing theory.

Finally, the theory of democratic weakness posits that in an asymmetric, "small" war involving a democracy, the democratic side is likely to lose if it suffers from the rise of middle-class opposition that constrains its military policy and prevents it from raising the level of violence needed to defeat the insurgency. From this perspective, the French *won* the Dahomean War because they were able to fend off domestic antiwar forces. Yet the theory's validity is questionable because it is possible that France would have won without raising the level of violence necessary to win it. The level of French violence in the war did not precipitously rise relative to that of Dahomey, but France still won the war. Instead, relative military effectiveness might have *declined* for France because of Dahomey's interwar rearmament, Dahomey's expectation of better performance, and a decline in the French commitment to the war. In other words, French victory came without the necessary variable that this theory generates.

CHAPTER 5

The Primitive Model: Malayan Emergency (1948–1960)

In this chapter, I apply the primitive model of sequencing theory to show that insurgent rebels are likely to lose war if they fight solely using a guerrilla strategy. Between 1948 and 1960, insurgents belonging to the Malayan Communist Party (MCP) fought British and local forces for the independence of a communist state in Malaya in a protracted guerrilla war until they were defeated. There were many factors that led to their defeat, including balance of power, minority status, and lack of external support, but here I demonstrate that the main reason is that they failed to move the fight beyond guerrilla warfare. Although they managed to stretch the war for over a decade by capitalizing on Malaysia's vast jungle geography and evasive maneuvers, the MCP ultimately failed to move out of a guerrilla war phase and gain popular support to sustain its operations. The lack of strategic adaptation cost MCP victory. Much research has already been done on the Malayan Emergency, but it is a good case study for this book for two reasons. First, it is one of the best exemplary cases from which to draw "lessons" in modern COIN studies. It has garnered great scholarly attention in recent years since the beginning of COIN missions in Afghanistan and Iraq. Scholars have looked back to Malaya to show how British forces succeeded in reducing communist guerrillas to a shambles and what U.S. forces in other areas can learn from that experience.[1] The other reason is that the case study dispels the myth that guerrilla forces are invincible in modern conflict. While the likelihood that guerrillas will defeat states has increased in the post–World War II era, I show that not all insurgent forces are successful. In fact, the Malaya case demonstrates what can go wrong on the part of the insurgents who would otherwise have a good chance of protracting the war and why they still fail in their endeavor. Thus, the case study answers an intriguing puzzle as to what the MCP did wrong and what that means for the future of extrasystemic war.

The chapter consists of five sections. In the next section, I show how the war began, proceeded, and ended. Then, I apply the primitive model to explain the war outcome. I show that both the British and communist sides failed to fathom the nature of guerrilla conflict for the first few years of war, resulting in a protracted conflict. But the British "learning" to fight COIN led to their success in gaining local support, which proved critical in the later years of the emergency. On the other hand, increasingly cognizant of its own errors, the MCP failed to win the hearts and minds of the people and develop institutions and military power in a sequential fashion. Without extensive local support, the MCP gradually faded into obscurity. In the final section, I show that three theories of asymmetric war do not adequately explain the war.

Background

The emergency erupted in the aftermath of World War II. The departure of the Japanese occupation reignited local desires for autonomy and rekindled a largely ethnic Chinese movement to establish a state independent of the returning British forces. The emergency was asymmetric in that Britain had clear advantages in military power as one of the remaining colonial states in Southeast Asia. Britain's armed forces ranked among the most capable in the world and had deep knowledge of Malay cultures and society through colonial control. After all, British presence in the country for over a century had enabled them to build long-term relationships and cultural awareness that proved crucial for winning over popular support. The result was the relative ease with which the British forces drew attention and support from the population.[2] Britain had numerous active and reserve forces in Malaya, whose numbers increased once Japanese troops departed the region in 1945, centered on the Southeast Asian command and joined by Commonwealth forces from Australia, New Zealand, Nepal, Kenya, Fiji, and Rhodesia. In Malaya Britain had thirteen battalions, composed of four infantry, seven Gurkha infantry, and two Malay regiment battalions, with each battalion having about seven hundred men, and one British field regiment, one Squadron Royal Air Force regiment, and some four thousand Gurkhas.[3] Soon London began to invest in manpower and firepower. In 1951, forces grew to eighteen battalions, composed of seven infantry, eight Gurkha, and three colonial battalions, in addition to two Royal Armored Corps Regiments, four battalions in the Malay Regiment, Malayan Police Forces, Malayan Scouts of four squadrons, ten Royal Air Force squadrons with 114 aircraft, two Australian Air Force Squadrons with 14 aircraft, one frigate, six minesweepers, and two motor launchers.[4] In 1952, Britain added its nuclear arsenal.

On the other hand, the MCP had fewer warriors and weapons and less ammunition. Originally formed in the early 1940s to counter imperial Japan's invasion, the MCP consisted mostly of ethnic Chinese who turned against British colonialism. It relied on a loose network of informal Chinese supporters known as Min Yuen, organized secretly cell by cell and located in squatter villages to offer food, medicine, information, recruits, and money. It was not composed of MCP members but linked to them through lower branch committees.[5] The MCP sought to establish an independent communist state in Malaya by waging guerrilla war and sabotaging key industries to reduce British profits and interest in the war. It relied on a small portion of Chinese immigrant groups who constituted 38 percent of the country's population. The Chinese were underrepresented in the colonial administration, and about half a million of them lived outside Malay society. In addition, despite the Chinese presence, China never supported the MCP during the conflict. With scarce weapons left by the Japanese, the MCP armed 12,000 men and women. MCP head Chin Peng organized his fighters into eight regiments of some 3,000 men in the jungle and another 7,000 in the Self-Protection Corps outside the jungle. The asymmetry in power became more pronounced as British deployments grew. By 1950, about 2,840 of the 12,000 guerrillas had been killed in the field, had been captured, or had surrendered, and an estimated 540 had died of wounds or disease. A Malayan estimate put the guerrillas' strength in 1951 at about 8,000.[6] In 1952 a British estimate set the number of active, organized, and working members of the guerrilla logistical structure at about 11,000, of whom 3,500 to 4,000 were armed.[7] Thus By the mid-1950s the military aspects of the war were largely over. By 1953 Britain had 40,000 troops, 45,000 regular and special police, and 350,000 Home Guards, summing to a total of 435,000 troops. In that year, the ratio of full-time British forces to active guerrilla irregulars was seventeen to one. Comparing the overall force sizes of 435,000 to 14.000, this ratio increases to an impressive thirty-one to one.[8]

The asymmetry in power shaped the choice of guerrilla war. The MCP chose guerrilla war because they realized that they would not match the British in direct confrontation. The MCP believed that the campaign of terror would force the Westerners to flee and complain about the lack of safety. The MCP also took advantage of topographic features that favored hit-and-run operations around jungle bases, which lessened the impact of resource limitations, an arms embargo, and the lack of Chinese support. Britain faced pressure from European communities in Malaya to act against a growing insurgency and the prospect of long-term destruction. It had little choice but to take a strong position to eradicate the communists rapidly and reinstall order. Thus British forces acted forcefully against the MCP by responding often aggressively to innocent supporters of the insurgents.

The outcome of the war is clear. Britain achieved its objective of defeating the MCP and restoring the political stability in the government through Malaysia's 1957 independence. The MCP was shattered, with two-thirds of its members eliminated. In 1960, as the MCP leader Chin Peng fled to China, the Malaysian government declared the emergency over. The MCP failed to achieve its goal of communist independence. Yet scholars have yet to agree on the cause of the outcome. Some believe Britain won by preventing the MCP from dominating Malay politics and dominating the region with the United States. This view contradicts that of those who argue that victory was won through British mastery of small-unit operations and separation of civilian populations into safe zones. Robert Thompson, for instance, raised six COIN "principles," including the recognition of the need for political action, civil-military cooperation, coordination of intelligence, separation of the insurgent from the population through the winning of hearts and minds, controlled use of military force, and implementing lasting political reform.[9] For Deborah Avant, Britain was successful in Malaya because it had united political institutions in which the army remained sensitive to civilian leaders who rewarded its adaptive behavior.[10] Still others argue that British victory had more to do with the MCP's mistakes to adopt measures to coerce villagers into cooperation. Civilian terrorism had the negative effect of reducing morale among the combatants and populations alike and consequently encouraging defection. Such measures were coupled with the MCP's offering of diplomatic concessions to the Malays that they might surrender in exchange for greater political representation in post-emergency Malay politics.[11]

The Primitive Model and the Emergency

The primitive model shows that the MCP lost the war because it failed to evolve beyond a guerrilla war phase. Some contend that the MCP lacked innovation beyond guerrilla strategy and Chin Peng did not learn from history and assess their performance to make adjustments in the middle of the war. Evidence shows, however, that the MCP actually sought to fight the war in a three-phase sequence that he borrowed from Mao. In the first phase, the MCP would gain control of target areas through a mix of terror and persuasion. In the second phase, however, it would coerce, rather than befriend, the inhabitants of these areas into joining military units. In the final phase, these units, now trained and armed, would move out of the occupied areas, which would serve as their bases. Ultimately, this plan would culminate with a union of operational units forming an army to retake the country. The MCP sought to set up four such base areas in Malaya and one across the Thai border, although such a plan was prevented in advance by the British forces.

Sequential analysis was not limited to the MCP side. Huw Bennett analyzes three different periods in British activities in Malaya. From 1948 to 1950, British security forces deliberately coerced the Chinese population into supporting the government with threats of mass arrests and property destruction. From 1950 to 1952, Britain forcibly placed the population into resettlement projects and applied punitive measures to restrain them. From 1952 to 1960, however, Britain adopted the so-called hearts-and-minds strategy to win over the masses.[12] The key factor for the change in British COIN missions rested with the strategic innovation that London engineered through the appointment of leaders such as Gerald Templer. The introduction of policies designed to provide for the population played a central role in turning the political tide in favor of British forces.

Britain's initial inclination to punish the population was influenced by a combination of bad planning, lack of organization, poor leadership, and shortage of analysis.[13] In this organizational vacuum, undisciplined soldiers occasionally killed civilians and destroyed their property. These problems were exacerbated by policy makers involved in this problem. Bennett argues that the British army aggressively intimidated the Chinese into supporting it until 1949 and even when the consequences became clear.[14] British action was driven by a belief that the war would be overcome by a less conventional solution that placed the main emphasis on the economic and political factors that sustained the will of the insurgents.[15] This belief reflected British military performance. Operation Warbler, for instance, involved eight infantry battalions and four police jungle companies, with a task of destroying the MCP in Johore, a southern province in Malaya. This was where the bandits sought to strike back at enemy patrols and "show the flag" to the civilians, following their considerable success against the patrols in North Johore. General Bruce Lochhart concluded that "no major result can be expected inside six months." It was strategically meaningless to send large units into the jungle looking for guerrillas who could easily evade them. Captured MCP documents indicate that jungle sweeps did not worry the insurgents; they feared only surprise raids and ambushes.[16] These operations would not be strategically effective unless they were employed in combination with other operations and in ways that led to accumulated victories over a long time. At the same time, British commanders had little grasp of the political nature of the war.[17] Furthermore, communist victory in China in 1949 boosted MCP morale and caused the Chinese in Malaya to believe that British control would be over soon. Therefore, the first few years left the British forces frustrated and tied down to the absence of strategic breakthrough. Moreover, they were made dependent on police intelligence that was nevertheless inadequate.[18] Richard Clutterbuck wrote that "there was a growing danger that the police and the civilian population would lose confidence in the government and conclude that the guerrillas in the end must win. . . . The guerrillas could get all the support they needed—food, clothing, information, and recruits—from the squatters. It was quite impossible to police and protect them. . . . Thus, the communists were fast building up their strength and their support, and at the same time, stocking up arms and ammunition by raiding or corrupting the village police posts."[19]

The MCP took advantage of these British mistakes to continue to sabotage mine equipment, slash rubber trees, and murder planters. The MCP implemented the "two-base/liberated area zones program" to conduct hit-and-run operations between two bases. It also gradually increased the unit size in each mission from a platoon of ten to twenty troops to a company size of over one hundred in hopes of increasing a kill rate during raids and shooting uncooperative civilians and hostile forces.[20] Guerrilla incidents, which had dropped from over two hundred monthly in 1948 to fewer than one hundred in 1949, increased to over four hundred by the middle of 1950.[21] The combination of jungle war, indiscriminate targeting of the population, and lack of strategic innovation put London on its heels in the first years of the emergency.

Changes to British strategy came with General Harold Briggs, who arrived in 1950 with plans to use a number of small patrols and ambushes so as to prevent the guerrillas from gathering power for large-scale operations and to cut off their food supply. He saw winning popular support as more important than forcibly defeating the insurgents and ensured that armed operations would come after the police and that they would not only operate in conjunction with the police but also cover populated areas. He also ordered forces to protect populated areas, emphasized the value of human intelligence, and strengthened joint police and military operations.[22] For the first time since the beginning of the emergency Britain began to take initiative in the art of fighting guerrillas.[23]

Briggs's reorganization and resettlement program was followed by the arrival of General Gerald Templer, who lifted the importance of the population in guerrilla war, a point echoed by other leaders of the British administration.[24] War Secretary John Strachey stated that the struggle centered on the support of Chinese squatters, and called for avoiding punitive and indiscriminate measures in favor of protecting the population.[25] Colonial Secretary Oliver Lyttelton added that one could not win this sort of war without the help of the population.[26] Further recognition of the centrality of popular support was reflected by the British publications on COIN. Among them was the "Conduct of Anti-Terrorist Operations in Malaya," which discussed a range of issues including enemy organizations, combat regulations, search methods, platoon formation, equipment, patrolling and ambushing, intelligence, and training.[27] It influenced the British army deeply; Templer acknowledged that much of the army's strategy relied on the handbook.[28] Furthermore, Britain established the Far East Land Force Training Centre to train officers on tactics, techniques, and procedures.[29] These views reflected some innovations made on the operational level. Britain departed from concentrated movement of large-scale battalions to "distribute" operations and disperse highly decentralized small-unit formations.[30] The army learned that sweeping operations were counterproductive; instead of massing troops and relying on firepower, it developed small patrols that used native scouts and intelligence provided by surrendered insurgents. In addition, Britain placed greater emphasis on capturing and deterring the insurgents than on killing them. This helped it gather intelligence and encouraged the rebels to surrender. As such, Major General Edward de

Fontlangue expanded the Home Guard paramilitary forces, also known as the "Chinese Army," from 79,000 men in 1951 to 250,000 by 1943 and assigned the Home Guard to defend seventy-two new villages.[31]

British success came as a result of multiple factors. R. W. Komer, for instance, raises a few successful programs, including registration, travel control, curfews, and ID card checks, food and drug controls in "black" areas to deny the guerrillas access to supplies, social and economic development, and information programs designed to keep the population abreast, all of which were implemented within the framework of a rule of law.[32] Here I raise a few others that I consider to be key to British success. First, Britain moved the population into protected zones called New Villages. Known as the Briggs Plan, this had the effect of isolating the bandit forces from the Min Yuen and noncombatant Chinese and cut enemy lines of supply. It resettled around 530,000 people—over a tenth of the entire population—into some 557 villages, most of which consisted of ethnic Chinese squatters and plantation workers. In sum, resettlement served local development, reestablished colonial authority in the countryside, and integrated rural people into the society.[33] Historians have widely embraced the plan as one of the most effective policies. After the war, Chin Peng, too, admitted the plan's long-term damage to his recruitment efforts.[34]

The second factor was the MCP's failure to win hearts and minds. Of course, the MCP recognized its mistake in its 1951 "October Directive," in which it ordered its branch committees to adopt a conciliatory approach toward people.[35] The directive was a direct testament to the MCP's admission that reform was needed to stop the decline in popular support. A senior party official in Johore, Siew Lau, for instance, complained that the MCP used coercion against civilians, thoughts that were echoed by his comrades.[36] As Lucien Pye noted before the emergency ended, complaints reflected a high degree of internal discord over policy, producing various logistical problems for their lines of communication and supply and slowing down jungle operations.[37] Thus the MCP carried out a series of internal evaluations of their performance. To work with various segments of Malay society, the MCP elevated Malay and Indian minority politicians to high positions in the Central Committee and expanded its school systems as a recruitment base.[38] These attempts nevertheless fell short of winning over popular support beyond areas of its control.

Third, the MCP lost their influence over the Chinese as postwar urbanization of Malayan society encouraged them to move into cities. Guerrilla activity in the jungle deprived their economic prosperity. Economic development in urban areas, therefore, undermined communist bases, while it benefited a rival organization—Malayan Chinese Association (MCA)—which had its own base from Chinese businesses. The MCP sought to respond by cultivating its contact with political parties in Singapore, labor unions, and Chinese language schools, but its base was largely lost by 1952 when Chin Peng withdrew to the relative safety of southern Thailand.[39] Finally, British forces managed to win local support by working with national leaders and providing resources and directions for the population. Templer is widely credited with bringing about a major strategic shift in British policy to win hearts and minds from Chinese populations; "the answer (to counterinsurgency) lies not in pouring more troops into the jungle, but in the hearts and minds of the people."[40] Britain's decision to transfer power in 1957 was consistent with its policy to gauge a support base through the gradual empowerment of local parties. John Nagl has argued that British victory had much to do with the learning nature of its military institution. Templer switched course to soft-line policies and sought to win hearts and minds.[41] The strategy of winning over hearts led the civilians to be convinced that their future was in an independent Malaya rather than one subordinate to the guerrillas.

Necessary Conditions for Sequencing Theory

The Malayan Emergency is a good example of the primitive model because it shows that the main reason for MCP failure lies with the absence of insurgent evolution through the periods of state building and conventional war. Examination of state building during the emergency requires us to look at various institutions both on the MCP and British sides. There are many works in the literature on the growth of the Malay state during this period, including the development of the state apparatus and revenue system and the enlargement of its administrative capacity, designed to reinforce the peasant allegiance and structure the postcolonial order.[42] But the major problem stemmed from a lack of support for the MCP, which made it difficult for the MCP to build institutions, authority, and legitimacy.[43] This also deprived the MCP of its ability to generate leaders who could serve as commanders and administrators on the national stage. In particular, the absence of statesmen capable of uniting the movement across different ethnic lines was devastating.

The MCP's effort to build a state in Malaya was curtailed by Britain's effort to strengthen institutions. One such institution was industry. In 1950, Britain built the Rural Industrial Development Authority to oversee small-scale development projects in the countryside, assist locals in building infrastructure, extend loans for crops, and sponsor cooperatives in agricultural, fisheries, and infrastructural industries.[44] Briggs recognized that political development could not be achieved immediately but would have to be achieved in small economic sections of the country first. Development of its institutions was strengthened during the Briggs era, when policies regarding food control, road control, and curfews were also introduced.[45] Political development rested also with a distribution of rights to minority groups. In 1952, the Legislative Council extended Malayan citizenship requirements for ethnic Chinese, which served growing demands for democratic governance from the bottom up. In 1953, Malay statesmen introduced elections for the Legislative Council. Growth in political participation was evident in the fact that 78 percent of eligible voters turned out for the first town council election.[46] The birth of electoral system was a significant step toward the consolidation of popular votes at the national level.[47]

British-led state-building efforts came with discussions about the eventual transfer of sovereignty to the Malays.[48] Templer worked with national leaders willing to negotiate settlements while making sure that the process would be stable and fair to competing loyalties on multiple ethnicities.[49] The United Malays National Organisation (UMNO), the MCA, and the Malayan Indian Congress (MIC) formed a coalition called the Alliance, led by UMNO chairman Tunku Abdul Rahman, which extended land reservation and quotas in employment and public positions, published electoral manifestos for self-rule, and secured communal compromise.[50] The establishment of this coalition that spoke with authority for multiple ethnic groups accommodated the communal interests with British COIN strategies.[51] In 1954, Alliance candidates swept the state-level elections and similarly the 1955 national elections, placing the Tunku in control of Malaya. The British effort became merged with much of Malay domestic politics and social conditions. The Malay government provided civics courses for the public, who would discuss matters of statecraft with civil servants. With development in the educational system and the corresponding rise in public consciousness about the direction of the country, public expectations for independence naturally grew during this stage, igniting debates about the sense of national direction.[52]

The other missing factor was conventional war. The MCP was such a disorganized insurgency that it had difficulty fielding an army to fight effectively. The MCP was a pyramid-shaped guerilla organization composed of a central committee all the way from the regional, state, and district levels to branch committees. The hierarchy between command and units did not function because of weak social foundations and lack of technology, such as radio, which forced the guerrillas to rely on couriers, which meant a time lag and unreliable communication. The organizational style had an effect of distancing itself from society. Weak social institutions led to shortages of supplies, disruptions in communications and intelligence, and lack places to rest, sleep, and regroup. Coercion of Chinese civilians challenged MCP recruitment. On the other hand, the absence of national representation prevented it from addressing policy, which was necessary to bring about popular votes to consolidate its legitimacy and attract cash and military advisement from China.

Britain's military activities were supported by regional security arrangements it made. The British Defense Coordination Committee—Far East created a civil coordinating officer, directly under the high commissioner, responsible for conducting the war. Heavy reinforcements of firepower and airpower would bring about tactical improvement only if they were followed up by vigorous action on the civil side. In 1950, Briggs built the Federal Joint Intelligence Advisory Committee, which coordinated the collection, analysis, and distribution of intelligence on insurgent locations, activities, and plans. In fact, Briggs launched what Harry Miller calls "joint thinking," combining three branches—the army, police, and civil administration—which led to the creation of the Federal War Council to coordinate all civil, police, and military efforts. The council, whose members included the chief secretary of the Malay federation, general and air force commanders, the police commissioner, and the secretary of defense, was replicated in the District and State War Executive Committees throughout Malaya. These committees controlled the police, and soldiers superimposed a military striking force. Each included a senior civil official for the area, a senior soldier, and a senior policeman, each responsible for day-to-day operations.[53]

In addition, Britain's innovations boosted military operations. A shift in the mobilization of strategic resources came with the arrival of Templer, who carried out multiple tasks: reorganizing the military operations, changing army doctrine to suit local conditions, and beefing up the intelligence apparatus, police forces, the Home Guard, and the overall coordination among them.[54] The midwar rectification of British strategy exemplified Templer's efforts to adopt the much-needed disciplined use of force, civil-military and interagency coordination, and flexibility in tactics through a highly decentralized approach. As a result, military incidents and casualties diminished after he arrived.[55] To consolidate local support, Britain strengthened the Psychological Warfare section of the Information Services and built the Intelligence Training School to train agents on intelligence collection and analysis.[56] Britain added civilian officials to expand local political activities in order to "Malayanize" the war. Geoffrey Bourne, director of operations, added Malay politicians to his operations committee and some more to district and state commands. Integration of Malay leaders into the war process escalated, as Bourne's committee was revamped in 1956 as the Emergency Operations Council with the Tunku as its chair, and all war executive committees were placed in Malay hands.

Most important, the MCP failed to adapt and put the war into sequence by using an effective political organization to create adequate military capability. A lack of material power could be seen in the organization of committees and the armed branch. On the one hand, the MCP had hardly evolved from the so-called 1948 structure, which was composed of a central committee, three regional bureaus, ten provincial committees, and fifty district committees.[57] One notable difference was the creation in the early 1950s of a more specialized group of Armed Work Force sections, which had their roots in the recruitment of MCP guerrillas as well as in the Min Yuen. They grew larger than the MCP itself, a concern to be addressed to ensure the latter's survival, but ultimately they failed to exhibit influence across ethnic lines.[58] The MRLA, the main military arm of the MCP, was informally organized into what could be considered regiments, with operations carried out according to geographic needs. These regiments had prematurely developed to have specialized units responsible for maintaining lecture courses on ideological issues and propaganda publications. These regiments, however, did not represent what we today consider an army, as the MCP continued to place its focus on jungle war that restricted its maneuver in urban areas without a central command and control structure for the purpose of fighting the emergency.

The Malaya case shows that the MCP satisfied none of the necessary conditions that I have identified. The MCP had access to the Min Yuen, local intelligence sources, knowledge about jungle terrain, and stable supply lines to do better in the guerrilla phase. But it made a series of mistakes concerning the fundamentals of guerrilla war by coercing the population from which it needed to draw support and allowing the British to take advantage of the political void to establish an alternative authority. Of course, the assumption that the MCP would fare well by coercing the population was clearly misplaced. Indeed, MCP failure to gain popular support was one of the most critical factors that prevented the MCP from carrying the war beyond the guerrilla phase. Undermined by the Briggs Plan, it also fought in territory where it was an ethnic and political minority. East Asian leaders gave Britain additional support when they created the Southeast Asia Treaty Organization in 1954 to provide a military shield to the north of Malaya. Long before 1955 when the MCP asked the Alliance to allow the MCP to survive, it had lost the military and political grounds to sustain operations.[59] In sum, the MCP lost the war because it failed to evolve as a guerrilla organization. It neglected the civilian role and failed to establish a base area to sustain its efforts. The sequencing theory is more persuasive than the theories of balance of resolve, strategic interaction, or democratic weakness.

Other Explanations

The literature on asymmetric war provides a host of explanations for how the weak beat the strong. While these works are not designed to address the Malayan Emergency per se, they offer basic ideas regarding the outcome. I select here three theories that challenge sequencing theory. First, the theory of balance of resolve posits that in war between unequal powers the weaker side is likely to win if it has greater resolve to withstand the cost of war. From this perspective, the communist insurgents *lost* the war because they were not as determined as the British. However, there are two problems with this view. First, the theory does not capture the fact that the level of resolve may change over time and leaves it unclear how such a change affects the war. Because motivation a priori resides with the weaker side of the outbreak, it may be the case that resolve is a better explanatory factor to explain why the MCP was able to *sustain* the war rather than to lose it. It also remains an open question whether resolve is the best factor to explain the timing of war termination. The premium that the theory places on the role of resolve has a danger of looking away from other variables that play significant roles. Such is especially the case for this twelve-year war, not only that the war may end as a result of omitted variables but also, even if resolve *did* play a decisive role, it would likely be a function of other variables. In Malaya's case, British strategy of winning hearts is the main determinant, coupled by the MCP's coercion of civilians and the Briggs Plan.

The theory of strategic interaction posits that the weaker side is likely to win a war if it adopts a military strategy that is different from that of the strong. According to this theory, the insurgents lost the war because they used the same conventional military strategy as the British. To put it another way, the insurgents would have *won* the war had they adopted a conventional strategy *and* the British used a guerrilla strategy. Alternatively, the insurgents would have won the war had they adopted a guerrilla strategy *and* the British had used a conventional strategy. There are two issues with this approach. First, it assumes that there is only one strategy used at one time. Because of this assumption, the theory overlooks the fact that one side can use different military strategies at the same time. Mixing strategies is a common practice designed to maximize effects and hedge against problems. Britain's switch from conventional to guerrilla war is a case in point, which dramatically improved its military effectiveness. If both sides engage in the same military strategy of guerrilla war, then the victor would be the British. Yet the emergency did not end at that point; instead, the war *went on* to stalemate. If the theory is correct, the war should have ended with an early British victory. Instead, the best explanation for the stalemate can be Britain's failure to realize the importance of popular support and the MCP's failure to take advantage of it.

Finally, the theory of democratic weakness posits that in an asymmetric war, it is the insurgent side that is likely to win because the democratic side suffers from the rise of middle-class opposition movement in a way that constrains the government's policy and prevents it from raising the level of violence needed to defeat the insurgency. From this perspective, the British won the war because they were able to fend off domestic antiwar rivals who threatened to destabilize internal politics as a means of their opposition to the war. The MCP lost the war because the British public did not embrace a strong resistance in ways that constrained the use of firepower. This theory suffers from two critical shortcomings. First, the assumption that the democratic government must overcome domestic social pressure against the continuation of war in order to have enough military capability to defeat guerrillas is infirm. It works in conventional war, where combat power plays a central role, but in small wars against guerrillas it is not only untrue but also counterproductive. In unconventional war, the less power one uses, the better one performs. This is why in Malaya, Britain began to do better once it began to use physical violence more judiciously. In the initial years when Britain had a hard time using a firepower-intense strategy and heavy weapons in large units in the jungle, results were suboptimal. Therefore, restraint in the use of force is a luxury for democracy at war. The state-society relationship is a positive constraint for democracies to use force wisely. The other problem with the theory is that it ignores the development on the MCP. It is concerned solely with how the "war" shapes politics inside democracies. But the reality is that it also affects insurgents. It is also the case that decision making in democracies reflects what the enemy does. Thus the theory suffers from the understatement of MCP development in the outcome and complete ignorance of how British strategy was affected by the way the MCP reacted. This one-sided analysis also indicates the possibility of overstatement of the democratic performance. This is why, in 1951, although Britain had a change of government from the Labour Party's Clement Attlee as prime minister to the Conservatives' Winston Churchill and Anthony Eden, this did not fundamentally change Malaya policy.[60] Yet the defeat of the MCP in 1960 cannot be explained without consideration of the MCP's failure to win popular support, build institutions of self-rule, and build military capability to support it, as this chapter has shown.

CHAPTER 6

The Degenerative Model: The Iraq War (2003–2011)

There have been few research projects to date explaining why the military operation in Iraq ended the way it did in 2011. Here I present the degenerative model of the sequencing theory to do so. The model deploys a useful framework to deconstruct the complex war, which involved a number of states, armed forces, tribes, insurgents, and warlords as well as a diverse set of competing political, economic, and military interests, aims, and interpretations. The importance of this case study is obvious, as scholars acknowledge the significance of the war, although not always its necessity, in terms of its repercussions reaching beyond the immediate Iraqi theater, spilling over to other countries, and causing civil wars in the Middle East. They include cross-border refugees, terrorism, radicalization of neighboring populations, regional secessionism, economic losses, and neighborly interventions.[1] The war has far-reaching strategic implications not only across the region but also for the future of international politics.

The primary purpose of this chapter is to use the degenerative model to explain the seeming defeat of Iraqi insurgents in two phases. I show that the key factor for this outcome rests with the fact that the destruction of Iraq's armed forces in early 2003 set the stage for a number of challenges for insurgent groups who emerged out of the mess to fight guerrilla war. The degenerative model illuminates why the transition from conventional war to guerrilla war is such a bad course of actions for insurgents. The sequence is critical here because the brief conventional war phase of 2003 shaped the way the insurgents fought in the second phase and because the collapse of guerrilla operations cannot be explained without it. More precisely, the destruction of the Iraqi army in the first phase paved the way for the creation of disorganized groups, who for a while were able to operate relatively effectively against unaccustomed American forces but eventually failed to win popular support in the postconflict phase. The insurgents were resilient and tactically innovative throughout the duration, but they failed to neutralize some of the major strategic innovations by the U.S. side, including the so-called Awakening movement, the surge, and the general improvement in American COIN operations.

Because of space limitations, I focus on events that matter directly to the sequencing theory, including the nature of military balance, popular support in Iraq, as well as some of the major political and military programs, as they profoundly affected the process and outcome of the conventional and guerrilla operations. Unlike in the other chapters of this book, I start by defining who the main actors were and examining which side won the war. I then trace the war using the degenerative model, covering both the conventional war phase and the guerrilla war phase in turn. I also deal with some of the existing explanations that challenge sequencing theory.

Who Fought and Won the War?

Because of the complexity associated with the Iraq War, I need to use some space to settle on a few things, even though scholars have not reached consensus on them. To begin with, who were the warriors? The "state" side consisted of the so-called coalition of the willing composed of nearly fifty nations who contributed close to three hundred thousand troops for the U.S.-led Operation Iraqi Freedom (OIF). But because the United States dominated in the making of financial and military contributions and conduct of the war,[2] I consider the United States to be the main actor on the state side. At the same time, I refrain from claiming across-the-board implications of this war for the other coalition members. The insurgent side was more complex. When the war started, the primary enemy of the United States was the Iraqi ground forces. With their destruction early in 2003 emerged a host of violent groups, many of which once served in the organized forces in various capacities. These groups became quite diverse in the short run, some of them developing hostility toward the occupying U.S. forces. Mohammed Hafez points to two types of such groups: Islamic nationalists and ideological Ba'athists and Sunni extremists. The Islamic nationalists were those who used Islam as a religious instrument to throw out the coalition forces and bring the Sunni

Arabs back into the political process. The other group was Sunni extremists who were Ba'athist and Sunni ideologues, also known as jihadi Salafis, and sought to oust the coalition out of Iraq while seeking to create a failed state because they believed that only then would they be able to survive.[3] Because the Sunni insurgency was a highly decentralized organization made of multiple hybrid groups across overlapping tribal interests, these groups can be further divided into subsets, such as Ba'athists, nationalist Islamists, Salafists, and transnational Salafi jihadists associated with al-Qaeda. Not coincidentally, all of them had different histories of growth, ideologies, objectives, strategies, and levels of impact on policy and the war.[4] There were some Sunni groups who were part of the Anbar Awakening movement and Sons of Iraq (SOI), so only a few Sunni groups were adversarial. But because the Sunni extremist groups constituted the most hostile insurgency fighting the coalition forces, in this chapter I consider America's adversaries to be a constellation of highly hostile Sunni groups who sought to force the United States to withdraw from Iraq. In this book I call them the Sunni insurgents.

A problem here is the fact that neither the American nor the insurgent side had ever clearly defined their objectives of the war. The insurgents' overarching goal was to "defeat the United States and establish a base of operations in Iraq for waging war against the West."[5] American objectives were more difficult to discern, in part because many actors were involved in the decision-making process. The decision to wage war was shaped not just by President George W. Bush but by various civilian and military leaders, political contexts, domestic and congressional forces, intelligence agencies, public opinion, and the United Nations, all of which had different goals to pursue.[6] Vagueness about goals stemmed from the presence of multiple objectives, ranging from short-term objectives, such as the removal of the Saddam regime and alleged weapons of mass destruction (WMD) programs, to medium-term objectives such as the destruction of al-Qaeda operations in Iraq and defeat of the Iraqi insurgency, to the long-term goals such as democratization and installation of a democratically elected, pro-Western leadership in Iraq. Brendan O'Leary assesses three "formal" American objectives: disarmament of Iraqi forces, demolishing of terrorist infrastructures, and the liberation of the Iraqi people, in addition to regional stabilization and prevention of Iranian domination in the region.[7] To make our analysis difficult, these goals were never static. America's prewar definition of victory was repeatedly compromised during the war and changed to suit its domestic exigencies. If destroying the WMD programs was the objective, then the United States had won the war even before it started. If the goal was the promotion of democracy in the Middle East, then victory would require the United States to have successfully coped with Kurdish, Shia, and Sunni interests and acquiesced to a partitioned Iraq.[8] The plausibility of a democratic Iraq was in question, too, because the Sunni population in postwar Iraq would oppose democracy in fear of being outnumbered and outvoted by a majority Shia.[9] Prevention of Iranian domination in Iraq was also part of U.S. interest, but it was hardly the main purpose of the invasion. Acknowledging that it is difficult to settle on a single objective, I define the U.S. objective in terms of withdrawing from Iraq with a stable government in place to control the civil war there. While this definition is not perfect, I believe that the United States achieved this objective when it terminated its operations in 2011.

Another question is whether the war was worth the cost that the United States paid. It is unclear how much the American blood and treasure were worth, but clearly the war was awfully expensive in both financial and human terms, with over four thousand deaths and an alleged three trillion dollars in costs.[10] For the most expensive military enterprise to date in the history of human conflict, one would need a powerful justification to have spent so much on a war whose necessity was questioned.[11] A good way to assess the cost of the war is to see the public reaction to American policy because, as Ole Holsti argues, "even in the face of vigorous public relations efforts by administration officials, public opinion, in the aggregate, seemed to reflect a sensible appraisal of events on the ground."[12] From the public standpoint, the war appeared worth the effort. While American people may have punished the Republican Party for the Iraq debacle in the 2006 and 2008 elections, they largely considered the absolute cost as secondary to the objective sought through the use of force. In fact, Peter Feaver and Christopher Gelpi show that the American public is not necessarily afraid of casualties; they accept casualties if they are necessary to accomplish a declared mission.[13] This finding is largely consistent with others showing that the public will support casualties if the mission is being actively pushed by the nation's leadership.[14] Gelpi, Feaver, and Reifler further argue that when American people face the question of supporting an ongoing military mission as casualties increase, it is the expectations of success that matter the most. Many factors, such as the stakes, human and financial costs, as well as the "trustworthiness of the administration, the quality of public consensus on the foreign policy goal in question, affect the robustness of support."[15] The view that success matters applies to the Iraq War as well.[16]

The final question of relevance is who won the war after all. Of many ways to assess the question, I choose to divide the war into two phases to assess victory—2003 and 2003–2011—because these two phases were quite different and both sides fought the two periods quite distinctively. As William Martel writes, "The initial outcome of the U.S. invasion of Iraq generally aligned with strategic victory," but before long the United States found itself mired in an intense insurgency and the initial victory eroded. Eventually, however, the campaign generated the basis for a "limited strategic victory" for the United States in the latter part of the conflict.[17] While scholars have yet to agree on the final outcome, I see a Pyrrhic American victory in 2011 because the United States achieved its prewar goal at bearable costs of the actions reflected in terms of the significance and challenge, while the opposing insurgents failed to do so. This verdict is not perfect, in part because the insurgents themselves have claimed victory, but it is a reasonable one for this book's purposes. Of course, America's "limited strategic victory" does not mean the wholesale end of violence; it indicates a formal end of the coalition forces' military operations in the country in 2011. In the next three sections, I trace the development of the war in the order of the conventional war and guerrilla war phases.

Iraq War in Sequence

My analysis of the war expands upon some of the existing works that used a sequential framework. The U.S. government's prewar plan for OIF, OPLAN 1003V, outlined four phases: (1) enlisting international support and preparing for deployment, (2) shaping the battlefield, (3) major combat operations, and (4) postcombat operations.[18] The U.S. Central Command described the progress of this war in five phases: (1) planning and decision making, (2) complete posturing of initial force, (3) attack the Iraqi regime, (4) complete regime destruction, and (5) postwar stabilization and pacification.[19] These sequences were, however, so unique and entrenched to the Iraqi setting that they would not necessarily work elsewhere, so an ideal sequence with decent applicability would instead be a generic one. In *On Point II*, Donald Wright and Timothy Reese strike the core of the sequencing theory when they write that "most commanders and units expected to *transition to a new phase of the conflict* in which stability and support operations would briefly dominate and would resemble recent experiences in Bosnia and Kosovo. This *phase of the conflict* would require only a limited commitment by the U.S. military and would be relatively peaceful and short as Iraqis quickly assumed responsibility. In this mind-set, full spectrum operations would occur *sequentially* over time as one type of operation finished and another began."[20] While underdeveloped, this statement nicely captures the fundamentally chronological nature of sequential analysis. As such, I use the sequencing framework to show how the most fitting explanation of the sequencing theory—the degenerative model—demonstrates that the main reason for insurgent failure was the shift from conventional war to guerrilla war without a state-building phase.

The war followed the degenerative path as it turned from a conventional war into guerrilla war in 2003 and lasted as such until it ended in 2011. Critics may object that the war was not precisely degenerative because the conventional war phase was so short in duration and because the insurgency and the Iraqi army were not the same group of people; instead we should count guerrilla war only, they may claim, which would make the war a primitive model. However, looking at guerrilla operations only would make an incomplete analysis of the war. Conventional war and guerrilla war were inseparable in this case because the insurgency was a direct fallout from the destruction of the Iraqi army and Republican Guards. Buttressed partly by the sudden inflows of foreign fighters, most insurgents had served in the Ba'athist, regular and even Republican armies and had fought the coalition forces in March and April 2003. Irregular militias had long supported the regular army and provided the very individuals who would later fight as insurgents. The guerrilla phase would not have been what it was without the conventional war phase. Thus it is possible to see the insurgency as a different stage of conflict because many in the Iraqi army made a conscious effort to expel the coalition forces by creating a paramilitary insurgent force of its own. As such, Keith Shimko argues that U.S. performance in the guerrilla war phase "will always stand in stark contrast to the stunningly successful campaign that deposed Saddam. But rather than view success and failure in isolation, common wisdom treats both as results of the same war plan, a plan brilliantly crafted and executed for our task but completely inadequate for another."[21]

The first phase was conventional, in which the U.S.-led forces quickly destroyed the Iraqi army under a "shock-and-awe" strategy, a military doctrine that promoted the use of overwhelming power to gain rapid dominance of battlefield. But failure to stabilize the country once the initial campaign ended allowed the insurgent forces and various warlords to regroup as guerrilla forces. Thus the insurgents replaced the Iraqi army as the main adversary of the U.S. forces in the second phase. Their use of urban tactics and improvised explosive devices (IEDs) proved to be highly effective against conventionally minded modern forces, who nevertheless made strategic adjustments in the following years to increase troop numbers in the surge and punch out a more population-centered COIN strategy under Army General David Petraeus. With both sides struggling to dominate the military and political scenes in Iraq, but with U.S. forces more successful with popular support, the military mission ended in 2011.

First Phase: Conventional War (2003)

America's preparation for OIF began with a smaller and agile force in the mind of national security leaders in the Bush administration, especially Donald Rumsfeld, Paul Wolfowitz, and Dick Cheney, as well as some of their military commanders like General Tommy Franks of the Central Command. The preparation was driven in part by the doctrine of transformation and revolution in military affairs (RMA), which saw advanced technology play a central role in the conduct of war. With another war going on in Afghanistan, this doctrine served America's interest in keeping resource balance within the military. In effect, Franks sought a "rapidly deployable small invasion force that would exploit American advantages in information, speed, precision, and air power to bring about a rapid collapse of Saddam Hussein's regime with a minimum of casualties and collateral damage."[22] Needless to say, the application of this small-unit mind-set to the Iraqi theater came without serious consideration for the postconflict stability operations, which was embodied by the congressional testimony of Army General Eric Shinseki, who received stern admonishment from the top officials like Wolfowitz. The key point about this preparation phase was that the United States sought to fight a conventional war and win it decisively in Iraq with limited forces and pervasive military technology.

In a semblance of the Gulf War of 1991,[23] Saddam Hussein's troops confronted the joint and multinational coalition forces in the air, in the desert, and in urban areas when the war broke out on March 20, 2003. Iraq's ground forces comprised three elements: the regular army, the Fedayeen, and the elite Republican Guard. Saddam split the Republican Guard for southern

Iraq to confront several mechanized divisions charging north and amassed other elite forces around Baghdad to protect the capital in a circle. Like during the Gulf War, the Republican Guard's powerhouse depended on four main divisions—Hammurabi, Medina, Baghdad, and Al Nida. These units were supported by irregular militias, although most of them were generally trained soldiers armed with modern weapons. On the other hand, American forces were configured to fight a desert war in a rush toward Baghdad, with soldiers and marines looking for fixed targets and aircraft seeking to destroy radar, missile, and other high-value communication facilities. Given the balance of power favoring the coalition of 300,000 men, the shock-and-awe strategy led quickly to the destruction of the 375,000-man Iraqi forces. Even though the numerical gap was relatively narrow, OIF proved to be overwhelming and ended in less than three weeks. The coalition had an apparent edge in airpower, firepower, and technologies such as night-vision goggles, precision bombing, and satellite-guided munitions.[24] This part of the war ended with the declaration of the end of hostilities on May 1, 2003, followed by actions of L. Paul Bremer, head of the Coalition Provisional Authority, to disband what remained of the Iraqi military. Importantly, the decisive rout of the Iraqi forces was consistent with the expectation of the conventional model of the sequencing theory; insurgents are likely to lose wars if they fight states conventionally. Not surprisingly, Martin Van Creveld calls the campaign a "walkover" for the United States.[25]

There are many reasons why the Iraqi forces lost this phase of the war so decisively. Among the strongest explanations is the centralized nature of Iraq's political and military decision-making process. Saddam was the only decision maker who mattered, and he made a series of errors regarding the intentions and capabilities of the coalition forces, which, coupled with Americans' own misperceptions about Saddam, generated a spiral of misperceptions on both sides that became difficult to overcome.[26] According to Stephen Hosmer, problems emerged when his decisions were distorted by optimism and overconfidence. He had a limited grasp of international and military affairs, while his advisers were uninformed and timid and routinely provided false information about enemy and Iraqi forces. These problems forced him to make a series of strategic misjudgments. In particular, he thought that Iraq would not be invaded in the first place and that, even if it was, coalition forces would be afraid of casualties and limit their operations to air attacks. His regime would also survive an invasion because his forces would be strong enough to force the coalition into signing a negotiated settlement short of defeat. At the same time, he was so concerned with Iraq's internal issues that he organized his forces to prevent coups and the infiltration of insurgents from Iran rather than to fight external threats, which meant that regular forces remained in prewar deployment areas when the invasion took place, making them vulnerable to an invasion from Kuwait. On top of this, Iraqis had a substandard situational awareness. They were poorly trained and positioned for defense and operated with old equipment. Once the war broke out, the coalition forces attacked Iraqi forces at standoff distances and at night, with armor units dominating ground combat and air power destroying much of Iraqi capability and resolve. These offensive operations quickly demoralized the Iraqi soldiers and forced mass desertion and the collapse of Saddam's regime. As a result, the vast majority of Iraqi forces failed to mount effective resistance.[27] While U.S. advantage in speed, precision, and airpower was overwhelming, Iraqi weaknesses also served as a necessary condition for their defeat. According to Stephen Biddle, the key factor rested with "the interaction between the Coalition's strengths and the Iraqis' military weaknesses . . . [and the] skilled use of modern Coalition technology interacted synergistically with Iraqi errors to produce extreme lethality and a radically one-sided military confrontation."[28]

The key implication of this phase is that the destruction of Iraqi ground forces and the elimination of the Saddam regime generated a host of rebel insurgents, both Sunni and Shia, who would operate without structure and were able to wage guerrilla operations effectively—for some time. A problem emerged soon with the insurgents, however, in terms of control of their organization, which made it difficult for them to plan and carry out their operations and protect the population. As the conventional war phase closed, therefore, what followed was the amalgamation of insurgent groups fighting nearly aimlessly against foreign troops without unified, long-term political objectives and sufficient resources to engage with the population. In other words, the first phase of the Iraq War vividly demonstrated the key reason for insurgent failure, which we would see grow in the second phase. In the next section, I show that the Sunni insurgents would enjoy a series of successful guerrilla operations for a short period of time before the United States developed a series of schemes to carry out COIN operations more effectively. The key factor for the strategic turnaround rested with the fact that the United States managed to execute a number of policies to effectively counter the insurgents—the introduction of the surge, the arrival of the Awakening movement, adaptation of U.S. forces to the changing battle environment, and ultimately gaining support from the population.

Second Phase: Guerrilla War (2003–2011)

The second phase was full of guerrilla operations, hit-and-run movements, car bombings, ambushes, IEDs, assassinations, and counterguerrilla/stabilization efforts led by the U.S. forces. Clearly less conventional than in the early part of 2003, this phase required both sides to restrain the use of force against the population, while encouraging them to increase manpower, stealth, and innovation. The advent of guerrilla war resulted from a number of factors, such as the desertion of Iraqi soldiers, large-scale arming of insurgents, dispersal of illicit weapons, and release of criminals from prisons. The key factor, however, was the collapse of the Ba'ath regime of the first phase, which introduced the flight of a number of motivated foreign jihadists. The destruction of the Iraqi Army led to the release of numerous security and intelligence personnel and militia fighters able and willing to continue resistance. The elimination of support bases further meant that they were forced to become submilitary units and join the expanding Fedayeen militias. Organized crime increased after the collapse of the regime.[29]

Once in the guerrilla war phase, the coalition forces found themselves devoid of resources with which to restore security forces and civilian officials needed to help stabilize Iraq. Because there are many academic works on why this phase turned so ugly for American troops, especially in light of the fact that the Bush administration had had so much time to prepare for Iraq (compared to the relatively expedient mission in Afghanistan in 2001), I do not repeat their claims here. But it is fair to say that the calamity started with the prewar intelligence failure, which caused the Bush administration to launch this war with a rosy picture of how it would deal with the alleged WMD presence without proper assurances from the intelligence branch. Intelligence errors arose from a series of conditions, as Robert Jervis argues, including the shortage of attention to the ways the intelligence community (IC) collects and interprets information, the presence of quite plausible inferences about the program,, and an organizational culture inside the IC that failed to explore alternatives in the face of what was considered to be compelling evidence.[30] A doctrinal preference for the transformational army and RMA inherent in the Bush administration's mind-set also contradicted the need for a more population-centric approach necessary for the stabilization operations. As Shimko contends, the "transformation, and the underlying vision of the RMA that drove it, left the American military unprepared for the challenges it faced after the fall of Saddam."[31] Furthermore, the United States saw irregular military operations as a means of managing international stability and made civilian agencies overdependent on the military.[32] All these biases meant the presence of inherent challenges with the U.S. forces to engage guerrillas and protect the local population. Absent active assistance of organized local forces, the U.S. troops found it nearly impossible to stabilize the country, which in turn undermined public support for them.[33] These challenges rested in part with the claimed institutional resistance of the U.S. armed forces to changes necessary for combating insurgents.[34] Peter Mansour, who led the Army's 1st Brigade, 1st Division, also known as the "Ready First Combat Team," blamed the post-Vietnam neglect of COIN and made a strong case for integrating military forces with civilian experts who can aid reconstruction in COIN operations.[35] Calls for increased troop deployments ensued among policy makers and scholars alike.[36]

One of the most difficult tasks of this phase—and the key to success in guerrilla war—was the protection of the population. American operations suffered from their inability and unwillingness to maintain a military presence to protect Iraqi citizens and embrace a legitimate political process.[37] Dangers in urban fighting were so great between 2003 and 2006 that American troops had to focus almost exclusively on roadside bombs at the negligence of suicide bombings of the population.[38] In Baghdad, combat took place heavily against and among the population in what Biddle called a communal civil war.[39] As a result, the population arguably shed the most blood. For some time, media reports portrayed American troop behavior toward Iraqi civilians to be substandard, a fact that was underscored most notably by the Abu Ghraib atrocity and charges of indiscriminate civilian killings by Blackwater mercenaries. This attitude resonated with Washington's often-cited propensity toward the use of violence as one of the main solutions to insurgency problems. For instance, the Quadrennial Defense Review's draft study on "irregular warfare" in 2005 stated that "there was a strong focus on raiding, cordon, and search and sweep operations throughout: the one day brigade raid is the preferred tactic . . . [and the] focus is on killing insurgents, not protecting the population."[40] Difficulty with the protection of population led to a reduction of support for the U.S. forces.

A number of other challenges exacerbated the problem with popular support. Edelstein's "duration dilemma" meant that while U.S. forces sought to protect local citizens, the "welcome" initially afforded to the forces obsolesced over time. This took place as populations sought to regain control over their territory and began to press the U.S. forces to withdraw. The United States also had a "footprint dilemma" when it faced incentives to provide a large military presence for security, but this intrusiveness increased the risk of stimulating nationalist resistance against it.[41] Resistance in the form of suicide terrorism increased as a result, which may have stemmed from the presence of U.S. forces in the first place, because suicide terrorism was used to coerce foreign governments to cease their occupation of the territory.[42] Ultimately, America faced what Edelstein calls an "occupational dilemma," a situation in which both prolonging the occupation of Iraq and ending it became unattractive options. That is, a successful withdrawal required returning sovereignty to Iraqis and ensuring their security after withdrawal. The problem was, however, that neither was desirable or politically feasible.[43] These dilemmas turned into challenges to bridge the gap between the Iraqi government and society. Eric Herring and Glen Rangwala argue that the United States "resorted to a range of strategies that trade off control and effectiveness, such as sectarian balancing, creating institutions but limiting their authority, playing off the center against the periphery, playing off political parties against tribes," and allowing the Iraqis to "remilitarize."[44] In addition, there were simply too many warlords for the United States to handle and coalesce into a functional COIN strategy. While the United States made a choice to work with *moderate* Sunni groups through the Awakening and the SOI program, Washington never managed to garner Iraqi consensus on these programs because the number of competing groups and interests overwhelmed its ability to deal with the complexity.[45] The net result was a lack of an integrative process that subordinated the local politics, which "promoted neo-patrimonialism, sectarianism, and unregulated national-local political conflict."[46]

Yet beginning in late 2007, the Iraqi situation improved for the United States. Figure 5 shows that while attacks against coalition forces continued to rise through 2007, after 2007 the rate declined. This improvement stemmed from many factors, but the important point here is that they all pointed to the need to secure popular support one way or another. Among them was the 2006 publication of the Army/Marine Corps Field Manual 3–24, which indicated that the key to success lay in the protection of the

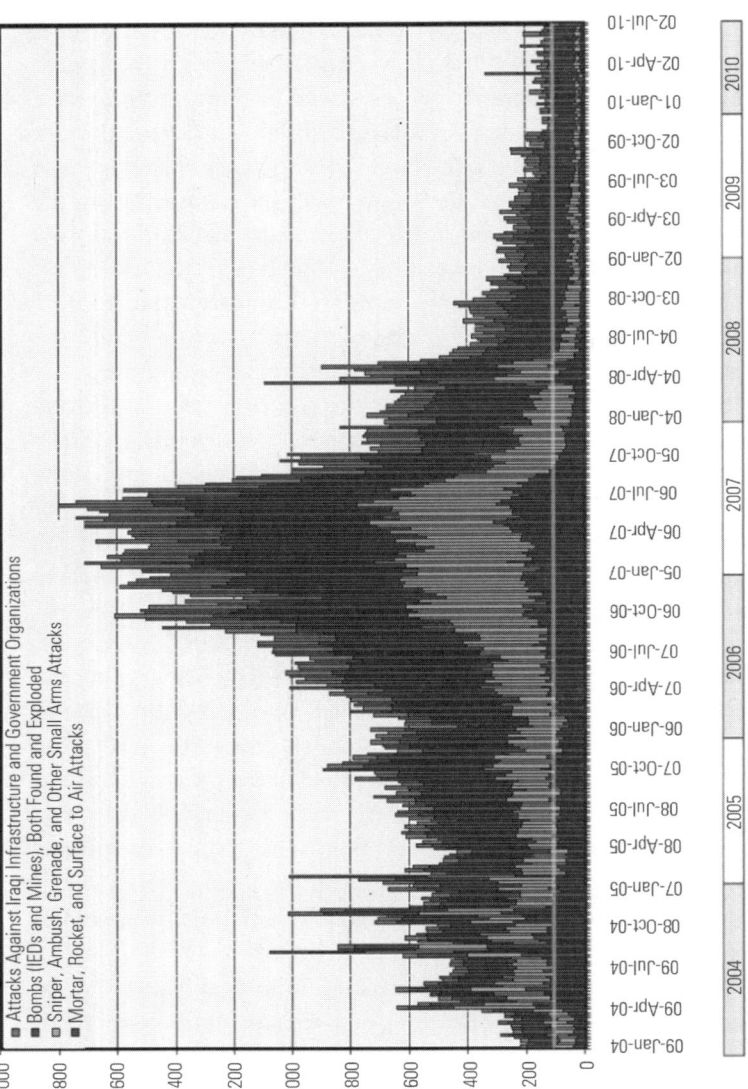

Figure 5. Enemy-initiated attacks against the coalition and its partners, by week. Source: Brookings Institution, "Iraq Index: Enemy-Initiated Attacks Against the Coalition and Its Partners, by Week," http://www.brookings.edu/~/media/Centers/saban/iraq%20index/index20120131.PDF, p. 4. Reproduced, by permission, from "Iraq Index: Tracking Variables of Reconstruction and Security in Post-Saddman Iraq" (Washington, D.C.: Brookings Institution, January 2011). Copyright 2011, The Brookings Institution.

population ensured by the establishment of various political and economic programs. The doctrine explicitly stated that "a successful COIN depends . . . on a legitimate and effective host nation justice program integrating law enforcement, the judiciary, and a penal system."[47] The rise of the Anbar Awakening movement in Iraq in 2006, which led to the decline of support for insurgents in various provinces, also vindicated the significance of local support gained through patient diplomatic efforts to bridge the gap between interested groups.[48] The movement's success hinged also on the Bush decision in 2007 to surge the number of deployed troops by thirty thousand and concentrate security forces within Baghdad to secure the local people and operate in small groups partnered with Iraqi units. This change departed from the earlier focus on "enemy-centric" (2003–2005) visions toward more training and supporting Iraqi forces with the coalition forces performing an "overwatch" role (2005–2006). This change also represented a major breakthrough in strategic innovation and adaptation. The U.S. military made a critical departure from the substandard operations and "learned COIN" when it embraced a small group of officers who advocated integration of armed forces in civilian environments.[49] For David Kilcullen, the United States "finally began to reflect counterinsurgency best practice as demonstrated over dozens of campaigns in the last several decades."[50]

Scholars have identified multiple causes behind the improvement of the U.S. performance in the 2007 to 2011 period, but they also were associated with the protection of the population. Kilcullen writes that America's improved performance in 2007 came through "a strategy of protecting the people from intimidation, forging genuine partnerships with local communities, co-opting 'accidentals' (including reconcilable Sunni insurgents and Shia communitarian militias), and killing or capturing the few on the extreme fringes . . . who proved themselves irreconcilable."[51] The surge strategy centered on the presumption that numerical increases in combat brigades and support troops would contribute to the closer network between the indigenous Iraqi security forces and coalition forces. Here the key was not in the number of the troops but the fact that the troops helped boost civilian protection. Recent research shows that manpower—based on the claim that COIN requires roughly twenty troops per one thousand inhabitants in the area of operations in order to be successful—has no discernible empirical support.[52] The end of sectarian cleansing and voluntary insurgent stand-downs of the Sunni Awakening have also been identified as background issues to the dramatic decrease in insurgent activities. But it was also the case that there was a synergistic interaction between the surge and the Awakening that was required for violence to drop.[53] Indeed, popular support was so important that even insurgents themselves learned to court civilians in order to fight their more powerful enemies.[54]

Popular support was not the only ingredient for success in guerrilla war, but it was the most important because much information about the enemy—its capability, location, and intentions—must be acquired by and from people. Thus, the United States paid great attention to ascertaining a sufficient degree of "footprint" for the protection of Iraqi people in ways that would sway the popular minds in favor of occupation forces.[55] Despite Abu Ghraib, the Blackwater incidents, and other alleged war crimes claims, American troops respected noncombatant immunity,[56] as seen in the Anbar Awakening and SOI and police auxiliary units associated with it. The SOI movement provoked a female counterpart to the policing program designed to mitigate the use of female suicide bombers, known as the Daughters of Iraq.[57] A number of Sunni tribes sought to ally with the American military and joined the Sunni Awakening.[58] The United States recovered from the problem in the prewar planning period to secure a semblance of emerging trust with locals who would have otherwise supported the insurgents. Indeed, James Russell contends that local security had *already* improved greatly in Anbar and Ninewah by 2007 when tribal leaders revolted against al-Qaeda. The leaders had transformed themselves from organizations structured and trained for conventional military operations into ones capable of conducting a wide range of combat operations.[59]

Of course, improvements in American performance in this phase were partly a function of the problems facing the Sunni insurgents, who were divided along ethnic and sectarian cleavages. Infusion of foreign fighters exacerbated the division as newcomers were not well incorporated. They shared the broad goal of pushing America out of Iraq, but many of them were in rival relationships. Troubled by sectarian violence and chronic division over territorial control, resources, and revenue distribution, they did not articulate a long-term strategy. Instead, they paid attention to the more operational, immediate aspects of the war and sought to expel the occupier without any further description of what exactly would replace the U.S.-sponsored process.[60] Moreover, accusations against insurgents' use of terror against people, led by prominent figures like Sayyid Imam Al-Sharif, better known as Dr. Fadl and a former member of al-Qaeda's top council, undermined their ability to unite.[61]

Indiscriminate killings of civilians as well as bombing of various important buildings, including one of the holiest Shiite sites, the Golden Mosque in Samarra, proved to be unpopular and countereffective. Sunni insurgents failed to turn resources into a cohesive set of political gains. Not surprisingly, many objected to the formation of a radical Islamic State of Iraq (ISI) in 2006 which symbolized the beginning of secession from Iraq. Instead, they sought to restore their position of power as a main objective, essentially becoming a *reactionary* insurgency seeking little more than the status quo ante. As a result they failed to establish a legitimate national polity.

For the United States, the strategic gains from March 2003 proved to be short-lived because it failed to execute the postconflict stabilization operations effectively. After a few years' struggle, however, it recovered through a series of programs including the Awakening, the surge, and the COIN doctrine that laid a foundation for ultimate military withdrawal. These projects were necessary factors for America's success in the guerrilla war phase, but we need to note that they were part of a sequence. That is, these projects worked because the first phase reduced the organizing power of the Sunni insurgents and American forces learned the importance of working with the Iraqi population around 2007. The war turned out to work for the U.S. forces, albeit with great challenges, because these events took place in a proper sequence for them. The most important reason for insurgent failure, however, rests with the fact that they failed to adapt. The destruction of the Iraqi army in the first phase paved the way for the creation of disorganized insurgencies, which were able to operate relatively freely against unaccustomed American forces for the time being, yet they failed to win popular support in this phase. The insurgents lost the war not just because of American COIN operations, their failure to win popular support, or the decentralized nature of their organization. The second phase followed largely by default; the guerrillas did not intend to fight that way because they thought of fighting conventionally a better option to take. Thus the war was not over after the first phase. Guerrilla warfare proved costly for the United States, but the degenerative model worked for it after all.

Necessary Conditions for Insurgent Victory

The Iraq War is a good example of the degenerative model because, while it had elements both of conventional war and guerrilla war, neither side made serious efforts toward state building. Generally identified as one of the major problems facing a number of weak states, most visible state-building efforts failed utterly in Iraq for both sides, even if the United States may have viewed them as key to establishing grounds for its departure. If executed properly, state building in Iraq would have entailed the establishment of a number of institutions necessary for the Iraqi government to function, including defense, law and order, poverty reduction, public health, education, utilities, and business regulation. Yet most new institutions built in this period were transitional, be they the Coalition Provisional Authority mandated under UN Security Council Resolution 1483 or a temporary "constitution-drafting" government in 2005. Although Washington demonstrated interest in reconciliation to facilitate the military drawdown and institutional capacity building, the net result of the political effort was Iraq's fragmentation, not reconstruction. The United States derailed the reform and reconciliation under the name of divide and rule. With Washington deeply entrenched in retaining control of the country, the fragmented Iraqi state wound up generating scores of militias claiming local power in southern and central Iraq.[62] Feldman argues that the main purpose of nation building in Iraq was nothing other than protecting America's interests through the creation of a legitimate, functioning democracy.[63] Likewise, emerging Iraqi leaders recognized the importance of reconstructing a state; thus they held national elections for transitional national assemblies, drafted a constitution, set up agencies to run a government, raised taxes, regulated trade, and provided services for the population. These efforts failed, however, "because of the combined pressures of the transnational informal economy and insurgency undermining the state's control of the economy from below, and the U.S.-led drive to open up the economy and configure the Iraqi state as an instrument of that process." As a result, there was no agreed-on procedure or unified agencies to resolve disputes between groups who claimed their own interests in the state-building process at the same time as they were tied to multiple nonstate networks and patrons.[64] While the new Iraqi state gained international support from foreign sympathizers in various forms, the end state was far weaker than it had been in 2003, and the rebuilding effort was highly unsuccessful.

The failure of state building on the part of U.S. and Iraqi leaders posed an opportunity for the insurgents to establish their positions, but they failed to capitalize. Rather than concentrating power at the center of insurgency, they deployed a diffuse, cell-based structure resembling a leaderless resistance, which helped increase their lethality. They used information technology and social networking to "cyber-mobilize" the population via transnational networks of sympathizers and sophisticated propaganda and foreign donations.[65] Yet the organizational structure was hardly conductive to state building. The insurgents had neither the ability nor the willingness to reestablish order on the public's behalf, rebuild institutions on a large scale, or unite various groups they purported to represent. They shied away from articulating goals and carrying them out at the same time. The loose nature of the network allowed foreign fighters and transnational Islamist groups to bring their own ideology into the conflict. The agendas of these groups extended beyond Iraq, hardly considered Iraq as a national entity, and made it difficult for them to create organized forces.[66] Problems like factionalism and rivalries emerged when subgroups developed independent wings, such as the Islamic Resistance Movement Hamas-Iraq, which split up from its main body and operated throughout Iraq. The decentralized system prevented the insurgents from uniting their goals under a single leadership.[67]

The emergence of warlords, defined here as individuals controlling territory though a combination of force and patronage,[68] was another impediment to the establishment of state institutions in Iraq. As Marten shows, four main characteristics of warlordism—small territories, personalistic rule, force-based patronage system, and warlords' function as middlemen—generated insecurity and instability that marred efforts to rebuild a state in Iraq. "Warlord territories are prime examples of economies operating without institutions. . . . Underdevelopment becomes chronic and subjects populations to ongoing long-term cycles of civil unrest and violence, even when temporary stability is possible."[69] One of the extended consequences of this situation was the rise of suicide bombings as a means of inciting public fear, attracting international news coverage, gaining support for their cause, and creating solidarity between disparate groups. Because it served the social and political motivations of the insurgents rather than inherently religious ones, however, suicide terrorism created contrary effects.[70] Mounting collateral damage provoked public scrutiny of insurgent behavior and legitimacy. As a result, much political power remained in the hands of the Maliki regime, which more or less controlled oil revenues, constitutional and electoral systems, and the social and economic infrastructure. For these reasons, there was no state-building phase in Iraq.

In short, the failure to build a stable Iraqi state made it difficult for the United States to claim a clear-cut victory in the war. The same can be said about the insurgents. From the sequencing perspective, lack of insurgents' ability and willingness to build a state further weakened their inability to reorganize the armed forces. The intensity of the post-2003 conflict and the prevalence of multiple tribal groups and competing insurgents and warlords made it impossible for the Sunni insurgency to provide for a collective action to enable the creation of organized armed forces. This meant that the Iraq War would end up as not just a guerrilla war but also an eventual failure of insurgency. Indeed, even if the insurgents had managed to build institutions, it would have taken years to build an army strong enough to challenge the U.S. forces in conventional settings. Counterfactually, it is possible to expect that state building and military power would have probably allowed the insurgents to fight the war better. Yet in reality, the insurgents never moved on to a next phase, with the Iraq War following a degenerative sequence.

Other Explanations

The literature of asymmetric war provides a host of explanations for how the weak beat the strong. While they are not designed to address the Iraq case per se, they offer basic ideas about why the war turned out the way it did. First, the theory of strategic interaction posits that particular configurations of military strategy affect the likelihood for insurgent success.[71] According to this logic, the Sunni insurgents lost the war because they adopted the same strategy—either a conventional strategy in the first phase or a guerrilla strategy in the second, or both—as the United States. Had they adopted a different strategy in either phase, the theory infers that they would have won the war. At a glance, the theory may seem to work because both sides appeared to have used a guerrilla strategy in the second phase. The problem, however, is that while the United States was *near defeat* in 2005 and 2006, it never was after 2006. In fact, the theory does not explain the stark difference in U.S. performance between the presurge and postsurge periods, a point of contention stressed by the closest observers of the U.S. COIN program, such as Nagl, Ucko, Taw, Davidson, and Serena.[72] In contrast, the Sunni insurgents consistently used a guerrilla war strategy throughout the conflict. The fact that the insurgents would have won the war in the presurge period of 2003 to 2007 severely undermines the theory. The theory, however, provides a useful framework. It explains that the U.S. forces suffered during the guerrilla war phase because they initially failed to adapt to the guerrilla environment when the insurgents used a guerrilla strategy. It also shows that the reason why the first phase of 2003 was so easy for the U.S. forces was because both sides fought a conventional war.

The second theory concerns the role of external support in that insurgent chances to beat powerful enemies rest with the availability of outside aid. Record argues that external aid is the key to underdog victory because historically foreign assistance has played a crucial role in war outcomes between the strong and weak.[73] However, the theory of external aid is empirically weak for the Iraqi case because the war outcome does not reflect the fact that the Sunni insurgents had benefited from foreign fighters and support and external sympathizers from Iraq's neighbors. The insurgents took advantage of weak border control, foreign import of weapons and fighters, and online financial contributions. External aid played a key role in allowing insurgents to gain weapons and expertise. Because this did not lead to insurgent victory, the theory of external aid is inadequate as an independent explanation. But it has some appeal to work with the sequencing theory because material resources were one of the key reasons for U.S. success. At the end of the day, the U.S. forces took advantage of enormous resources they had at their disposal to create an environment that was flawed but adequate for U.S. departure. Had the insurgents had an overwhelming advantage in material support, they might have performed better. Therefore, the sequencing theory agrees with this theory in that resource gap played a key role in the guerrilla war phase. For this reason, the theory of external support presents a pillar of the explanation of the Iraq War.

Finally, the logic of mechanization would posit that the U.S. forces won the war because they overcame the challenge of adopting a COIN strategy that relied less on mechanized forces than on manpower. The shift away from machine-intensive warfare allowed them to tell combatants from noncombatants, work more closely with civilians, collect intelligence from the population, and selectively apply rewards and punishments to the population, which proved to be critical in the latter part of the guerrilla phase. This change indeed crystallized a major departure from the standard interpretation of the "American way of war," a propensity to use massive force to destroy an adversary.[74] The problem with this theory, however, is that the United States did not lose the war despite its preference for mechanization earlier in the war. Shimko and Murray and Scales see the war as an extension of the Gulf War because the Iraq War, at least in its first phase, was heavily influenced by the Bush administration's preference to wage a machine-centered mission.⁷⁵ The Bush administration deployed forces that were fast and agile under the transformation and RMA concepts,⁷⁶ but that did not generate a catastrophic impact on the enemy. Thus, while the United States shifted focus to intelligence and manpower late in the game, earlier in the war U.S. leadership had failed to grasp the true nature of COIN, which according to the theory would have likely led to defeat. This theory, however, is not completely wrong. In fact, it works with sequencing theory because the focus on manpower in COIN operations, championed more in the postsurge period, is a key ingredient for success in the guerrilla war phase. The theory identifies one of the most important necessary conditions for the successful execution of guerrilla war—popular support—by means of using manpower to protect the civilians. For this reason, the theory would work with the degenerative model to explain the Iraq War.

CHAPTER 7

The Premature Model: The Anglo-Somali War (1900–1920)

The Anglo-Somali War of 1900 to 1920 is an example of the "premature" model. In this model, insurgents initially fight like guerrilla forces yet evolve into regular armies to fight conventional war. The model depicts the fighting in terms of a midwar transition from a period of guerrilla war into a phase of conventional war. The model is a "premature" one because, while insurgents do evolve from a guerrilla organization into a modern army, they do so without a critical phase of state building. The evolution ends up being premature because, while the transition into conventional war is a necessary factor for insurgent victory in extrasystemic war, it is not sufficient. It is not sufficient because the evolution comes without state building. The early twentieth-century war between Somali insurgents and British forces usefully characterizes the struggle on the part of the insurgents associated with this model. This case study shows that when the insurgents evolve this way, they are likely to lose the war.

The Anglo-Somali War carries two benefits as a case study. First, it displays a clear transformation of guerrilla war into conventional war, a process that is normally subtle but in this case is quite apparent, and demonstrates why this sequence of events does not work for insurgents. The war is an interesting case that shows that guerrilla movement can turn into a conventional military force without a strong role of a state. It shows that this kind of evolution produces nothing more than a halfhearted outcome for insurgents. Second, this chapter shows that the order of the sequence is crucial. Somalia's small territory and geographic location gave the movement only a modest resource to transform. As Saadia Touval pointed out, Somalia's location on the Horn of Africa has long attracted the attention of outside powers and often translated into strategic vulnerability,[1] which has in turn led to continuous struggle for autonomy in political, economic, social, and linguistic spheres. For these reasons, Somalia after the war developed a political entity that relied on the languages of foreign societies in its early years of independence, which lasted until people there accepted Somali as the official language.[2]

This chapter contains five sections. In the next section, I discuss the background to the outbreak of the war. In the third section, I use the premature model to capture the war's development over two decades. Specifically, I show that the insurgents' evasion strategy and British difficulties in chasing the Somalis allowed the insurgents to survive the first phase of the war, but the second phase—conventional war—saw the gradual decline of the insurgents' ability to fight a well-armed British force. I show how the war proceeded in two phases—guerrilla war and conventional war without state building—and how it ended with British victory, an outcome consistent with the expectation of the premature model. In the fourth section, I identify necessary conditions for insurgent victory and point out why state-building efforts did not occur in Somalia during the war. In the fifth section, I examine theories of asymmetric war and demonstrate why none of them adequately describes the cause of insurgent defeat in this case.

Background

The Anglo-Somali War originated from the colonial background of European exploitation in East Africa. In the second half of the nineteenth century, Somalia was partitioned by Britain, France, and Italy while Ethiopia, or Abyssinia as it was known then, invaded Somalia as an Italian proxy. These countries had different colonial policies in the Horn of Africa and therefore managed the territory differently. For instance, while France practiced an ambitious assimilation policy, Italy was more ambivalent between colonialism and minimalism that thwarted wholesale expansionism. In fact, as Robert Hess shows, Italy was so loath to expensive campaigns that it sought colonialism indirectly and financially through the use of chartered companies.[3] On the other hand, Britain extended a benevolent imperialism based on an indirect approach to much of its empire, including Somalia, where it controlled the most lucrative ports, imposed heavy tax on farmers, and recruited local levies to defend its protectorate. Most of seven principal tribes in the protectorate did not challenge British authority there,[4] which meant that British control was relatively firm. At the same time, Britain's projection of power came both from its naval supremacy around the globe and its advantage in military technology and firepower. One of the world's strongest powers at that time, Britain enjoyed advantages in virtually all aspects of military power over the Somalis. One exception, however, was in the number of combat troops. Britain only had twenty-four hundred troops when the war broke out, while the Somalis numbered over ten thousand, excluding local residents and nomads all over the territory. This numerical difference reflected a gap in terms of commitment to the war. While for the insurgents, Britain meant the major archenemy threatening their territorial integrity, Britain had other wars to fight simultaneously, so its deployment as well as commitment to this theater were relatively thinner. That did not, however, prevent Britain from collaborating with Ethiopians and Italians in its operations, with the latter two moving around the country so as to restrict the insurgency operations and serve as de facto British allies.

The insurgents were led by Muhamed Abdullah Hassan, also known as the Sayyid and the Mullah. His objective in the war was to establish suzerainty over the Somali country.[5] The insurgency was called Dervish, an Islamic term for the dedicated in service to God and community. According to Abdisalam Issa-Salwe, the Dervish movement contained four subgroups: (1) the ministerial council, which presided over state affairs, (2) bodyguards for senior members, (3) the regular army, and (4) the population. The bodyguards were mainly conscripts drawn from people the Sayyid trusted, such as former slaves he had adopted as sons. The regular army was organized into seven regiments, although, as I show below, in the first four years of the war its role in combat was kept to a minimum and it was overshadowed mostly by guerrilla operations. For the rest of the war, its combat capability gradually declined. Each regiment varied between one thousand and four thousand men, and had separate quarters, horses, and arms.[6] On the other hand, Britain's main goal was to defeat the Sayyid and secure the protectorate.[7] Like most extrasystemic wars I examine, the Anglo-Somali War saw some external actors; Italy and Ethiopia were occasional intruders. Because of the involvement of Ethiopians who exploited the territory and the Italians who colonized them, scholars such as David Laitin and Said Samatar see the war as a Dervish struggle against *Ethiopian* plunder of Somali livestock rather than against Britain.[8] Yet most of Somalia's main battles occurred against the British, who called the insurgents "tribes," or followers of the Sayyid's,[9] so in this book I treat the war as one between Britain and Somali insurgents.

The Premature Model and the Anglo-Somali War

The war followed the path of the premature model in that the insurgents evolved from a group of guerrilla fighters into an organized force. Specifically, the insurgency, as well as British forces, fought in a series of low-intensity skirmishes until the former realized that the strategy was not working. So they began to organize a military force and armed and trained soldiers to fight in open terrain just like a normal army in the second phase of the war. Thus the insurgents evolved, but without making serious efforts to build institutions as the means to keep the war going. Instead, efforts to build a state structure were left out of the process of military modernization. In short, the insurgents followed the premature model, a sequence of actions that expanded guerrilla war into conventional war short of state building. To use the language of sequencing theory, it was a primitive model followed by a conventional one, which reflected a form of strategic evolution in the direction opposite to the degenerative model discussed in the previous chapter.

At first glance it appears paradoxical that a military organization would evolve without a political structure. Military modernization requires a strong role of the state (and vice versa). Yet military modernization can take place without the state if there is sufficient ability and willingness on the part of the insurgents to organize a group of violent warriors and procure weapons domestically and from import. For this reason, the model focuses on state building as a main determinant of outcomes in extrasystemic war. Absent a sustained process of building a state, the insurgents are likely to lose the political drive for its armed operations, which in turn hinders their efforts to win the war. The premature model connotes the notion of prematurity in the sense that the insurgents evolve short of essential political process. Of the three past attempts to fight along this model, no insurgent group has ever been successful. There are two reasons.

First, the premature model is likely to fail because the population does not find much incentive to cooperate with insurgents. The population will lend support when there is a clear and achievable objective. They may not be motivated to fight a long war if they do not have a vision of stable governance in sight and if they fear that their lifeline is threatened. They will withhold support for the insurgency also if they receive from the state a set of assurances, material benefits, immunity from prosecution, and various sociopolitical packages such as education and constitutional protection. In other words, the absence of state building strengthens the enemy-civilian linkage at the exclusion of insurgents. Second, the premature model is likely to fail because the insurgents do not have a political drive to build a powerful army. Absent these, the premature pattern is doomed to failure. In the next sections, I use the premature model to show how the war proceeded. I divide this war into two phases—the first phase of guerrilla warfare between 1900 and 1904 and the second phase of conventional warfare from 1909 to 1920.

First Phase: Guerrilla War (1900–1904)

According to historical accounts, most of the war's first phase was characterized by guerrilla actions, major features of which include hit-and-run operations carried out by loosely structured groups of insurgents. Abdi Sheik-Abdi writes that the Sayyid "was not at the head of a military force with the cohesion and discipline of a regular army but an irregular tribal levy whose ranks could swell or shrink as quickly as the ephemeral streams of the monsoon season."[10] Laitin and Samatar write that "although inferior in firepower and organization, the Somali resisters managed to hold their own by engaging in guerrilla tactics strongly suited to the terrain of the country."[11] The Dervishes displayed signs of disinclination to get close to the British unless they could take them by surprise.[12] Instead, they would carry out light raids and surprise attacks. The Isa tribe, who had no horses and fought on foot, distinguished themselves in particular by their inclination toward night attacks.[13] It was in part intentional for the Dervishes; evidence shows that the Sayyid was concerned about the way his forces were perceived and cognizant of the need to influence the public opinion in Britain. On the other hand, British forces were armed with modern weapons and lined in unit formations for most of its combat operations. After all, although the British Empire encountered irregular tribal forces in various parts of its colonies, their main defense policy centered on continental military capability combining the use of ground and naval assets. Yet, for Somalia they had adjusted forces to fight enemies of irregular nature, by deploying forces drawn from the population. As Britain's War Office wrote in 1901, "There was at this time no regular organization of the lines of communication under a commander, ... nor was such an organization apparently considered necessary for the requirements of the expedition."[14]

Guerrilla operations were pervasive for two reasons. First, guerrilla warfare fit Britain's preference for light operations because the army had trouble justifying resources for initial deployments in Somalia. It had just fought a series of unpopular wars against Mahdist forces in Sudan (1893–1996), the Matabeles in South Africa (1896–1897), and Ashantis in Ghana for the fourth time (1994). When the war broke out in Somalia, the army was fighting the Boers (1899–1902) for a second time and Pashtuns in Afghanistan, among others. These campaigns took away resources for the mission in Somalia. The official British account acknowledged that in 1907 the army had not been ready for the Somali campaign because small wars of this kind generally broke out unexpectedly and because local military systems were often unsuitable to meet such a contingency.[15] Deployment would involve enormous resources for even a small place as Somalia. Mobilizing one army corps would require the government to buy eight thousand extra horses and fill gaps in various staff positions. To mobilize another would mean an additional eleven thousand horses, in addition to artillery, transport, and medical staff as well as camels to transport a large amount of men, animal rations, supply, and water. Thus the army had to save money by using levies and soldiers drawn from other parts of the empire, commanded by a small number of British officers. The army's resource constraint came in part from its position relative to the navy. Britain being a maritime power, the army had long been in a weaker position. As historian John Gooch shows, the army lost a budget war to the navy, failing to raise money for the operations in Somalia. The navy had put forward the so-called blue water theory, which posited that the need to protect the empire lay in the command of the sea. Imperial defense could be achieved only by a program of naval building based on a sufficiently mighty fleet. It also implied that the strategy would work by giving up localized ground operations in favor of overall command of the sea. The army had little intellectual argument against the then powerful idea of Alfred Thayer Mahan.[16]

Second, the choice of guerrilla warfare was influenced by the fact that the Sayyid was neither willing nor able to take on the British. From the beginning he aimed to rule the interior of the Somali country and leave the coast to the foreigners.[17] Aware that his forces would be outmanned if they fought face-to-face, he chose to strike at a time and place of his choosing.[18] As Issa-Salwe writes, the Dervish movement proved to be "a natural military organization that was ingenious in guerrilla warfare, drawing their enemy to ideal terrain and striking at will."[19] His unwillingness to engage was coupled with his inability to secure modern weapons. Because of an embargo, the insurgents acquired arms only secretly and in small numbers, not enough to build an army. Somali commerce in arms and slaves had been prohibited, as stipulated in the General Act of Brussels of 1890. In 1894, Italy, which was occupying part of today's Somalia territory, had forced some of the Somali Sultans, namely of the Mijertein and Obbia, to agree not to import firearms.[20] Arms embargoes restricted movement; every move to buy weapons had to be made covertly and risked exposure.

The first phase of the war was characterized by four major British expeditions carried out between 1900 and 1904. The first expedition was led by Lieutenant Colonel Eric Swayne of the Somaliland Field Force. Sent out from Burco, a city in northwest Somalia, it consisted of twenty-one British and Indian officers and a levy of fifteen hundred Somalis. The British fought major battles in Beerdhiga, Afbakayle, and Fardhidin with significantly smaller forces, outnumbered by six to one at times. Yet they did fairly well by taking advantage of speed and power. For instance, a corps drove back a set of Dervish soldiers by moving fast through the field and outflanking them. They then pursued the enemy toward their camps and, after destroying them, chased the Dervishes farther down the road to inflict additional damage. Nonetheless, for much of the early period, the British forces struggled with the insurgents' evasion tactics and stealth. The first expedition ended with no victor, but it was soon followed by a second expedition, starting in 1902. By this time Swayne had understood the fanaticism of the Dervishes and their ability to recuperate after battle. He ordered infantry and camel corps to garrison Burco, while disbanding the Somali levy and forming a militia not only to fight the guerrillas better but also to ascertain coastal protection and maintain a measure of operational flexibility. After a few skirmishes, however, Swayne's forces proved to be ineffective. His forces found their morale to decline in the face of what could now be seen as a long war, stemming in part from severe fighting in the bush in deep Somali interior, which was, according to Douglas Jardine, mostly "unexplored and unmapped, and the impossibility of determining accurately the numbers and dispositions" of the Dervishes.[21] To make it worse, the British government sought to cut defense expenditures and avoid further entanglements in the interior, forcing Swayne to curtail operations and limit resources to main strategic centers of its territory, such as Bohotle, Burco, and Berbera. This in turn allowed the Mullah to recuperate and regroup and Swayne to call for reinforcements, composed of a small number of King's African Rifles members and the warship *HMS Cossack*, from Aden to Berbera. The Battle of Awan Erigo occurred in 1902 in a manner that was indecisive for both sides.

By the time the third expedition started, in 1902, Swayne had been replaced by Brigadier General W. H. Manning, who had successfully suppressed insurrectionary movements in other parts of Africa. This time, the British were ready to carry out frontal attacks. During this expedition, fierce battles took place in Cagaar-weyne, Daratoole, and Ruugga in 1903 and Jidbaale in 1904. The Dervishes fought quite well against better-armed adversaries, occasionally defeating some of the best troops the British had on the field. This was partly because these "best" British troops included multiple ethnicities, such as Punjabis, Yaos (China and Laos), and Somalis. While the Indians and Yaos demonstrated good combat discipline, the Somalis were seen to have been intimidated by the Sayyid. Coordinating, motivating, and leading so many disparate nationalities turned out to be challenging. Malcolm McNeill points out that Somali levies had a serious problem of overexcitement, which reduced their combat focus. They were better, he concludes, as irregulars with assignments of scouting, raiding, and harassing the enemy than as combat troops.[22] Furthermore, the levies had trained for neither coordinated movements nor combined actions, and every man acted independently as he thought fit. Thus when combined into units under British officers, they found it difficult to adjust and to trust their leaders, which in turn reduced unit cohesion.[23] As Jardine writes, "There was a strange medley of men drawn from many different corners of the Empire: from the British Isles, from South Africa, from the frontier of India, from Kenya, from the Nile, from the uplands of Central Africa. Boers and Sikhs and Sudanese."[24] The use of a mixed-race army under the command of white officers meant substandard effectiveness. The British were indecisive against elusive enemy forces, so they replaced Manning with General Charles Egerton and gave him more control of operations.

The fourth expedition of 1904 was an extension of the antiguerrilla campaign. Away from the nearest enemy outposts by about 150 miles, the Dervishes awaited the advance of British troops to decide whether or not to fight because there was no overwhelming force to keep them where they were. On the British side, according to the official document, there was no forward operating base from which to deploy even a small force.²⁵ Thus in the first phase of the war, we witness an evolution of the Somali military strategy that was completed with the gradual convergence of guerrilla strategies by both sides. Yet the expedition slightly differed from previous operations in that the war became increasingly conventional. Britain made an attempt to put its forces into conventional combat, a move reciprocated by the Sayyid. Egerton sought to execute frontal attacks, moving around his forces to push the Sayyid northward again in order to get him exposed to naval fire. He also used an Ethiopian army to deny the Sayyid access to the west and southwest of the country and therefore geographically trap him. Threatened from all fronts, the Sayyid began to hold territory and build an outpost at Jibdali. By this stage both sides placed more focus on firepower than on movement. The official British account writes that "[w]hen the attack on the left front failed, two determined rushes [by British forces] were made on the front and right flank of the square, but they met with such terrific fire from rifles and Maxims that the charging enemy could not face it."²⁶

Despite the challenges, the Dervishes resisted the British expeditions by the use of guerrilla strategy and secured a limited number of notable victories. These victories were never decisive, but the strategy was effective enough to prolong the war. In late 1904 the Sayyid agreed to a cease-fire with the British, who decided to evacuate all posts in the interior and withdraw to the coast of Berbera. These four expeditions taken together, the first phase of the war was effectively a draw, which was followed by a five-year cease-fire during which the Sayyid and Italy signed the Illig Agreement, making Sayyid an Italian-protected subject. The Dervish adoption of guerrilla war helped them prolong a war that would have otherwise been brief. The guerrilla strategy allowed the insurgents to neutralize British superiority in firepower and fight the enemy on near parity. The cease-fire nevertheless broke down in 1909 when the second phase of the war began.²⁷

Second Phase: Conventional War (1909–1920)

Contrary to the first phase, the second phase was distinctively conventional. Because the transition between these two phases came slowly, it is difficult to pinpoint the precise moment of change. The year 1909 may not be perfect, as in 1903 British forces had already introduced square formations.²⁸ Yet it was during the interwar period when the Sayyid's forces adopted what can be considered an "orthodox" military strategy. Touval suggests the beginning of the third expedition as the transition to conventional war when he writes that while the "forces employed in the first two expeditions (1901–1902) consisted mainly of locally recruited levies, in the third and fourth expeditions (1903–1904), principal reliance was put on regular troops of the Indian army and the King's African Rifles."[29] As Abdisalam Issa-Salwe writes, "During the first period the Daraawiish won many battles because many factors such as their knowledge of guerrilla warfare, knowledge of the territory, their adaptability to the environment, their belief that they were fighting a jihad and their well organised military. However, after many successes over the intruders, they changed their tactics of guerrilla warfare to conventional warfare. This was a change of strategy that proved fatal for them."[30] In the meantime, Britain's armed forces gradually expanded in size, and their missions grew complicated and better coordinated. Well into the 1910s both sides confronted each other in a consistent manner. The British army marched within square formations with transport units placed in the center and cavalry following scouts, who moved forward while looking to the flanks and rear. The army also transported soldiers' guns and protected their camels, a key source of water, while fighting columns moved forward accompanied by the camels.[31]

There were two factors that facilitated the transition from guerrilla war to conventional war. First, the Sayyid sought for an organized force as a solution to divisions inside his insurgency. For instance, the Sayyid's loyal troops consisted primarily of the Ogaden branch of the Darod tribe, who were simultaneously interested in settling old feuds with their Ishak enemies and thus occasionally lapsed into a state of chaos. Aware that his followers joined the rebellion for the purpose of looting, which was inconsistent with his cause, the Sayyid disclosed that "most of the Dervishes have got beyond my control and frequently raid the people without my orders."[32] In this context, providing discipline and organization made sense for him as an instrument to solve the problem of internal instability. Yet the Sayyid imposed a hierarchical authority on the otherwise egalitarian society by compelling his followers to address him as "Father Master." Designed to place under his command people outside the traditional clan system, this leadership contradicted with the Somali authority system. Traditionally, the Somali people loathed totalitarian top-down control and suspected any form of centralized rule. His leadership became unpopular, which stemmed in part from some of his personal traits, including "his religious fanaticism, despotic rule over his followers, and bloody massacres of different tribes."[33] His indiscriminate raids and seizure of property of the clans he suspected were not popular. All these issues put his relations in trouble with the internal clans.[34] It created rivalry from nearly all clan leaders who followed him initially but soon began to question his legitimacy. Soon a challenge of leadership between the Sayyid and clan leaders followed, culminating in the so-called Tree of Bad Counsel crisis, where major leaders renounced the Sayyid, and the erosion of his moral credibility to lead the movement. Fighting erupted between his troops and the conspirators, turning into a bloody war that decimated several clans.[35]

The other factor that turned the insurgents into an organized force was the fact that the Dervish managed to gain weapons. The Illig Agreement gave the Sayyid access to some areas within the British protectorate and gave him freedom to trade, except in slaves and firearms.[36] He got by the arms control by arranging for German mechanics to join as armorers,[37] as well as through Sultan Osman Mahmud, who sold him rifles.[38] In addition, the Sayyid courted major clans, Warsangeli of the powerful Harti clan and Cumar Maxamuud and Ciisa Maxamuud, both of Mijjarten Harti clans, who sponsored the movement. He married and divorced children of powerful clan leaders frequently to expand his political network. The alliance he built helped him buy weapons and reconstruct his forces. As a result, these arms helped him to organize his force more conventionally. By the time the war resumed in 1909, the Dervishes had established an operational base in Caday-Dheero, amassed weapons, and attracted more recruits to the movement. Two years later they moved to Dameero and later to Taleex, the heart of the Nugaal Valley, and built garrisons. The location was strategic because it had abundant water and pasture. The Sayyid's forces gradually grew into a larger and more organized army. His followers reached twelve thousand troops, of whom ten thousand were mounted and one thousand had rifles. So as to facilitate the growing movement, he divided his forces into three main parties: the Haroun in the Italian Somaliland, the Warsangli in the northeast, and the Bagheri in the south. British occupation of the interior allowed him to operate in the hinterlands and halt British trade with the hinterlands, which in turn prompted the British to form the Somaliland Camel Corps in 1912 to police the hinterlands under Commander Richard Corfield.[39]

As the premature model predicts, however, the shift to conventional war did not help the Dervish much. British power was overwhelming, as the Camel Corps won battles at the Endow Pass and the Ok Pass, captured the Shimber Beris fort in 1915 to reduce the insurgents' supplies, and seized Taleex, which forced the Dervish troops to abandon their forts in the Nugaal

Valley. In Western Somaliland smallpox severely undermined the insurgents. Ultimately, the introduction of British warplanes in 1920 proved to be decisive against the insurgents. The combined-arms offensive used aircraft to provide close air support for ground forces and dispersed the Sayyid's forces, coupled with wireless telegraphy enabling cooperation between British columns and friendly tribes. The "joint" operation decimated the Dervish and drove them to Iimay in Ethiopia. After arriving in Iimay, the Sayid started to build new garrisons, but he soon died of influenza, ending the war.

Necessary Conditions for Sequencing Theory

The premature model identifies a state-building phase as the key variable for successful insurgency against state actors. This resonates with the fact that the Dervish embraced military modernization without making much effort to build a national movement. There was no identifiable government in Somalia at the time; a constitution, political parties, and fixed borders did not exist. The insurgency did not reflect an organization with a monopoly on the use of violence in the contested territory. As such, Somalia's geographic and political partition of the time undermined internal unity of the Sayyid's community. He failed to unite the Somalis, furthermore, because the traditional Somali society was too widely dispersed to form a political unit and because of the internal rivalries.[40] He never attempted to unify his territory within a national boundary. As I. M. Lewis writes, the Sayyid "allocated many civil and military responsibilities to the ablest of his adherents, remained always personal in quality and never developed into a strongly hierarchical organization."[41] Absence of political foundations for statehood hurt the insurgents. Their inability to govern the territory meant an inability to collect taxes, procure arms, and mobilize people. The movement failed to expand their armed forces beyond their immediate tribal linkages and provide functional command and control to solve the collective action problem stemming from the lack of coordination among various tribal groups that constituted the movement. Moreover, the movement remained unable to exert border control, which meant that they were constantly on the move to avoid confrontations with moving adversaries.

Second, the lack of state-building efforts generated heavy military costs. Although the Dervishes used fortresses to gain the appearance of positional supremacy in the area, they were strategically disadvantageous because they gave the British a fixed target.⁴² Even before the war began, it was the British who had a set of inherent advantages; they had not just more economic and military resources but also control and legitimacy established in the territory. This came in part from their historic control of British Somaliland and interactions with residents who traveled and traded with the Westerners. This posed a challenge for the Sayyid as he tried to build a counterauthority to the well-established government. Contrary to the foreign invaders, he had a weak institutional leadership that could not consolidate widespread social and military gains. At the same time, Britain enjoyed advantages in many combat operations. Specifically, Malcolm McNeill points to the fact that Britain used firearms skillfully. British Mauser rifles were more accurate and effective than other weapons of the day, serving a deterrent effect against rushing Dervishes.⁴³ In contrast, the Dervishes turned out to be inaccurate shooters. The widespread adoption of firearms meant that they used their spears less than guns when it was the very spears that made them competent combatants. Incidentally, as Jardine argues, the exchange of spears for rifles inspired the Somalis with a false confidence in their strength.⁴⁴ These skill and training deficits in each of major combat experiences added up to chronic deficiencies of Dervish military power that ran for much of the second phase. B. G. Martin describes the Dervish inability to fight in general: "By inactivity, by building masonry forts, the Sayyid's situation altered drastically. Stone forts are fixed, immobile, and the Sayyid himself had become overconfident. The roles of attackers and defenders were reversed; the British [with their superior military technology] could go over to the offensive whenever they wished. The Sayyid and Dervishes were not the targets—defending their flocks, forts, wells, and encampments. This was a style of warfare that did not suit them at all. When they abandoned their older basic strategy of guerrilla warfare their hopes for ultimate success evaporated."⁴⁵ Thus, the absence of state building on the insurgents' part translated into major political and military challenges. The premature evolution of the movement, from the guerrilla war phase to the conventional war phase without state building, demonstrates in Somalia that the failure to adapt and evolve led the Sayyid's war into disaster.

Other Explanations

The literature on asymmetric war provides a set of theories for how the weak beat the strong. While they are not designed to specifically address the

Somali insurgency, they do offer basic ideas about why the insurgency failed in the end. First, according to the theory of balance of resolve, the Dervishes lost the war because they were not determined to withstand the cost of the war. Conversely, they would have *won* the war if they had had greater resolve.[46] I find limited empirical support for this explanation. On the one hand, the Sayyid and his forces were determined to oust foreign colonial forces and tolerate moderate risks. The official British document readily acknowledged the Sayyid's tenacity of purpose that never broke down in the face of dangers. Even when he lost a large portion of his territory, he scorned peace terms that favored his movement and preferred to continue the war to regain what he had lost.[47] In contrast, the British did not demonstrate a high level of resolve. The British army was severely constrained in its budget for Somalia because the British navy had powerful say in foreign policy. The army was fighting wars outside Somalia that bore greater importance. Throughout the twenty-year war in Somalia, British forces were outnumbered, and London tried to seek an easy way out by substituting manpower with local forces. Britain also displayed signals of weakness. In 1908, there was a complaint about fighting in an "apparently worthless country." Due to the rising cost of operations and the inconclusive results of the first four expeditions, a new military venture was opposed and vetoed down by the government. Not surprisingly, the level of interest in the war reflected the defense budget. Britain's aid program to Somaliland declined precipitously after the first expedition. Although the program resurrected for the second campaign, there was debate over complete evacuation from Somalia.[48] As the second phase resumed in 1909, Sir Reginald Wingate, the secretary of state, declared that the continuation of the military occupation with a large expenditure was no longer justifiable.[49] After the war resumed, the killing of Camel Corp Commander Corfield put the British combat morale perhaps at the lowest point.[50] During World War I, too, Britain committed a great deal of resources to major operations on the European continent, which took away manpower and firepower from operations in Africa. If the theory is valid, the Somalis would have outrun the British resolve and won the war. Even if not, Somali resolve would have still remained strong enough during World War I to outwill the British. Historical evidence, however, shows otherwise.

Second, the theory of strategic interaction posits that the weaker side is likely to win the war if it adopts a military strategy that is different from that of the strong. According to this theory, the Dervishes lost the war because they used the same military strategy as the British. Specifically, the Dervishes lost the war because both sides used guerrilla war between 1900 and 1904 and conventional military strategy between 1909 and 1920. Conversely, the Dervish would have *won* the war had they adopted guerrilla war strategy and the British adopted conventional war strategy, or had they adopted a conventional war strategy and the British guerrilla war. Yet the theory is weak because evidence points to its inconsistencies. For instance, when the two sides fought guerrilla warfare in the first phase, Britain should have won but did not. When the war resumed in 1909, furthermore, the theory would expect that the British would have won at this point because both sides used the same strategy. But they did not, and Britain did not attain victory for more than a decade. What happened, instead, was a series of military operations carried out with substandard effects that prevented Britain from achieving a clear-cut victory. One of the major military debacles in fact took place when British camel operations failed to prevent the death of Commander Corfield. The theory is not completely falsified, however, because it does correctly state that when both sides fight using conventional military strategy, it is normally the stronger side that wins the war, which is consistent with this case study. The theory agrees with the sequencing theory about this outcome in that when both sides engage in combat in open terrain, it is the state side that comes out of the battlefield in good shape.

Finally, the theory of democratic weakness posits that in an asymmetric war involving a democracy, the democratic side is likely to lose the war when the democratic side suffers from middle-class opposition in ways that constrain its military policy and prevent it from raising the level of violence needed to defeat the insurgency. From this perspective, if there had been a strong force that constrained its budget and justification for the continuation of this war, chances were that Britain might have withdrawn at some point or significantly reduced its commitment. But Britain won the war because it did not suffer from the problem. This view has some credence in Somalia because Britain retained a high degree of political unity despite radical changes of personnel in the war government.[51] If British voters were pleased in any way possible with World War I, they would probably have been less so with the smaller war in Somalia. This means that British victory in Somalia was closely associated with British efforts to suppress public discontent. At the same time, domestic opposition to the war increased as wars elsewhere went bad. The Second Boer War of 1899 to 1902 turned out to be more costly than expected.[52] Opposition to World War I grew as the war lasted longer than initially thought. With the growth of powerful middle-class discontent with the war,⁵³ political liberals became modestly successful. The House of Lords lost power when the Parliament Act was passed in 1911, and the Fourth Reform Act, passed in 1918, boosted British democracy and saw the rise of the Labour Party. David Cannadine shows the widespread dissatisfaction among the populace with the existing domestic order, which, as R. H. Tawney had discerned in Britain on the eve of World War I, was nevertheless nothing compared with the greater anxieties that resulted from the physical destruction of the war itself and the sense of uncertainty in the future.⁵⁴ Some even feared that the war's end would overthrow the overarching social order by rebellious elements that belonged to an obscure worldwide conspiracy.⁵⁵ The ensuing sales of land and property in the postwar period implied the emerging end of the traditional rural life in Britain.⁵⁶ After World War I, the Somali war lingered on, forcing Colonial Secretary Alfred Milner to advise Prime Minister Lloyd George to choose from abandoning Somaliland or mounting yet another army expedition. The latter choice was taken, but it was financially prohibitive, as the War Office admitted that the campaign would require at least two more divisions to be deployed and take a year to reach any conclusion. Moreover, the war exhausted the British so much that in 1919 they passed the Ten Year Rule not to fight any major war for a decade, a government guideline that effectively tightened the Exchequer's hold on the military budget.⁵⁷ In sum, all this meant that while there was powerful opposition to the war, Britain managed to hold together its war effort. Despite the rise of antiwar forces and the perceived decline of domestic stability, Britain managed to raise camel missions, increase firepower, and mobilize airpower in the Somaliland. Historical observation of British social behavior during the 1920s shows that when World War I was over, there was a significant debate about whether fighting in Somalia would be worth the effort when the country should have focused on recovery and reconstruction in the homeland and armed services. Despite these challenges, Britain chose to fight on in the remote horn of Africa and persevered to win the war. Thus one may rule out drawing a direct line between the rise of antiwar forces at home and the war outcome.

CHAPTER 8

The Maoist Model: The Guinean War of Independence (1963–1974)

> Enormous modern armies, of a general toughness and tenacity never in doubt until near the end, were outpaced and outmatched by fighters who came from populations whose technological level, whose initial command of any aspect of "modern science," whose "starting point" of literacy or any other educational preparation, were always far lower.... In technological terms, these were very "backward" populations. And yet they were able to win.
>
> —Basil Davidson[1]

The Guinean War of independence receives little attention from the public today, but it reveals quite a drama about extrasystemic war. A tiny political group called the African Party for the Independence of Guinea and Cape Verde (Partido Africano da Independência da Guiné e Cabo Verde, or PAIGC) in this little-known place began a rural guerrilla insurgency before it became one of the best organized armed forces of West African history and defeated the colonial power in a highly contested war of attrition. In one of the recent revisits to the war, Mustafah Dhada writes that "the war was a much more complex contest between the PAIGC and the Portuguese army, each seeking to outwit the other's countermoves, than most accounts would have us believe."[2] The PAIGC managing to overcome enormous disadvantages to defeat Portuguese forces came as a surprise, as Basil Davidson acknowledges above. In this chapter, I show how this conflict turned into a set of surprising events that ended up working well for the insurgent group. I use sequencing theory and focus on a set of key events in sequence that caused the insurgency to succeed.

The central proposition of this chapter is that the PAIGC's victory is best explained by the Maoist model. According to the model, the PAIGC's evolution through three stages played a decisive role. In the first phase of state building, the insurgents outdid the Portuguese in building political and economic institutions and laid a foundation for a protracted war. The second phase began when the war turned into guerrilla combat, in which the Portuguese forces launched a powerful "Africanization" effort and the insurgents responded with strong popular campaigns. The PAIGC's success in "winning hearts and minds" in this phase allowed it to take the war to the final phase of conventional war, in which it developed a modern army to match the Portuguese in combat, before Lisbon succumbed to domestic turmoil to withdraw its troops from the territory. While each of the three phases had an important impact on the war, none of them was independently decisive. Instead, it was the insurgent evolution that had the strongest influence on the outcome.

The Guinean War fits the book's purpose in two ways. First, it offers the best empirical support for the Maoist model. Of the four former Portuguese colonies in Africa—the others being Mozambique, Angola, and São Tomé and Príncipe—Guinea was the last to launch its decolonization missions but it became the first to be sovereign. The Guinean War impacted the other movements more than vice versa.[3] Thus, Joao Paulo Borges Coelho writes it was none other than Guinea "where the nationalist war effort was the most advanced, and where the new Portuguese authorities thus had less room to negotiate."[4] In contrast, the Angolan and Mozambican wars are not suited for this case study because they were more like civil wars. Angola had three anticolonial movements fighting *each other* at the time they each fought Portugal. Although they were loosely led by the Popular Movement for the Liberation of Angola (MPLA), they were divided in a multiparty competition for leadership. Mozambique had two groups under the Liberation Front of Mozambique (FRELIMO), which nevertheless split into various groups and demoralized an otherwise unified front.[5] Independence movements in Cape Verde and São Tomé and Príncipe were hardly violent. Second, events in Guinea provide one of the most dramatic examples of extrasystemic war. Under Portuguese control since 1446, the Guineans should have lost the war thoroughly. The prospect for Portugal's easy victory was widely shared among military leaders, including General Kaulza de Arriaga, commander in chief of Portuguese troops in Mozambique, who stated that "subversion is a war above all of intelligence. One must be highly intelligent to carry on subversion, not everyone can do it. Now the black people are not highly intelligent, on the contrary, they are the least intelligent of all the peoples in the world."[6] As Davidson's statement above implies, few had predicted Guinean triumph, but Guinea became "the most successful nationalist movement in Black Africa and the first to achieve independence through armed struggle."[7] The war turned out to be so important that, as Kenneth Maxwell writes, events in Guinea were directly instrumental in what happened to Portugal in 1974. Nothing else more strongly marked the end of Portugal's imperialism.[8]

This chapter contains four sections. In the next section, in a brief background to the war, I show that various leaders on both sides of the war saw sequences as a means of fighting. While they differed radically from the Maoist model that I use, they demonstrate the centrality of innovation and gradualist approach in their strategic thinking. In the third section, I trace the war from the late 1950s to 1974 and describe how the PAIGC evolved across three phases. Specifically I use the Maoist model to analyze the trajectory of the war and the evolution of military engagements and political institutions in Guinea. In the fourth section, I examine three existing theories of asymmetric war.

Centrality of Sequence in the Guinean War

Central to PAIGC victory was the art of sequencing. Scholars see it as critical, although most of them simply stress how the war changed over time rather than how the phases reinforced or weakened each other as they constituted a sequence. Examples abound in the works of historians, starting with Dhada, who calls the war "a fight in which factors beyond the PAIGC's control determined some of the initial phases, the subsequent pace of the evolution of the war, and the final outcome of the PAIGC's armed struggle to free Guinea-Bissau."[9] Lars Rudebeck writes that "flexibility and adjustment to the shifting conditions of the struggle are key expressions in the organizational vocabulary of the PAIGC.... The general institutional and structural framework of the revolution in Guinea-Bissau is characterized by a combination of continuity and frequent pragmatic adjustment."[10] Before the war ended, fundamental changes had taken place in the social structure of the Guinean colony, changes that did not obtain in countries that decolonized peacefully. So argued Amílcar Cabral, the PAIGC's central figure, whose vision of advancing African society was made of three phases. The first phase was the formation of a communal agricultural society without a state structure and social classes. This phase morphed into an agrarian feudal society, constituting a second phase, with social structures becoming what would essentially look like a state. The final phase was the emergence of a socialist society in which the state and class struggle disappeared. This development relied on the growth of the society's productive forces, governance, and class systems.[11] *No Pintcha*, a major PAIGC publication, noted that "in the *beginning* of the struggle, our young militants were poorly equipped and for the majority military training was very elementary.... *After years* of struggle and victories, our Liberation Army has become powerful, well-equipped and technically advanced.... *[T]oday* our new combatants can receive political and military training in accordance with the actual demands of our struggle."[12]

More systematic analysis has been conducted by scholars and practitioners of Guinean insurgency, each of whom has revealed different sequences. Aristides Pereira, the first president of Cape Verde who fought along the PAIGC, broke the war into seven sections: (1) from the founding of the PAIGC to an incident called the Pidjiguiti massacre, (2) preparation of the masses for armed struggle, (3) liberation struggle, (4) the Congress of Cassaca and the Battle of Como in 1964, which led to the restructuring of the party and armed forces, (5) decisions to consolidate the state and create the National Popular Assembly (NPA), (6) the first NPA meeting and proclamation of independence, and (7) negotiations with Portugal for independence and the Third Party Congress in 1977. Patrick Chabal saw three phases: (1) peasant mobilization, (2) group consolidation, and (3) political maturation. During mobilization, people would join an alliance of radical petite bourgeoisies in order to promote socialism. During consolidation, this alliance would formulate policies and embrace an ideology to bring about revolutionary change via a people's war. Finally, the party would lead political, social, and economic changes in liberated areas by upgrading collective production and distribution systems and setting socioeconomic priorities for an independent polity to be followed by a social revolution.[13] Basil Davidson presented four phases. The first phase of "guerrilla operations" saw the formation of the working class. The second phase of "diffusion of peasant support" began in 1962 when the workers pledged support for the rebellion en masse. The third phase started in 1964 when the PAIGC transformed itself into a regular army and waged a coordinated mobile war. The final phase saw the PAIGC's direct assaults on Portuguese-held towns coupled with simultaneous insurrections in Bissau, Cape Verde, and the Bissagos Islands, which culminated in the elimination of Portuguese power.[14] Last, Gerard Chaliand presented a four-stage assessment. In the "preparation" and "reconnoitering" phase, the PAIGC mobilized the peasantry through political agitation and propaganda, followed by the second phase in which commando members provided for villages to be accepted as a legitimate group. In the third phase, the PAIGC persuaded the peasantry to control greater areas and more people. In the final phase, which the war had not reached because Chaliand's book was published at a later time, the PAIGC would achieve a relatively substantial concentration of material means and defeat the Portuguese.[15]

Cabral, too, showed a great deal of thinking through phases. His ideas started with new independent states in Africa, which he argued evolved through armed struggle toward the freedom of the suppressed. Although Cabral did not believe that there was a universal model of social revolution applicable everywhere,[16] he promised a better life in a new statehood through gradual change. He argued, for instance, "the main characteristic of the *present phase* of our liberation struggle is the *progressive reversal* of the relative positions of the two forces [Guinean and Portuguese forces]."[17] Sequential thinking addressed the future of Africa, too. For Cabral, the African revolution meant an opportunity for the "*transformation* of the present economic life in accordance with the movement of *progress*."[18] The PAIGC needed to be at the center of this revolution in Guinea and beyond. At home, it must (1) mobilize and organize popular masses in urban areas and unify all nationalists, (2) mobilize and organize *rural* masses, and (3) train them for the final phase of the struggle. Overseas, it must (1) denounce colonialism and mobilize international opinion, (2) fight colonialists from within international organizations, especially the United Nations, and isolate colonialists, and (3) organize patriotic forces abroad, coordinate their activity, and consolidate unity. The PAIGC would train "*for the final phase of the struggle* for liberation and the consolidation of the unity among forces existing abroad and their collaboration with the party."[19] Similarly, in economic policy Guinea started with a system of private appropriation of the means of production short of class struggle. The second phase was expected to see the level of production rise and the mode of production develop, in which class struggle might be provoked. In the final phase, people would eliminate the private appropriation of the means of production.[20] It turned out that these phases were more rhetorical than substantive, but the important point here is that many scholars and participants in the war saw the sequencing approach as central to PAIGC victory.

Evolutionary thinking was professed in the realm of colonial policy. Cabral's strategic calculation began with his analysis of Portuguese COIN strategy that developed in three phases. In the first phase, he argued, Portugal would attempt to replace the PAIGC with a party more sympathetic to its domination, promise Guinean people autonomy short of independence, and, with this puppet regime installed, assassinate nationalist leaders. In the second phase, Portugal would revise its constitution and allow greater autonomy for Angola and Mozambique, which had a large population of European origin. In the final phase, Portugal would seek diplomatic relations between African countries and colonialists of southern Africa, while conducting military operations to eliminate the PAIGC.[21] More specifically, Portugal would seek to destroy the PAIGC from within through three subphases. First, it would send in agents trained in the techniques of sabotage, provocation, and instigation of confusion. The agents would then penetrate the PAIGC ranks and cause an internal rift. If accomplished, the final phase would see the agents stir more confusion and turn it into permanent instability.[22] Although Portugal's actions deviated from these expectations, they did reflect Cabral's thought on how Portugal would attempt to undermine his insurgency.

Naturally, Cabral's effort to counter Portugal consisted of three phases. In the first phase of "pacification," the Guineans would begin to resist Portugal's effort to pacify their society. In the second phase he called the "golden age of triumphant colonialism," the PAIGC would mount a passive resistance, which involved only the occasional instigation of minor revolts. In the final phase of liberation struggle, the masses would employ armed resistance to destroy foreign dominance.[23] Because this plan was made in such a vague and broad expression, it was not particularly off the mark. From this one can infer what he imagined Portugal's COIN strategy would look like as a way of developing his strategy. Few good strategists would publicize details. The problems identified here must be understood within this constraint and evaluated through a study of what actually took place. Yet Cabral's plan was nowhere near completion as an elaborate insurgency strategy; it made no mention, for instance, of the need to modernize armed forces or of how to build a nation and carry out diplomacy with other decolonization movements. Cabral's strategy was so full of inconsistency and missing factors that Henry Bienen calls Cabral "not a comprehensive thinker about revolutionary processes."[24] My job here, then, is to deconstruct the seemingly abstract series of conceptual phases of conflict as consistent with the way the war unfolded in Guinea.

The Maoist Model and the Guinean War

Using sequencing theory, I argue that the Maoist model is the best fit for the Guinean War, which proceeded in three phases. The first phase (1956–1963) began with state building on both sides of conflict. The PAIGC first built base areas in countryside where Portugal's influence was infirm. While Portugal controlled major cities, the PAIGC built constitutional, educational, and tax systems in rural areas and gained some foreign recognition. The success in building what began to look like an early stage of a national government about to replace the crumbling colonial administration gave the PAIGC more time to continue the war into the second phase (1963–1968). Then the focus of the war shifted to guerrilla operations as the war began to be highly contested. The PAIGC and Portugal engaged in intense guerrilla war, seeking to win hearts and minds of the population and wearing down the will of each other. A brief stalemate followed, yet before long the PAIGC found itself outperforming the Portuguese and continued into the final stage. During this period (1968–1973), each side strove to defeat the other in conventional combat, but by now the PAIGC had developed a modern armed force called Armed Forces of the People (FARP), which grew into an even more mature armed service called the National Armed Forces (FAN). At this point, the Portuguese army revolted and brought down its own regime, giving independence to the PAIGC.[25] In sum, sequencing theory demonstrates that the primary reason for PAIGC victory lay not in war materiel or willpower of the party but in the way the party adapted and evolved. The PAIGC adopted a sequence of actions that made the party stronger each phase.

First Phase: State Building (1956–1963)

During Portuguese colonialism, Guinea was among the few fortunate territories that evaded heavy-handed foreign exploitation because it offered few minerals and natural resources. For Guineans, however, colonialism meant not only psychological subjugation but also imposition of inferiority in every respect, which inevitably yielded a small number of local movements constantly vying for independence. The end of World War II in 1945 introduced an era of decolonization movements in Africa, energized by a revolutionary ideological discourse, often grounded in variations of Marxist-Leninism and Wilsonian self-determination that bolstered Guinea's own. For these groups, outright victory against a better-equipped colonial army was unlikely at best, but sustaining guerrilla violence in the global marketplace of ideas was seen to increase that likelihood. Studies about "African guerrillas" flourished, forcing Western governments to investigate ways to undermine these movements.[26]

Sandwiched between the former French colony of Senegal to the north and French Guinea to the east and south, Guinea had a population of about 520,000 in 1960, composed of seven major ethnic groups. The Balantes were 30 percent of the population and provided most PAIGC members who rivaled Fula minorities, most of whom supported the Portuguese. This ethnic diversity challenged the PAIGC to hold together leaders from varied backgrounds as the PAIGC consolidated its position as the sole independence movement in Guinea. Aside from the ethnic composition of the PAIGC, military and economic balance of power favored Portugal for much of the conflict. At the beginning, the PAIGC's manpower stood at just one-eighth Portugal's, with four to five thousand fighters and fifteen hundred militias against forty to fifty thousand Portuguese forces.[27] Portugal's GDP was $2.88 billion, while the PAIGC's operational budget was close to nil.[28] Outlawed in 1959, the PAIGC began with nineteen members, none of whom had experience with armed struggle and many international contacts.[29] In the context of this highly adversarial condition, the PAIGC achieved quite a lot, as it originally sought "(1) immediate conquest of national independence in Guinea and the Cape Verde Islands, (2) democratization and emancipation of the African populations of these countries, exploited for centuries by Portuguese colonialism, and (3) achievement of rapid economic progress and true social and cultural advancement for the peoples of Guinea and the Cape Verde Islands."[30] For the PAIGC, fighting its colonial master was needlessly a daunting task, which naturally called for the centralization of power within the party, especially at the planning stage of the war. Popular participation gave the rural-based party a powerful impetus in boosting institutions in reflection of its growing ability to persuade their followers. Thus

Cabral sought to consolidate its position as the sole challenger to Portugal in the initial process of state building. Bienen writes, "It has been Amílcar Cabral, more than any other leader, who explicitly takes up the development of the state, and what it means to build a revolutionary movement in a very undeveloped and ethnically heterogeneous colonial society."[31] Prewar state building also allowed the PAIGC to avoid major combat it was sure to lose and to preserve itself. Examination of Cabral's works points to his meticulous attention to state building without relying too much on the ideologies of Marxism and Leninism.[32]

State building under the PAIGC entailed two subprocesses—internal and external. First, with the Portuguese forces in control of major cities, the PAIGC sought to consolidate its position in the countryside. In doing so, the PAIGC faced challenges from six rivals: Union of the Peoples of Portuguese Guinea, National Union of Portuguese Guinea, Movement for the Liberation of Guinea, African Democratic Union of Guinea, Union for the Liberation of Guinea, and the Front for the Liberation of an Independent Nationalist Guinea. Rivalry with these groups endangered the PAIGC's internal unity and took away resources to fight Portugal. The Fulas did not hide their support for the Portuguese, which caused Cabral's efforts to establish base areas in places like Senegal to fail.[33] Portugal tried to kindle tribal and ethnic tensions between ethnic Balante and Mandjaques and Fulas and also sought to entice PAIGC members with high positions and honors and rewarded defectors.[34]

Cabral equated the party with a state. He stated that "as the party functions as a state in the process of development . . . the fundamental characteristic of a state is its ability to suppress those who act against the interests of that state. Our interests, the interests of our party which directs our state, these interests are also the interests of our people."[35] The PAIGC managed to stay intact by building economic systems in rural areas, setting up regional bureaus, and drafting a constitution in 1960. It sent out party members to rural towns, created village committees to supervise agriculture, marketed agricultural produce through its outlets, and built schools to train recruits.[36] Behind this development was a firm belief in the Leninist revolutionary party led by elites. Cabral stated that "only a revolutionary vanguard, generally an active minority, can be aware of this distinction from the start and make it known through the struggle, to the popular masses. This explains the fundamentally political nature of the national liberation struggle and to a certain extent makes the form of the struggle important in the final result of the phenomenon of national liberation."[37] The PAIGC kept popular support with revolutionary democracy, a rhetorically attractive description of work ethics that illustrated an idealized behavior of people for successful revolution. Hence Cabral's slogan was "struggle of the people, by the people, for the people." The revolutionary democracy concept discouraged demagogy, lies, and exploitation of the people and encouraged responsibility, mutual respect, tolerance toward errors, and honesty. In Cabral's words, revolutionary democracy meant a collective effort to make the party "an effective instrument for the construction of freedom, peace, progress, and happiness of its people in Guinea and Cape Verde."[38] This process of democratization allowed working-class groups, youth groups, women's groups, and labor unions to join the party. For Cabral the liberation of rural areas symbolized a first step forward and had "all the instruments of the state."[39] Although the PAIGC's political development would continue through the war, the most important thrust of institutions was built in the early phase. The PAIGC replaced the economic infrastructure with its own industries centered on rice, peanuts, and timber.[40]

The other part of state building was drawing international support. Portugal gained support from its Western allies and worked with the United Nations to maintain its colonial possessions. Cabral sought balance by networking at overseas conferences, including at the Afro-Asian Solidarity Conference, in Moscow seeking communist support and Belgrade seeking nonaligned support, and in other colonial territories seeking to establish the Anti-Colonial Movement, which later became the African Revolutionary Front for National Independence. The discourse included voices from members of the Conference of Nationalist Organizations of the Portuguese Colonies. Most African movements recognized the PAIGC and extended support, as well as members of the Organization of African Unity, the Vatican, the UNESCO, the Soviet Union, China, and Cuba.[41] These groups became international backers of the PAIGC and lent moral support and legitimacy, although it mostly came from the communist world. In sum, the PAIGC began its preparation for war through the creation of institutions in the countryside and mobilized support both internally and externally. The period of political institutionalization addressed indigenous resentment at foreign rule and constructed networks needed to consolidate support. Such efforts gradually gained in intensity as the PAIGC moved its base of operations from rural to urban areas and expanded its propaganda program. These efforts won international support from similar-minded groups outside Guinea.

From the standpoint of sequencing theory, the phase of state building was effective because it was followed and complemented by successful execution of guerrilla warfare. The shift to guerrilla warfare proved to be effective because the first phase set the stage for the second phase.

In 1961, an uprising in Angola prompted violent reactions from Portugal, setting a stage for independence movements across the lusophone communities. The war began in 1963 when PAIGC operatives attacked Portuguese military headquarters south of capital Bissau, and violence spread out quickly. The PAIGC seized vast rural areas, while the Portuguese forces occupied major cities. The war grew intense in the mid-1960s as PAIGC firepower increased thanks to weapons import. In 1964, the PAIGC adopted insurgency strategy and won a series of combat victories in places like Gabu and Como, an island strategically important for supply lines in the south.[42] PAIGC growth increasingly put Portugal on the defensive and in 1966 controlled about half of the countryside. With the PAIGC's destruction of the Bissau airport, the war became so bad for Portugal that in 1968 Spinola replaced Schultz as commander and sought a negotiated settlement with the PAIGC.

Second Phase: Guerrilla War (1963–1968)

In the period following the initial years of the war, both sides found themselves in intense struggle for popular support. Central to their efforts to gain popular support was the creation of institutions that provided policy direction, infrastructure, and public goods to elevate local quality of life. The PAIGC chose to fight guerrilla war because fighting conventionally was deemed too difficult at this point and because party members had to constantly move. Armament was grossly in short supply to launch a serious mobile war. A PAIGC commander stated that "at the very outset we had seven base camps, and three guns per base. It did not amount to much but we were getting there; one submachine gun and two revolvers. There were far more men than guns. Only the real specialists in guerrilla warfare were given the firearms. We promptly got busy ambushing convoys in order to seize more guns."[43] So it had to "go underground" by deploying forces that avoided face-to-face combat in open terrain. Popular support was a necessary factor for the PAIGC's success in guerrilla warfare. Cabral stated that "national liberation, the struggle against colonialism, working for peace and progress, independence—all these will be empty words without significance for the people.... It is useless to liberate a region, if the people of the region are then left without the elementary necessities of life."[44]

The Portuguese displayed an absence of cohesion between government policy and its behavior toward the population, although the military leadership did recognize the importance of winning local support. Spinola, the governor and commander in chief of Portuguese forces in Guinea, argued that "the policy of a government can never be real if it is not oriented toward the aspirations of the *popular masses*" and "the *popular mobilization by means of persuasion* and the consequent *winning over the masses of people* were the real weapons in this war."[45] To build a "better Guinea" entailed the provision of a variety of incentives, such as education, socioeconomic development, infrastructure, civic action programs, political assemblies, and literacy campaigns.[46] John Cann has published key works on Portuguese COIN performance. According to him, the Portuguese reacted with a COIN strategy based on an army document called "Guerra Revolucionaria." It anticipated that this war would evolve in five phases: (1) preparation, (2) agitation and creation of subversion, (3) terrorism and guerrilla action, (4) creation of bases, a rebel government, and pseudo-regular forces, and (5) regular war.[47] In 1961, the army stationed a few COIN units in cities and gave them a surveillance function. In 1963, it published a manual called "The Army in Subversive War" that emphasized the use of minimum force, small-unit operations, and popular support while elite units carried out combat operations, elements consistent with emerging COIN principles of the time. In addition, Portugal used some American manuals, such as Field Manual 31-20, "Operations Against Guerrilla Forces," and Field Manual 31-15, "Operations Against Airborne Attack, Guerrilla Action and Infiltration."[48]

Portugal's COIN strategy had three components. The first was resettlement, a program quite similar to the Briggs Plan during the Malayan Emergency; Portugal built villages where it protected the population, trained, armed, and organized the villagers into self-defense forces, and set up mines to prevent concentrated enemy charges.[49] Second, Portuguese forces expanded their operations into the countryside, the traditional safe haven for the PAIGC, to cut off PAIGC access to the population. In doing so, the army gave more power and autonomy to the regional commanders and brought theater commanders closer to the people. The PAIGC also increased the number of special forces and expanded local paramilitary forces. Finally, Portugal systematically recruited local soldiers to form "second-line units" in the so-called Africanization program and work as guides, civil militia, and auxiliary forces. This program reflected the so-called same element theory, which mixed an old colonial tradition with local forces to mirror their organization, weaponry, and knowledge of terrain and language and increase combat efficiency.[50] The Africanization program helped free the expeditionary army for offensive operations, defended vulnerable villagers, and moved other troops away from the metropole. The population size from which to draw was small, but African troops were cheap and familiar with terrain and provided good intelligence.

In the next few years, the PAIGC survived the guerrilla war phase because it countered Portuguese COIN operations with insurgency plans that worked, which were designed to (1) mobilize the masses into the main combat force, (2) carry out clandestine activities in towns, (3) reinforce unity across ethnic groups, origins, and social strata, (4) give members leadership training, (5) elicit foreign fighters, and (6) acquire resources.[51] The PAIGC extended activities to include education, health, and social services, in a hierarchical manner. The 1964 Cassaca Congress led to the reorganization of various command posts designed to improve educational, agricultural, and medical facilities. The PAIGC's survival in the guerrilla phase generated a new phase. Popular support was critical for the consolidation of state building and allowed the PAIGC only to *continue* fighting. While popular support was necessary for success in insurgency, it was not a sufficient factor for outright victory at this point. In fact, the Portuguese side, too, survived this phase. In the end, the war would culminate in a phase where military power played a decisive role.

Final Phase: Conventional War (1968–1973)

The war culminated in a period featuring the mass use of firepower and territorial competition in which military power played a key part. Portugal's army had expanded after the debacle at Como Island in 1964 killed nine hundred men. In 1968, Spinola replaced Schultz as commander and stressed the use of mobile units and fixed infrastructure. He restructured African troops to fit better with the modernizing Portuguese army, and the air force deployed a squadron of modern aircraft, such as Fiat G-91 light jet fighter-bombers and Alouette helicopters, although they faced hostile antiaircraft guns that kept them at high altitude and rendered their bombing inaccurate.[52] Portugal's operational coverage broadened as combat lethality grew.

Control of violence became essential for PAIGC victory. As Rudebeck writes, "The struggle for national liberation made a military organization necessary."[53] In the words of Cabral, national liberation could not occur "without the use of liberating violence by the nationalist forces."[54] As such, the PAIGC modernized its forces in ways that supplemented the previous phases of state building and guerrilla war. The growth of Guinean military capability closely reflected Cabral's realization that modern technology and combat formation would be effective. Cabral wrote that "in order to dominate a given zone, the enemy is obliged to disperse his forces. In dispersing his forces, he weakens himself and we can defeat him. Then in order to defend himself against us, he has to concentrate his forces. When he does that, we can occupy the zones that he leaves free, and work in them politically so as to hinder his return there."[55] Along with this came his call for military professionalism for his forces, coupled with his growing skepticism of peasant revolution as central to decolonization. The case in point was his statement that "the peasantry is not a revolutionary force—which may seem strange, particularly as we have based the whole of our armed liberation struggle on the peasantry. A distinction must be drawn between a *physical force and revolutionary forces*; physically, the peasantry is a great force in Guinea: it is almost the whole of the population it controls the nation's wealth, it is the peasantry which produces; but we know from experience what trouble we had convincing the peasantry to fight."[56] He believed that the administrative body must control the revolutionary peasantry and tied the local defense forces closely to the civilian structure and professional military and placed them under the direction of zone committees.[57] What began to emerge, therefore, was the crystallization of professional coordination between local civilian sectors and the centralized military structure. In 1968, Cabral acknowledged that the PAIGC was in a stage of mobile war, with a diminished focus on guerrilla strategy. The essential feature of the struggle became the systematic attacking of Portuguese fortified camps and fortresses.[58] As it turned out, conventional war fit well.

Thus in the span of five years (1968–1973), Guinean forces managed to bolster their military capability in every aspect. Weapons importation increased in the 1960s along with the reorganization of command zones, creation of a supreme council of war, and formation of special units. The result of the reorganization was the birth of a three-tiered structure. At the top was the modern FARP of five to ten thousand troops, which consisted of infantry, tank, and engineer units, placed directly under the PAIGC interregional committee command via the regional or special committees. A peasant militia was composed of three elements of part-time farmer-guerrillas: (1) the Exercito Popular, which consisted of the most experienced fighters; (2) the Guerrilla Popular, drawn from the population; and (3) the Militia Popular, which consisted of trusted villagers.[59] Along with the reorganization came the development of mobile war. The PAIGC used the "Armed Action and the Military Methods," accompanied by the development of military organization called Acção Revolucionária Armada, which advised the military to seize territory and destroy enemy lines of communication.[60] The final phase of war saw the PAIGC gain the upper hand. In 1970, the PAIGC evolved further when the more flexible, better coordinated FAN replaced the FARP and spread out into three services—the army, navy, and air force. These forces combined attacks with mortars and cannons, striking from a distance before attacking from close range. The PAIGC changed the way it used its bases. During the guerrilla war phase, Cabral had built large bases where supplies and ammunition could be stored and guerrillas trained. Supply and transport of military equipment was unreliable and had to be carried by hand and on foot. As offensive actions became intense, the PAIGC made these bases mobile in order to avoid reconnaissance and loss of equipment and ammunition.[61]

Guinea's military modernization opened a crack in Portuguese forces who suffered poor training, declining combat effectiveness, and tensions between officers and conscripts. Conscripts were trained in large camps by young officers and noncommissioned officers just out of training themselves, rather than by experienced soldiers.[62] Portugal's air operations were severely limited; there was only one squadron of twelve helicopters ready to use, and they were relatively slow, expensive, and in short supply and had little natural terrain to shield their movement, making surprise difficult to achieve.[63] Furthermore, various obstacles, such as enemy sentries, mines, and booby traps, hindered helicopter assault, which reduced combat performance.[64] Portugal's army suffered nearly eight thousand deaths by 1972.[65] Cabral was assassinated in 1973, but the PAIGC kept control of much of territory and declared independence from Portugal.[66] According to John Keegan, "Portuguese forces [were] ejected in the final phase of a protracted insurgency campaign at times more resembling a limited conventional war."[67] At this point, Prime Minister Marcelo Caetano told Spinola that he preferred that Portugal leave Guinea with an honorable defeat, which angered Spinoza and his army subordinates. In 1974, a group of frustrated young officers staged a coup in Lisbon and toppled the Caetano regime in the so-called Carnation Revolution. With Spinola assuming the presidency, Portugal recognized Guinea-Bissau's independence for the first time.[68]

Scholars have offered competing explanations for the outcome of the war. Hugo Gil Ferreira and Michael Marshall indicate that it was conscript army officers, drawn from college graduates influenced by Marxist social revolution, who played a key role in mobilizing the officer corps in toppling the regime.[69] According to this view, Caetano blamed them for infecting regular officers with left-wing ideas; "because of this constant infusion of university graduates, the armed forces absorbed the ideas that agitated the younger generation and circulated in the schools."[70] However, the officers' role was minimal because universities in those years were rather conservative. In fact, Douglas Porch indicates the role of young officers who, opposing the wars in Africa on the basis of weak public support and inherent risks, were most involved in combat and suffered the most. They developed collective objections to the war, which weakened morale and combat effectiveness and increased the desertion rate, which was 7,857 per year on average between 1961 and 1974.[71] The other view is that as cohesion and strength within the armed forces declined and as a military coup appeared likely, the power of the secret police increased. This shift of power from security to police forces led to the division of Portuguese domestic politics.[72] In contrast, Aquino de Braganca and Basil Davidson give credit to the PAIGC side and contend that the PAIGC won independence thanks to its successful offensives in early 1973. The PAIGC controlled two-thirds of the country and achieved equilibrium in the other third. Radical elements in the Portuguese forces campaigned to end a war that they saw as hopeless and morally wrong and emboldened others to form the MFA to overthrow the regime.[73] Taken together, these views contend that PAIGC victory was shaped primarily by Portuguese politics and civil military relations. The problem, however, is that they assume that the revolt resulted apart from the broader strategic context of extrasystemic war and overstate the role of revolutionary officers who were part of the long conflict that involved a far larger number of soldiers. They also undervalue the complicated development of the war.

In contrast, sequencing theory offers a more comprehensive framework to analyze the development of the extrasystemic war. Specifically, the Maoist model sees the revolution as but a culmination of the fifteen-year sequence that included periods of intense guerrilla war and conventional war and efforts at building an independent state structure. Counterfactually, even if the 1974 revolution did not occur when it did, the Guineans were likely to have won the war, probably at a later time, because they had evolved through the accumulation of so much social, political, and military capital. As Gibson argues, examination of the development of the anticolonial struggle in Africa in the second half of the twentieth century must begin with its specific geographical, economic, social, and historical context.[74] In addition, Portuguese Admiral Rose Coutinho said that "we should not forget that it was the colonial problem that triggered the Portuguese revolution."[75] The Carnation Revolution was necessary but insufficient for Guinean victory.

Other Explanations

The literature of asymmetric war offers good insights into why the PAIGC managed to beat the Portuguese forces. As in the other chapters, here I select the theories that I consider pose the greatest challenge to sequencing theory. First, the balance of resolve theory posits that the weaker side in war, facing a greater chance of failure than the stronger side, makes more effort to withstand the pain of difficult war and thus is more likely to out-will the opponent. While the theory describes the PAIGC's commitment to the war and survival in the balance of resolve as key to the its victory, the theory stands on a weak empirical ground. If the PAIGC fought better as Portugal's military morale declined, as the theory suggests, then it is incongruous with the fact that Portugal continued to raise its military budget and troop size even though it continued to suffer growing casualties and equipment losses. There were many battles that Portuguese forces lost, which in turn increased the PAIGC's resolve.[76] Indeed, as Dhada shows, Portugal became committed to maintaining its control in Guinea when national security spending rose. In 1963, Lisbon increased the budget when the initial figure turned out to be short of meeting war requirements in order to augment conventional forces and enhance logistic capabilities. This increase allowed the army to grow its size by ten thousand men, which doubled the size of the army operations two years earlier.[77] This increase in the budget and operational size underscores an increasing resolve of the Portuguese. Thus, even if the PAIGC did retain a higher resolve to fight the war than the Portuguese did, resolve may not have been the main determinant of the war outcome.

Second, the theory of strategic interaction posits that particular configurations of military strategy affect the likelihood for insurgent success. According to this logic, the insurgents won the war because they adopted a different strategy—either conventional strategy or guerrilla strategy. Had they adopted the same strategy, the theory would predict that the insurgents would have lost the war.[78] The theory has three problems. First, it assumes that military strategies do not change, but the Guinean War evolved over time; military strategy, operations, tactics, main battlefields, leaders, and organizations all changed. The PAIGC's evolution from a small guerrilla group into the FARP and FAN has especially indicated the change. The theory falsely assumes the initial combat model locked in for the war's duration. It makes a better fit for wars that are shorter than the Guinean War. Second, the theory is inconsistent with two instances. First, between 1961 and the late 1960s, both sides were fighting guerrilla war. According to the theory, at this point the Portuguese should have won the war. The war, however, did not end. And in the final years of the war modern forces were at center stage. Again, if the theory was valid, the Portuguese should have capitalized on their conventional military advantage and defeated the Guineans, but they failed. For these reasons, there is strong doubt about the validity of the theory.

The last theory concerns the role of external support. This theory centers on two related arguments about the role of foreign support to the insurgents. Record argues that external aid is the key to underdog victory because foreign assistance has played a crucial role in war outcomes,[79] while Staniland maintains that insurgent cohesion and performance generally revolve around the nature of external networks that groups have.[80] Indeed, external support did play some role in the PAIGC's effort to increase its capability. Cabral's well-kept network of African allies and Russian, Chinese, Cuban, and Eastern European countries as well as various international conferences favoring decolonization movements provided key financial, moral, and military support. Sekou Toure's socialist regime in French Guinea allowed the PAIGC to set up base camps there and receive training in Cuba.[81] Kenneth Maxwell argues in a similar vein that PAIGC victory was the result of skillful political and diplomatic actions of its allies in Mozambique and its sister movements. From this perspective, Guinea's success was a triumph for African nonaligned movement across Algeria and Zambia and all the diplomatic efforts to help each other's decolonization movements.[82] Yet the theory falls short. External support was unlikely the sole determinant of

PAIGC victory. Cabral sought international support as one way to enhance the PAIGC in Guinea. Yet it was the strategic shift that took place from the political institutionalization period of the PAIGC to the guerrilla and conventional war phases that became the center of analysis. While regional and international actors sympathetic to the PAIGC movement did offer various forms of aid, they played mostly a contextual role in the strengthening of the party.

CHAPTER 9

The Progressive Model: The Indochina War (1946–1954)

> The great victory of the Vietnamese people at Dien Bien Phu is no longer, strictly speaking, a Vietnamese victory. Since July 1954, the question which the colonized peoples have asked themselves has been, "What must be done to bring about another Dien Bien Phu? How can we manage it?"
>
> —Franz Fanon[1]

The progressive model of the sequencing theory is best represented by the Indochina War, a war between Vietminh and French forces between 1946 and 1954. Apparently an underdog in this war, the Vietminh fought face-to-face with the colonial master that had dominated Indochina for decades.[2] The drama about this case was that after years of struggle, the Vietminh achieved the unthinkable. Vietminh leader Ho Chi Minh, who led the independence movement, proudly stated that "against the enemy's airplanes and artillery we had only bamboo sticks. . . . The discrepancy between our forces and the enemy's was so great that at the time some people likened our war of resistance to a fight between 'a grasshopper and an elephant.'"[3] Like that outlined in the previous chapter, this case study reveals a great amount of surprise about what the underdog insurgents can do to upset their colonial master in an environment quite hostile to their presence, although through a different sequence than the Guineans adopted.

Of the five wars that I classify as a progressive model in the extrasystemic war dataset, I select this war as a case study because it underscores one of the two best ways for insurgents to fight in extrasystemic war. Furthermore, it reinforces the power of the sequencing theory as an alternative framework to other explanations of the war. Indeed, the progressive model runs counter to one of the most popular explanations for Vietminh victory, which is embedded in the strategic ideas of Mao Zedong. Another powerful thesis in the literature revisits the role of antiwar forces in France as the key to French defeat because these opposition movements shifted resources to suppress decolonization movements in North Africa and therefore made only limited commitment to the war in Indochina. Theories of asymmetric war provide still different explanations, such as the balance of resolve (that the Vietminh were more determined to fight than France because survival was at stake), strategic interaction (the Vietminh won the war because they adopted a different strategy than France), and external aid (the Vietminh won the war because Chinese support proved decisive). Against these interpretations, sequencing theory deploys the progressive model, which posits that the primary reason for Vietminh victory rested with the fact that they adapted and evolved.

This chapter includes four sections. In the next section, I briefly discuss the background of the Indochina War. In the third section, I use the progressive model in the Indochinese context to argue that the Vietminh won the war because they evolved, and I review each of the three phases: guerrilla war, state building, and conventional war. In the final section, I examine three extant theories of asymmetric war.

Background

The Indochina War originated from the extension of French colonialism in Southeast Asia and the birth of a movement for Vietnam's independence in the 1940s. France's overtures in Indonesia had been part of its broader strategic interest in the region that was shaped by the desire to maintain empire as well as the consideration of political elites in France and its strategic culture.[4] Placed under French colonialism for decades, Vietnamese leaders had long sought a negotiated settlement for self-rule. After World War II, the Vietminh took advantage of the strategic void left by Japan's departure and France's weakening grip of the territory. The August Revolution of 1945 set the stage for a national revolution and opened the gate for Indochinese autonomy, which allowed the Vietminh to take up arms and declare independence the following year. The French were not ready to invest in the war when hostilities broke out in 1946. For the French, the timing of war was suboptimal. The Vietminh were motivated to fight, but they were not prepared for the war at the time of political upheaval. When Ho declared independence, he had little idea how many people were armed and willing to defend the movement. His Liberation Army unit numbered no more than twelve hundred trained soldiers. For some time, the Vietminh's priority was to tell all these armed men not to provoke conflict with Chinese troops crossing the border. Other Vietminh groups skirmished with the Vietnam Nationalist Party and Vietnam Revolutionary League that accompanied Chinese forces.

The Indochina War was asymmetric as well as extrasystemic. Between the French and Vietminh, military balance broadly favored the French, who had advantages in firepower, supplies, and manpower. French troops totaled nearly 75,000 men in 1946, with more than half of them located in the south. Later they deployed an additional 500,000 troops, 160,000 of whom were in the Vietnamese armed forces, including 120,000 regular troops and 47,000 auxiliary forces. There were also 70,000 Vietnamese serving in the French Expeditionary Corps and another 20,000 in intelligence.[5] Of these, 420,000 were drawn from the French Union and composed of the expeditionary corps, including Africans, Legionnaires, and Indochinese. By 1954, they had grown into a corps of 605,000 troops. In contrast, Vietminh forces were much smaller. They totaled only 82,000 men, composed of thirty-two regiments and eleven battalions for northern and central Vietnam. The regular army had somewhere between 40,000 and 45,000 men, soon growing up to 291,000, of whom 185,000 were classified as regulars.[6] These regular troops were nevertheless drawn from the peasantry and were largely underorganized, underarmed, and poorly led.[7] Training was minimal, maps were scarce, and illiteracy was widespread.

The Progressive Model and the Indochina War

The progressive model best captures the way the war developed in a sequence. It is in part because Vietminh leaders themselves treated it as such. Indeed, ideas central to the sequencing approach came from those leaders who sought to learn from foreign experiences. As I mentioned in Chapter 2, ideas of sequencing had proliferated from Lenin and Mao to various parts of the world where respective leaders of extrasystemic war strove to marinate with local conditions. The Vietminh's case underscores this learning trend of the time. Ho Chi Minh, for instance, had spent years as a communist in France, studied revolutionary movements, translated Chinese publications into Vietnamese, and published such works as *Experience of Chinese Guerrillas*, *Experience of French Guerrillas*, and *Experience of Russian Guerrillas*.[8] In particular, he praised the Soviet Red Army for "know[ing] the method of fighting" and agreed with the Soviet emphasis on the need to avoid excessive stress on modern armaments, the need for good education and a good relationship with people, and the role of virtue and talent among commanders.[9] These strategic thoughts were shaped by Mao's ideas as well, which considered sequences as central to victory and that Vietminh strategy would coalesce around them for the most part. William Duiker points out that the Vietminh's revolutionary inspiration came from China and they *intended* to follow the Maoist course.[10] Thus a national conference in 1950 called for the general mobilization of the population in liberated areas under the new slogan "All for the front, all for *people's war*, all for victory." Their faith was buoyed by the timely beginning of China's aid in northern Vietnam in the same year. After the war, Giap wrote that *his* victory was due to people's war of liberation.[11] Yet some remained skeptical of the Chinese role. Truong Chinh, in fact, cautioned that Chinese assistance would not guarantee success in Indochina and that Vietnamese people must rely on themselves to bring about victory.[12]

The Vietminh's view differed from the Chinese one in two ways. First, Chinh rejected Mao's analysis of the role of terrain in a revolutionary environment. Mao had argued that people's war in China hinged on two factors: (1) its semicolonial status that confined imperialist rule to cities and left rural areas relatively alone and (2) its large territory that allowed revolutionary forces to wage guerrilla operations in areas far away from enemy authority. In contrast, Vietminh forces had a limited physical environment in the north where they could create a base area and launch movements to expand their control down to the south. Yet while territorial size mattered in creating space for guerrilla retreat, what mattered more was the *suitability* of terrain to provide for peasant living, recruitment, and mobilization. Consequently, while Chinh admitted to the strategic importance of terrain in protracted war, he asserted that the combination of "popular support and a disciplined people's army could overcome the limitations of geography and colonial status." Despite territorial constraints on their operations, the Vietminh still had a chance to establish base areas.[13] Second, Vietminh strategists departed from the Chinese model in terms of their worldview. While the

Chinese model did not incorporate foreign affairs in its mechanism, the Vietnamese version assumed that external affairs shaped the resistance. Although reliance on foreign support was only a part of entire strategy, they recognized the linkage between foreign aid and victory; the more material they received, the more likely they were to win the war. Thus central to victory was the need to take advantage of the changes in the "world balance of power."[14] Developments in other parts of the region—notably independence in Cambodia and Laos in 1953 and 1954, respectively—provided a powerful impetus for them to follow suit.[15]

Not surprisingly, the Vietminh's strategic sequence differed from the Chinese one. Ho Chi Minh envisaged the war developing in three phases. In the first phase, spanning from the outbreak of the war in 1946 to the end of operations in northern Vietnam in the winter of 1947, he sought to preserve the main force. The second phase was when his forces would contend with the enemy, in the period between the northern campaign and 1954. The third phase was one of general counteroffensive that culminated at the Battle of Dien Bien Phu.[16] On the other hand, Giap advocated his own three phases. The first phase was guerrilla war against the French occupation. The second phase would set a strategic equilibrium with the French, by achieving a succession of small victories. The final phase was that of strategic counteroffensive, which, as Mao imagined, was to bring the battle in open terrain with Vietminh forces face-to-face with the retreating French.[17] Other leaders had different sequences in mind. Colonel Ly Ban argued that, without much elaboration, the war would be won in three different phases: (1) passive resistance, (2) active resistance and preparation for the counteroffensive, and (3) general counteroffensive. Hoang Van Thai, chief of the general staff, suggested (1) partial offensives, (2) raising a tactical force superior to the enemy, and (3) general counteroffensive.[18] The Vietminh made no effort to reconcile these differences, which underscored the variation in ways that these leaders planned the war. More important, none of these views precisely captured the war as it unfolded.

Indeed, the war followed the progressive model. The war began as a low-intensity conflict between the motivated Vietminh and French forces that hurriedly arrived in the region to suppress the independence movement. The early years of the war were characterized by a number of skirmishes between lightly armed insurgents and mobile French units. Although the Vietminh made some progress during this phase of guerrilla war, because they realized that they were not fighting decisively enough, they proceeded to strengthen their political institutions as they made a transition to a second phase of state building. Such a shift was supported by the growing success on the part of the Vietminh to exploit their position and power in combat through military modernization. In sum, the progressive model shows that the war went well for the Vietnamese because each of the three phases of the model was properly followed in sequence in a mutually reinforcing manner. Thanks to the Ho-Giap leadership, a relatively small guerrilla command with a weak political foundation evolved over the years to become a complex military organization and a proud sovereign. The army, despite limits on its powers of control during the first years of struggle, ultimately developed into a highly influential organization.[19]

First Phase: Guerrilla War (1946–1950)

The first phase of the war was characterized by guerrilla war. Vietminh preparation for this phase had begun years earlier. As mentioned earlier, Ho Chi Minh had studied guerrilla warfare in the Chinese context and irregular strategy from Soviet experiences. In his 1942 work on "Guerrilla Tactics," Ho wrote that "guerrilla war means making a surprise attack, attacking the enemy when they are unguarded. It is a war carried out by a people oppressed by imperialists with good weapons and a regular army." Guerrilla war consisted, according to Ho, of the following actions: hold the initiative, move rapidly, maintain the offensive, and plan appropriately and carefully. On the other hand, he defined guerrilla *maneuvers* as follows: avoid the enemy's stronger positions and attack the softer ones, avoid engaging in face-to-face battles and do not try to hold land at any cost, and attack the enemy in movement or when tired.[20] The best organizational form of the insurgency was that of decentralization. Thus the Vietminh dissolved the Indochinese Communist Party in 1945, formed a clandestine national front, and used it to lead the war effort on behalf of the party.

As expected, the Vietminh suffered multiple challenges associated with guerrilla strategy, decentralized organization, and voluntary withdrawal of territory even though the strategy suited them at this point. Absence of infrastructure and coordination mechanisms between branches meant that orders, instructions, and plans were slow to disperse. Troops were not organized into specialized units, commanders lacked experience, leaders failed to mobilize ethnic minorities, main units had been squandered, the strategic initiative had been lost, there was inadequate preparation for offensive operations, and units had not been formed properly to seize enemy weapons.[21] Each rural community followed its own development project. Communication beyond base areas was more challenging, with limited coordination between the disorganized units. Some use was made of wireless radios, but it was more common for them to transmit important operational decisions by couriers who took two or three months to go between Hanoi and Saigon by boat and on foot. Not surprisingly, the first few years of the war proved to be disappointing, with French forces seizing Hanoi and occupying most of the lowland regions in northern and central Vietnam. Second, French pacification operations proved to be powerful and spread French control further out into Vietminh territory. The forces were mostly infantry but included engineers, artillery, and cavalry. As George Tanham shows, they operated in columns that fit into the operational requirements that centered on the use of roads and dikes and generate firepower to be used in some of the battles of the early years. For instance, French forces launched a major offensive in the mountains north of the Red River Delta and put Vietminh safe zones near Hanoi in jeopardy, which prompted Ho to order reinforcements, only to be overwhelmed by French reinforcements attacking their rear.[22]

These challenges forced the Vietminh to innovate in ways that would boost their operations. The Vietminh disbanded regiments, built self-defense militia units in villages, and trained them in guerrilla war. The expectation was that when an opportunity arose they could recombine, accomplish the mission, and disperse again.[23] Consensus emerged that they would put up a better fight in a guerrilla than in a conventional war. Domestic weapons manufacturing also increased with new factories and arms workshops. The Vietminh carried out land reform and literacy campaigns to sustain popular support.[24] The Vietminh also launched campaigns to encourage the public to increase overall agricultural production and donate supplies to patriotic funds. They limited taxes on farmers in order to prevent starvation during the food shortage until the rice harvest in 1946 and keep popular support.[25] Revolutionary forces steadily rose in size. In 1948, Vietminh leaders recognized that the stage of strategic withdrawal had been completed and paved the way for a new stage of "strategic equilibrium."[26] The Vietminh extended operations into the south and into Laos and Cambodia in an attempt to force the enemy to disperse and increase their vulnerability. The Vietminh emerged from the initial phase of the war in better shape than the statistics demonstrated.

The key reason why the Vietminh survived the guerrilla war phase was that they succeeded in winning popular support to maintain the nationalist movement. They did so by reinforcing their organization and internal cohesion. They delegated to rising leaders from the peasantry a good deal of decisionmaking authority to plan and execute operations. Using a guerrilla command inherited from World War II, Ho and Giap paid meticulous attention to logistics, administration, and stockpiling of materials.[27] Their policies widely appealed to the peasants. They tapped into the vast peasantry and built irregular troops and trained them in logistics and on sabotaging, harassing, and ambushing. They won popular support and gained a key source of supply, recruitment, lodging, and intelligence. The need for popular support was keenly recognized by the Vietminh leadership. Giap stated that he needed "to educate, mobilize, organize, and arm the whole people in order that they might take part in the Resistance was a crucial question."[28] Vietnam scholars see that popular support was integral to the success of the resistance.[29] In the words of Ellen Hammer, the popular support "came also from techniques of control and organization through which the Ho government reached down into the peasantry, rally support and eliminating opposition."[30]

Second Phase: State Building (1950–1952)

After the first few years of guerrilla operations, the war turned better for the Vietminh, but they realized that their chances of victory had not tremendously improved. Indeed, success in guerrilla war alone was not sufficient to win the war. The war was going well enough for them to transition to another phase. They realized that they needed new institutions, public services, and infrastructure across the north in order to consolidate their base areas and buttress their political claims for independence. The birth of a modern Vietnamese state proved to be a crucial phase of the war, which was based on Lenin's call for the vanguard party leadership to take the role of leading the peasantry into self-determination. Thus the second phase of the war was characterized by the growing realization on the part of the Vietminh of the need to strengthen their institutions, especially their political party, as a means of winning the war. According to William Duiker,

> Vietminh strategy had become an amalgam of the ideas of Mao Zedong and Ho Chi Ming with a dash of Lenin thrown in. Lenin had laid the political groundwork with his concept of a four-class alliance, while Maoism weighed in with the strategy of people's war. Ho Chi Minh adjusted Leninism in order to broaden the base of the movement and then contributed his own views on the importance of political struggle and diplomacy.... What had been created then was not so much a new model of revolutionary war as a patchwork of ideas designed to meet the particular circumstances of the Vietnamese revolution. As an exercise in pragmatism, it was vintage Ho Chi Minh.[31]

The institutional engine of Vietminh state building was the Indochinese Communist Party (ICP), augmented by various political and economic programs designed to enforce its policy. The centrality of the party system in the effort of independence stemmed from their realization in the late 1940s that political development must sustain popular support. They resuscitated the ICP in 1951 to boost the state-building framework and added an elaborate network of educational and tax systems, military courts, and communication and liaison offices. In addition, they upheld the 1946 constitution, built a cabinet of advisors the same year, and ordered the draft. Ho Chi Minh became minister of foreign affairs and president, replacing a fugitive leader of the Vietnamese Nationalist Party, another independence-minded movement later to be eliminated by the Vietminh. Giap was chairman of the National Military Council as well as minister of national defense responsible for overseeing the war. These functions generated a number of follow-up projects, including lowered rent, new bureaus, and cooperatives in cotton, silkworms, and textiles. As Hammer shows, they sought economic self-sufficiency by promoting various projects, including ones to encourage peasants to collectively produce rice and promote competition between administrative districts in fighting famine and illiteracy.[32] An idea emerged gradually that the Vietminh should move to the next stage of conventional war provided they had advantage over the enemy in battlefields even though the French had material superiority. But Ho expressed his reservations about the plan, calling for patience to wait for a proper time for another offensive. Chinese military observers, too, cited a number of deficiencies in the Vietminh forces, including problems of unit cohesion, discipline, battlefield inexperience, and command and control issues.[33] So instead, they remained committed to filling the critical political void by building a series of organizations in the military, such as the General Staff Headquarters, Arms Office, and Vietnam Military-Political Academy designed to train young officers.[34]

There were two factors that facilitated Vietminh state building. First, the success in building state apparatus was in part a function of the strong popular support won in the first phase of the war. Ho had emphasized a powerful role played by party leadership in revolutionary guerrilla war. A slogan emerged in the concept that "National Insurrection (must be) carried out by the entire population under Party leadership"; this became an intellectual pillar of the movement. State-building projects encompassed efforts to reform land on the principle of "land to the tiller" and expanded the branches of small industries, which supplied consumer goods. The Vietminh tirelessly invested in the production of rice and subsistence crops, irrigation, and transportation systems. These efforts gradually paid off, as they began to draw profits from the operations of national banks, educational systems, and literacy campaigns.[35] Ho later stated, "The people of a small colonial country can carry the day if they are led by the proletariat and its party, rely on the broad masses of the people, first on the peasantry, unite with all patriotic sections of the people in a united front and win the approval and support of the revolutionary movement of the world, first and foremost of the powerful socialist camp."[36] In fact, his role was vital in gaining legitimacy from his allies in the Soviet Union and China and from nationalist leaders in South Asia and Southeast Asia who opposed French imperialism.[37] The state-building phase was possible because it followed the first phase of guerrilla war in that the insurgency combat had brought about support from non-communist, nonpolitical elements in Vietnamese society. As Hammer argues, a wide popular base was provided by the *noncommunist* intellectuals.[38] These changes took place across the emerging nationalist sentiment within the Vietminh. Without this political phase of state building, the sequence of actions that propelled the Vietminh to victory would not have been completed. The political factor was connected with social factors that exercised mutually reinforcing effectiveness to make the war easier for the Vietminh.

The other reason was that French efforts at state building fell short. They failed to have a succession of cooperative local leaders in the proxy government. In 1949, France extended nominal independence to Bao Dai's south in the French Union, in which France controlled all defense issues and foreign policy. Unpopular from the beginning, Bao Dai built no national assembly or political parties to support his authority. What existed were groups of self-interested religious individuals who carved out and defended their own zones of influence. A growing gap between French and Vietnamese interests discouraged the former from making unnecessary overtures in Vietnam. Bao Dai built a national army and held national elections. This French mismanagement of colonial politics left French officers unhappy. General Georges Revers, chief of staff of the French Army, stated that "if Ho Chi Minh has been able to hold off French intervention for so long, it is because the Vietminh leader has surrounded himself with a group of men of incontestable worth.... [Bao Dai, by contrast, had] a government composed of twenty representatives of phantom parties, the best organized of which would have difficulty in rally twenty-five adherents."[39] Bao Dai himself did little to make his government either more representative or efficient. He promised elections only in areas under his control, which constituted merely 25 percent of the national territory and roughly 50 percent of its citizenry. Only one village out of three was regarded as sufficiently "pacified" to vote. The right to vote was given only to those who had registered in the census taken in 1951 for purposes of mobilization, which led to a reduction in the entire constituency by targeting people who had already given in some measure to the authority.[40] What was worse, the voters preferred the Vietminh manifesto. Therefore, the elections repudiated the French policy of divide and conquer as well as the Bao Dai solution. The French conception of Bao Dai rule failed to provide a powerful counterweight to the Vietminh. Bao Dai spent much time visiting resort towns and essentially remained outside the process of government. A similar attitude was seen in the nationalist behavior of Prime Minister Nguyen Phuc Buu Loc, Bao Dai's cousin, who pushed France to recognize independence. Nationalist elements within the Bao Dai regime saw the enemy in France, not the Vietminh. His successor, Nguyen Phan Long, also failed to win over the people, even though they refused to collaborate with the French. Long sought to build an independent army, claiming that the people wanted their own troops, a gesture that alienated the French authorities. Nguyen Van Tam, prime minister between 1952 and 1953, also proved to be unpopular and supported only by politicians with no party affiliation. The Bao Dai zone was based on feudal units divided by major regional groups. These so-called Big Five groups sought to strengthen control over their own territories and to consolidate their private armies against Bao Dai and each other.[41]

Final Phase: Conventional War (1952–1954)

The Vietminh's survival in the second phase did not mean victory; instead, it ignited the perception that the movement must evolve. It was at this point that Giap wrote that the revolution "had to move" to a third stage and that they needed to field adequate military capability to defeat the French. The so-called general law of a revolutionary war dictated that this war would be won only when they could liberate land and defeat much of enemy power. It also assumed that mobile war would become the main part of the final phase in which they would capitalize on the spiritual advantage they had to gain an upper hand, while they would avoid positional war in which France had material superiority.[42] The final two years of the war would thus be dominated by combat operations, although preparation for it had begun some time earlier. Ho had founded a regular army in 1944 and begun to modernize their forces in 1945 when they formed the Liberation Army of Viet Nam. In 1947, he had formed mobile battalions. In 1950, Giap had already proposed to fight conventional war, believing that they had acquired sufficient capability to launch counteroffensives and that international factors were turning in their favor. Resistance forces would thus move toward a final showdown with the enemy, moving from limited attacks on smaller positions to large-scale attacks on major garrisons. The imbalance of military capability began to disappear two years after 1950 when China began sending massive amounts of military materiel and advisers across the border. The beginning of Chinese aid was matched by British and American aid to the French, but the Vietminh used resources effectively to modernize their armed forces, starting with the reorganization of his irregular forces into five full infantry divisions. Naturally this period of the war saw the intensification of armed combat. The Vietminh improved their defenses and fortified their villages so as to make French clearing operations more difficult. They issued manuals on counterencirclement tactics, following the general rules that Vietminh combatants caught in a tight encirclement would target weak points of the encircling forces. Some successful operations in the Tonkin area forced the French into a defensive posture.

The determinant of Vietminh success in the second phase was the creation of institutions intended to bolster the political integrity of emerging statehood. By this point the institutions began to extend to the military side. The Vietminh rebuilt their battle corps, increasing the number of units and giving these organizations power to ensure proper command and control and indoctrinate soldiers. They also restricted the power of commissars in order to clarify that military and political commanders would decide issues of their responsibility. They reformed the General Staff in 1953, which would have four staff divisions: local security, communications, accounting, and administration. They placed the army's regular units directly under the High Command, which controlled a number of territorially divided units and subunits below that controlled regional and popular forces, and organized the lowest echelon at the level of villages. These organizational changes took place concurrently with changes in the rules of conduct for soldiers through a number of directives to minimize conflict between political and military officers. In addition, they developed an extensive school system in the armed forces.[43]

When General Jean de Lattre took command, the French sought to create a stronghold in the Tonkin Delta region that would enable them to cut off Vietminh supply lines. He built a fortified line to hold the Vietminh in place and used his troops to smash them against this barricade, which became known as the de Lattre Line. Yet the line provoked a series of strong Vietminh efforts to penetrate it and caused the war to intensify. The French soon shifted attention away from de Lattre Line and resorted to the ideas of his predecessor, General Marcel Carpentier, who had established a system of posts and forts to protect the delta and provide a secure base and relied on mobile units for offensive operations. The general therefore regrouped his troops and organized mobile groups to act as his main force.[44] However, French gains from this interaction were negated by the increasing opposition in France to the war. Although all of their forces were volunteers, their officers were being killed faster than they could train new ones. The growing frustration at home impeded any further attempts to increase the troop size.

The Vietminh then managed to cut French supply lines. They continued raids, ambushes, and skirmishes, but in 1954 both sides retreated briefly to prepare for larger operations. Giap attacked French garrisons and extended control over much of Tonkin beyond the de Lattre Line. De Lattre's successor, General Raoul Salan, launched Operation Lorraine along the Clear River to force Giap to relieve pressure from Nghia Lo outposts. This allowed thirty thousand French soldiers to get out of the de Lattre Line to attack Vietminh positions at Phu Yen. Unable to react to these offensives, Giap waited for the French supply network to be overextended, cut them off from the Red River Delta, and ambushed the French at Chang Muong. These intense operations were never decisive, with the French fending off the growing enemy and the Vietminh searching for ways to defeat the French in the open. They sketched their plan for a general offensive that would extend the conflict into Laos and Cambodia, make French outposts more vulnerable, and cause a political problem for the French. They hoped that this move would weaken the French resolve to fight on and force a cease-fire, although the French were not interested in negotiated settlements at that moment. At the same time the Vietminh remained strongly keen to the need to make a military breakthrough and capitalize on the French vulnerability in the north. The chosen place was Dien Bien Phu, where Vietminh forces had a clear geographical advantage in logistics and access to Chinese equipment.[45]

On the French side, Salan and later his successor, General Henri Navarre, became more tied to the defense and faced declining morale among their troops. Faced with supply problems and heavy losses, they reluctantly abandoned one of their strategic outposts in Hoa Binh, a village not far from Dien Bien Phu. As the Vietminh occupied the areas around Dien Bien Phu, Navarre launched a series of raids into enemy rear areas, which in turn prompted Giap to invade Laos to extend the front line and force the French to disperse their forces. This move prompted Navarre to shift troops toward the Laotian border in an attempt to block Giap from going further north. He announced plans for a "general counteroffensive" designed to bring the strategic tide back in favor by deploying an additional ten battalions. Navarre ordered his troops to prepare for pitched battles, for which the Vietminh prepared five divisions to attack and opened a siege in March 1954. Giap outsmarted Navarre by equipping his men with artillery and trapping the French garrison in the valley. Once the battle began in Dien Bien Phu, French supply lines were interrupted while hard rain made dropping supplies and reinforcements difficult. Giap's troops overrunning most French garrisons, the Vietminh emerged victorious. Vietminh victory soon led to the Geneva Conference, where both sides agreed on a provisional division of Vietnam at the seventeenth parallel, giving the north to the Vietminh and the south to Bao Dai and Vietnam's independence, ending French colonialism in Indochina. The rebel group's primary objective of independence was achieved, albeit geographically compromised, whereas France's insistence that Vietnam not be partitioned, let alone be taken away from its empire, was denied. France lost significant economic interests around the Tonkin Delta region and a base of political influence in Southeast Asia. It is clear that the French were decisively defeated.[46]

In part, Vietminh infantry operations were effective as they fit the topography. They relied on foot soldiers until a large amount of artillery and tanks became available. Infantry tactics carried a few benefits. For example, the units avoided entanglements with tanks or artillery, which freed many of them from combat trains and heavy units. They had minimal dependence on road transport and thus were able to make long treks whenever necessary and stay mobile and flexible. Operational simplicity also enabled them to avoid detection and attack with surprise and gave them enormous advantages in speed and flexibility.[47] In addition, because the French could not deploy air support in the darkness and their artillery support was unreliable, the Vietminh took advantage of their mobility and knowledge of geography to carry out night operations, which normally involved four groups. The first group would man heavy weapons so as to neutralize enemy positions. The second group mostly consisted of assault engineers who would infiltrate enemy lines to make way for breakthrough and exploitation points. The third group—shock troops and assault infantry—would move forward to overwhelm enemy posts. The fourth group—reserves—would provide fire support when needed. Combined with ambush operations, these specialized forces offered highly reliable combat capability based on deception and mobility.[48]

These military advantages of the Vietminh left catastrophic damage on the French. At Dien Bien Phu, the French suffered a combination of several problems, such as shortage of airpower, poorly planned counterattacks, poorly built fortifications, intelligence, and supply shortages. France did not provide adequate air support and deployed only transport planes and a few fighter squadrons. Its mission within NATO in the early 1950s assigned France a small role in tactical support and air defense, and thus France did not possess aircraft capable of training heavy-bomber crews. There was no airfield capable of providing for the maintenance. General Charles Lauzin, commander of the Far Eastern Air Force, had reported a series of deficiencies in aircraft and personnel, calling for airfield construction and improvement.[49] Lauzin's predecessor, General Lucien-Max Chassin, had pointed out that it was the French army, rather than the air force, that dominated the French command and regularly ordered air missions on a "piecemeal basis," which forced ground commanders to operate with their own arsenal, and there was a shortage of coordination between the air force and navy. The focus on the army helped neither service contribute to strategic planning.[50] In addition, the Vietminh had more artillery pieces than the French in Dien Bien Phu, outgunning the French three to one on the ground. Logistical problems rendered the French incapable of interdicting weapons flows to Dien Bien Phu. The French also had irreconcilable problems between troop size and supply; the maintenance of larger supply drop zones would have required more reserves of troops, but the presence of larger troop reserves would have required more supplies.[51]

Two factors explain why the Vietminh ended up outperforming the French in the final phase of the war. First, the institutions built earlier increased Vietminh military capability. As soon as the guerrillas were incorporated in the regular army, they received proper combat suits, arms, supplies, and training. Integration of soldiers had begun in the late 1940s and led to improved war planning, more frequent offensives, and larger units. It accompanied a shift of greater focus on the regular forces and mobility, while incorporating the regional and communal forces as well. These institutions played a key role in allowing armed forces to fight with proper and clear instructions about what to do in combat zones. This worked along with lessons learned from an earlier failure to fight the French. Giap noted the dangers inherent in conventional combat: "For an army relatively feeble and poorly equipped, the classical concept of war is extremely dangerous and ought to be resolutely rejected." As Tanham argues, main failures had stemmed from the premature departure from guerrilla tactics. Failure in the south was due to the shortage of political guidance given to the troops, while in the north mistakes were made in military training.[52] In the third phase, the war matured with the modernization of the Vietnamese forces. Chinese material and advisory support expanded clandestine operations in Burma, Singapore, and the Philippines. The Vietminh maintained irregular forces during the process of reorganization, but these forces underwent organizational modifications to support regular forces. These groups grew in size in the early 1950s.[53]

Scholars disagree over whether the third phase was completed when the war ended. According to Giap, the second phase of the war began in 1947, and it was still in effect in 1950. According to Tanham, however, "most observers place the beginning of the second phase in 1949, as the French retained the initiative during 1947–1949 and the first Vietminh attacks of a formal nature did not occur until 1949."[54] According to Hammer, the Vietminh were still at the second phase in 1953, "close to its guerrilla origins, based on the population, fighting a war of movement and maneuver." Giap's troops rarely sought large-scale encounters in the late phases.[55] Of course, the difficulty of pointing out the timing of transition is associated with the challenge of overlaps between phases. It is true that even Mao had recognized the challenge of selecting the right moment for launching the general counteroffensive. Likewise, Giap found it difficult to pinpoint an exact date of the beginning of the second phase.[56] He felt uncertain about the possibility of drawing clear divisions between the phases. In fact, he was careful about the timing of moving to the third phase, as he conceived of this phase as having several subdivisions. Before progressing to the third phase, according to Tanham, the Vietminh sought to be certain that (1) they had moral superiority in the revolutions, (2) they would make improvement in material power, (3) they had a favorable international environment, and (4) they had a clear sense of direction.[57]

Other Explanations

The literature on asymmetric war provides a host of reasons why the insurgency succeeded in the end. First, balance of resolve theory posits that the weaker side in war is more likely to out-will the opponent because it faces a greater chance of failure than the stronger side and makes more efforts to withstand the pain. Indeed, the ICP stressed the role of resolve to fight. For the Vietminh, the war was a matter of survival, since defeat meant the destruction of freedom and another decade of imperial subjugation. Ho Chi Minh said, "We are ready to sacrifice some millions of combatants and to go on struggling many years to safeguard Viet Nam's independence and to save our children from a fate of slavery. The Vietnamese people are determined to fight and believe firmly that the resistance will win."[58] For the French, at stake was not only the stabilization of their colonial control in Southeast Asia but also their reputation as a recovering imperial power capable of keeping their other colonies, such as Algeria, intact. Although survival of a national polity was not the main concern, the entire French imperial system had been shaken by its defeat in 1945.[59] According to Charles Kupchan, Indochina reflected the dynamics of defense priorities in the eyes of French leaders, so they shifted a significant portion of resources from continental defense to Southeast Asia at the risk of potential German rearmament and Soviet attacks.[60]

However, there were two instances where the theory is inconsistent with evidence. The first instance was when the war started. At this point the Vietminh were likely to have had a higher level of resolve than the French, who were militarily and financially shattered by World War II. If the theory was valid, the war would have ended here, but the war went on. The French continued to fight despite low morale among French politicians and soldiers. This contradiction provides a ground to question the validity of the theory. The second instance was at the end of the war. In 1954, Vietnam's military capability played more significant roles than resolve. French concern about negative consequences of Indochina's fall on other parts of their empire likely made Paris more resolute. This high level of resolve was seen especially in the adaptation of the Navarre Plan of 1954, which called for massive increases in the Vietnamese army, French reinforcements, and offensive operations.[61] Furthermore, as Kupchan argues, French leaders arguably had incentives to make public their commitment to the war in order to avoid signaling weakness to their local allies and the Vietminh.[62] Thus, the balance of resolve was not against France in 1954, but unlike what the theory would predict, France still lost the war. The role of resolve in the outcome was indecisive.

Second, the theory of strategic interaction posits that the weaker side is likely to win the war if it adopts military strategy that is different from that of the strong. According to this theory, the Vietminh won the war because they adopted a military strategy that was different from that of the French. Specifically, they adopted guerrilla war as France fought a conventional war. Alternatively, they won the war because they adopted conventional war and France fought a guerrilla war. However, the theory fails to show that the Vietminh guerrillas won the war despite the fact that they used the same strategy of conventional war as the French in the final years of the war. Instead, what happened was that by 1954, the Vietminh had evolved into a modern army and had managed to trap the French in terrain most vulnerable to artillery fire and infantry attack. The theory also fails to explain the fact that although in the initial years the Vietminh primarily used guerrilla strategies and the French forces were designed to fight conventionally, the former did not win the war. If the theory is right that the regular-irregular strategic dyads favor the weaker side, then the insurgents should have won the war. Because France had conventional military strategy throughout the war, this theory posits that the Vietminh had a guerrilla strategy at Dien Bien Phu when the war concluded, which is historically inaccurate.

Finally, external material support has played a role in determining the outcome of war between unequal powers. Record suggests that no war ends in favor of the weak without external support.[63] From 1950 on, China sent weapons and advisors into Vietnam to help modernize Vietminh forces because it saw the war as a key battleground in the regional extension of the Cold War.[64] The Vietminh benefited significantly from Chinese aid, and it is hard to consider in hindsight that the Vietminh had much chance to win the war without Chinese assistance. The problem with this view, however, is that external aid is largely a contextual factor *indirectly* related to the outcome. There is little evidence that it was foreign aid that put the Vietminh disproportionately in favor. At the height of the Korean War, Washington's support of France grew and the United States paid a majority of France's costs. American aircraft dropped bombs at Dien Bien Phu. Mark Lawrence and Fredrik Logevall argue that Western aid to the French in fact was more substantial than Chinese aid to the Vietnamese, and the gap grew as the war went on.[65] Chinese aid was not enough to build a state apparatus that Ho and Giap did. While external support was necessary for the Vietminh to win the war, it was not a sufficient factor.

Conclusion

The central argument of this book is that insurgent groups are likely to defeat foreign states in war when they achieve an orderly combination of three phases: state building, guerrilla war, and conventional war. Adaptation and evolution through the right sequences, therefore, are central to success and failure. Evolving in the "right" sequences, however, imposes a significant burden on those seeking to achieve it. Insurgents need to outperform their enemy in the first phase, whether in guerrilla battles or in state-building efforts, which allow them to move the war on to the second phase. The transition alone will not be sufficient until they repeat it to complete a three-phase sequence. In successful sequences, normally either guerrilla war or state building comes in one of the first two phases and culminates with conventional war. This is because it generally takes a significant amount of social, political, and economic resources to produce an effective army. The order of the phases is critical because they must be sequential to be effective. The integration of these elements in orderly fashion has saved many insurgent groups from destruction. The very aspect of using these sequences is essential to sequencing theory.

I conclude this book by drawing a broader set of implications for the study of international security. First, I offer a set of theoretical findings from earlier chapters and bring together the case studies to empirically reinforce sequencing theory. A discussion of strategic implications for the United States is in order. Second, I apply the theory to the war in Afghanistan briefly to make a preliminary assessment of this important war. The aim is not to provide a comprehensive picture of the war, since a growing number of recent accounts have been published elsewhere,[1] but to buttress the analytical usefulness of the theory as a tool to describe the process of the war. The war has so far unfolded on a path close to the primitive model, which means that, theoretically, the insurgents are likely to lose in the end. I present some evidence to support this theoretical expectation. Yet al-Qaeda and Taliban insurgents do remain potent forces, and it is still possible that they could launch a major state-building effort and establish a well-organized force, especially after the scheduled withdrawal of U.S. and International Security Assistance Forces (ISAF) coalition forces in 2014, to turn things around. Under the leadership of Mullah Omar and Abu Musab al-Zarqawi, respectively, the Taliban and al-Qaeda have on numerous occasions claimed victory on their own terms against Western powers involved in the mission, and they are likely to continue to spread their claim as the organization remains relatively intact, which undermines the primitive model's prediction. For these reasons, I suggest that the United States should not be overly optimistic. The war will continue to be messy and costly to the end. Yet because sequencing theory is not the only option we have in Afghanistan, I bring in three existing theories of asymmetric war to assess the war and compare the expectation of sequencing theory against them. In the final section, I offer broader implications of sequencing theory for political science. I discuss what this analysis means to nations engaged in other types of war, especially civil war. I conclude by identifying three ways by which the theory contributes to the study of international security.

Theoretical Implications from Empirical Evidence

The main theoretical finding of this book is that nonstate insurgents boost their chance of victory in extrasystemic war by fighting through proper sequences. These insurgents have over time slowly learned to do so through dissemination of information about conventional and guerrilla wars, and thus becoming more capable of outlasting their state adversaries through wars of attrition. This trend has become the norm only since the 1940s, however, which means that most insurgents failed until then, resulting in a mere 15 percent victory rate across the 1816 to 1945 time period (20 victories out of 130 wars fought). As a result, we see a remarkable difference in the probability of insurgent victory between the colonial/imperial and decolonization periods. That is, insurgent forces lost most extrasystemic wars in the colonial and imperial era because they fought mainly in simple conventional and guerrilla methods. In the post-1945 era of Third World independence, however, they have won many of these conflicts by fighting through more complicated sequences, as seen in Maoist and progressive models, combining various military strategies and embracing institution building as part of war efforts. Even when insurgents fight in primitive models (e.g.,

Algerians against the French in the 1950s and Afghan forces against the Soviets in the 1980s), the lethality and effectiveness of the very terrorism and guerrilla method itself have increased dramatically to generate terrible outcomes for the outsiders. The introduction of sequences, however, has not necessarily revolutionized extrasystemic war. Even in the post-1945 world, many insurgencies have had difficulty adopting the right sequences. In Iraq's case, even after the Iraqi soldiers morphed into a Sunni insurgency, the group never grew out of a primitive insurgency and never adopted visions for mature statehood. As I show below, even al-Qaeda has not moved beyond the current primitive phase, although it has not fared particularly badly, either, against the U.S. and ISAF forces. In short, sequencing theory points to the variation in the probability of insurgent success based on whether they succeed to evolve in wartime. The empirical chapters that followed the first three chapters of theory development captured the variation in the processes and outcomes of extrasystemic wars.

Individual case studies in Chapters 4 through 9 demonstrate this variation. In Chapter 4, I showed that the conventional model does not work for insurgents because they find themselves overwhelmed by enemy power even when, as the Dahomey war of the 1890s indicated, they were able to prolong the war through attrition. In Chapter 5, in order to demonstrate how guerrilla strategy does not always work for insurgents, as in the primitive model, I investigated the Malayan Emergency in which Chinese communists in Malaya adopted a guerrilla strategy against superior British forces in the 1960s. While they managed to stretch the war for over a decade, they failed to win popular support, build an independent communist state, or garner an organized force before they were decimated in the wake of an emerging independent Malaysia. Then, I explored the "premature" model in the Somali context in which local rebels fought the British forces first by adopting what seemed to be a guerrilla strategy. The strategy failed because of a large power gap, and the conventional war that ensued did not turn the situation for the better for the insurgents, and they were finally defeated by British firepower and airpower. In Chapter 7, I showed that hostile Sunni insurgents of twenty-first-century Iraq fought a "degenerative" conflict against the U.S.-led coalition forces. Recovering from the destruction of the Republican Guard and organized Ba'athist forces in 2003, the rebels managed to drag the war out for years, but in the end they failed to unify a movement, build a state, and establish a well-functioning force. In Chapter 8, I applied the Maoist model to the Guinean War of independence, a case in which a small number of insurgents ended up defeating state adversaries through years of struggle. The PAIGC managed to fight a rural guerrilla insurgency in ways that helped the party build up its urban base and strengthened the military branch of its party operations. The path to the conventional war phase was rough and time-consuming, but the PAIGC succeeded in pushing the Portuguese forces to the brink of defeat, before the adversary succumbed to a revolution that brought down the Portuguese empire. Finally, I used the progressive model to characterize the three-phase evolution of the Vietminh, who demonstrated the impressive power of an underdog, built up an organized state structure, and fought what they knew would be a difficult war of independence. The Indochina War saw initial state-building efforts through the Vietminh's party enhancement, followed by the intense insurgency in various parts of the northern Vietnam, which culminated with modern battles where the Vietminh fielded a mature armed force.

The variation captured in the empirical chapters reinforces the utility of sequencing theory to explain different patterns of extrasystemic war. With its theoretical inspiration drawn from the field of biological evolution and developed through the ideas of Lenin and Mao, the book posits that what is central to our understanding of why some insurgents succeed and others fail is whether they can fight the war in ways that allow them to evolve in proper sequences. The empirical case studies also underscore the usefulness of the sequencing framework to describe the fact that those insurgents who adapt properly are likely to have a good chance to accomplish their objective while those who fail to do so are most likely to see their chances dwindle. Ultimately, what makes a difference between "proper" sequences and those that are not is the presence of a state-building phase in the process of evolution. Insurgent leaders fighting in extrasystemic war, furthermore, must learn to fight through sequence from overseas experiences if not from their own past. Those who adopt those sequences and adjust them in local environments are most likely to come out of war successfully.

In the literature of asymmetric warfare, sequencing theory positions itself in competition with a few hypotheses. As I show below, the existing views, stressing the role of resolve, strategic interactions, strong actors' mechanization of armed forces, or weaknesses in democratic institutions, have improved our understanding of how weak actors fight and defeat strong actors. Arreguin-Toft's and Lyall and Wilson's approaches, for example, have correctly identified the growing likelihood of underdog success in modern times. The theories are not particularly suited to explain how nonstate insurgents fight as they evolve against powerful state adversaries in extrasystemic contexts. In many instances, however, these theories can work with sequencing theory to better account for why insurgencies defeat states when they do. In fact, sequencing theory often complements these works, while putting forth a theory of its own focused on the way extrasystemic wars develop over time.

At the broader level, the world of global security has been changing dramatically in the past few years, especially when we look at relations between major powers. The post–Cold War world has seen the relative decline of U.S.-led unipolarity and the gradual rise of China as a chief contender for leadership in the new world order. China has arguably replaced Russia as a global challenger to American power, with its influence challenging the security and prosperity of its neighbors and the stability of international politics, while presenting a new engine of regional economic development and financial opportunities around the globe. China's rise has great implications for U.S. policy and strategy at both global and regional levels, most recently prompting responses from the United States in the form of the "pivot," or strategic rebalancing, to the Asia Pacific region. Other key countries, especially the so-called BRIC nations—China along with Brazil, Russia, and India—have grown to become major economic powers in the past few years and significantly influence world affairs. In the meantime, the global financial crisis has hit hard the economic foundations of many Western powers, made it difficult for them to intervene in the affairs of foreign nations, and constrained their resources to fight small wars, even when it appears necessary. The ongoing defense contraction further challenges the American effort to conduct small wars effectively.

None of these changes, however, is going to make extrasystemic war obsolete. Advanced nations will continue to engage and confront rebel groups in foreign territories in what may look like an extrasystemic war. In recent years, extrasystemic war has grown intense and generated more insurgent winners. It will continue to affect political and military stability in many parts of the developing world, including the Middle East, South and Southeast Asia, and Africa, where mostly unregulated flows of local insurgency and foreign fighters provide causes for military intervention of major powers and where minority groups at risk of national suppression may call for external help. Extrasystemic war may also be prominent where local rebel groups operate actively in areas such as Mali and Syria and in networks associated with al-Qaeda in Pakistan, Algeria, Yemen, Indonesia, and the

Philippines. Recent developments in Libya and Syria, although neither is part of an extrasystemic war, may stand out as relevant. In 2013 alone, NATO found itself in Libya, and debate continues as to whether the international community, especially the United States, might be drawn into rebel uprisings in Syria to stop the fighting. Sequences of events in both instances remain important, and sequencing theory would posit that rebel groups would benefit from first identifying the role of political institutions in the creation of an independent statehood and planning a national discourse on the path of sequencing the process. The case studies, particularly those of Indochina and Guinea-Bissau, demonstrate that rebels need to organize their forces in ways to make it possible to confront enemy forces in open terrain. In Libya, NATO bombings complemented sustained rebellions with import of weapons, leading to the fall of Gaddafi, which provided a way for the rebels to begin to create institutions to replace the regime and rebuild their nation. In Syria, rebel groups are seeking to enlist foreign supporters to gain advanced weapons to destroy the national army and create foundations for their statehood in case President Assad falls from power. Enormous challenges remain, however, because the groups are resource-scarce and internally divided, most notably between the Western-aligned Free Syrian Army and the al-Qaeda-related Islamic State of Iraq and Al Sham, and because Russia continues to support the Assad regime.

This type of irregular war will continue to grab daily attention from experts and the media and challenge the security policy of major powers. The series of asymmetric conflicts have thus generated a new type of security dynamics. T. V. Paul, among others, argues that it has become more difficult and complex to deter nonstate insurgents in asymmetric environments. It is in part because these actors may have cognitive styles, emotions, and expectations for rational behavior that are different from state leaders. In addition, the more states believe in the need to deter nonstate violence by taking action, the more nonstate leaders may leverage this belief system and subsequent violent actions from the states and raise their social status. So the more state actors engage in deterring these groups, the more difficult it may become for them to deal with these groups in the future. Yet not responding would be a senseless policy because it would mean capitulation and invite more attacks. This is what Emanuel Adler calls the "deterrence trap": damned if you do, damned if you don't.[2] Furthermore, Arreguin-Toft argues that weak actors may take up a guerrilla strategy and use the threat of nationalism to deter foreign intervention by powerful states. In such "unconventional" conflict, major states tend to realize that it is counterproductive to have high-intensity force structures, doctrines, and technologies appropriate for the conventional combat for which they are mostly trained. Therefore, weak actors can leverage particular interactions of strategy with state adversaries desirous of facing enemies in more conventional settings. In addition, weak actors can deny their foes the ability to use technological advantage by enlisting social support, sanctuary, and collective self-sacrifice.[3] These new trends constitute the bedrock of what Paul calls "complex deterrence." As such, Washington will continue to confront a number of violent nonstate challenges and find it imperative to prepare resources for untraditional security contingencies. One of the key security implications for the United States is that, while wholesale avoidance of extrasystemic war in the future is neither practical nor plausible, Americans are likely to face difficulty dealing with insurgent forces who adopt various methods of fighting in unconventional settings and learn to adapt.

What can the United States do to deal with the challenges? As suggested in the introduction, the United States is likely to face a number of insurgent forces around the world that are increasingly versed in the practice of sequencing. Thus the United States is well positioned to consider wartime evolution of enemies as a central part of its strategy making in future engagements in irregular war. Sequencing theory provides a range of further implications for the country. First, the United States should adopt a variety of means to curb insurgent evolution. Dominic Johnson and Joshua Madin's point about the need to curtail insurgent development early in the process, as discussed in Chapter 3, is critical in disrupting insurgent resource bases and preventing them from sequencing operations. While it is difficult for the state to detect a credible threat from an incipient insurgency and justify the threat to their public, it is imperative to arrest the development before the insurgency presents a threat.[4] Second, the United States should cut weapons flows into insurgent hands, particularly weapons that can be used to modernize their forces and increase capability to fight conventional war. As British success in COIN in Malaya (Chapter 5) and Somalia (Chapter 7) shows, the United States should continue to marginalize the enemy's access to resources to fight on, which helps prevent the enemy from taking the war to the next stages. After all, one of the decisive factors for insurgent failure in the two examples was the drainage of insurgent resources for modern war. Third, the United States should curtail insurgents' efforts toward state development not only by weakening their resource bases but also by building a rival political structure and making it legitimate in order to undermine the insurgent base. A key lesson from France's Bao Dai solution during the Indochina War (Chapter 9) is that a regime replacing the government will not function unless it is legitimate. Fourth, the United States should practice a strategy of winning hearts and minds at every point of the guerrilla war phase in order to undermine enemy guerrillas, maximize support from local populations, and bring them into operational alliances. The U.S. experience in the Anbar Awakening in Iraq (Chapter 6) stands out as a good example. Finally, the United States should keep in mind that not every insurgency will evolve the way I have identified. Many of them end up being simply insurgents and have no ambition to develop a state of their own, except that they are simply interested in driving foreign forces out of their homeland. As seen in the Iraq chapter and in this conclusion, the Iraqi and Afghan insurgents have fought U.S. forces largely outside the evolutionary framework, although even then sequencing theory offers a set of useful approaches for dealing with them.

In the next section I briefly discuss the war in Afghanistan from the sequencing perspective and apply the primitive model to explain why it is possible to think that the insurgents are unlikely to prevail in the war. I argue that there are reasons for us to expect to see them grow weaker over time as they fail to move out of the current guerrilla gridlock. However, the Taliban and al-Qaeda have their own ways of defining and interpreting victory, and U.S. departure will give them more ammunition to claim victory based on the achievement of their ends of forcing the departure. Thus, while sequencing theory may predict an ultimate U.S. victory, the reality may be more of a draw in the end, with both sides claiming victory at the same time. I do not list this war in the appendices because it is still ongoing, but because it is an important war I provide a preliminary analysis below.

The War in Afghanistan

The extrasystemic war in Afghanistan, begun initially as an antiterror campaign in response to the September 11, 2001, attacks, is a highly complex undertaking that involves a number of actors in and out of the country. They include various political actors in Pakistan (especially in the Federally Administered Tribal Areas, the North-West Frontier Province, and Baluchistan), Iran, India, and Central Asian states; violent nonstate actors like the Haqqani network, foreign fighters, criminal groups, and tribal militias; and a host of ISAF members who create an extraordinarily complicated and dynamic strategic landscape and whose summary discussion in these pages can be done only at the risk of oversimplification. That is particularly true given, as Hy Rothstein and John Arquilla argue, the multiplicity of challenges the United States faces, the range of options it has to deal with those challenges, and the need to understand different cultures and languages and build legitimacy and stability.[5] Of course, if we focus on the "terrorism" dimension of this war, from the literature we know how generic terrorism ends—through leadership decapitation, negotiated transition into a political process, achievement of terrorists' objectives or failure, as well as counterterrorist repression and encouragement of terrorists to reorient themselves to another form of operation.[6] But the war on terrorism is a unique campaign because, while there are those who engage in what people consider to be terrorist acts, in most cases terrorism itself is not an objective entity a nation can battle but a tactic designed to bring about political gains for the utilizer.[7] Some even argue that terrorism is an idea that is impossible to get rid of because killing terrorists will not eliminate the idea itself, so there is no victory against it and thus no need to "fight" it. Furthermore, it is a challenge to analyze this war as an extrasystemic war because it involves insurgencies that seem to have little interest in governance but instead seek an absence of governance as a means of achieving their ends. As William Martel puts it succinctly, "Classic measures of victory are inadequate" in this kind of war.[8] I acknowledge the limitations of the existing definitions of victory and defeat in the literature. Partly for this reason, in this book I treat the war more broadly as an extrasystemic war than as one on terrorism.

I proceed by designating the two sides of war. The insurgent side is represented by the Taliban and al-Qaeda and the state side by the United States.[9] It may be the case that the Taliban is a more principal part of the Afghan insurgency than al-Qaeda because the few hundred insurgents al-Qaeda has in the theater pale in terms of capability and size to the Taliban, or even Haqqani network members. This designation of combatants certainly comes at the cost of oversimplification, but it is the most reasonable one for this book because the Taliban and al-Qaeda continue to be the main protagonists on the insurgency side while the United States dominates the military and political aspects of the state side. At the same time, I do not argue that sequencing theory is the solution to the end of terrorism in its entirety; instead, the theory serves the limited purpose of understanding how the war between the U.S. forces and the insurgent forces can end up. Furthermore, U.S. victory against al-Qaeda in Afghanistan, if achieved at all, would not necessarily mean the end of al-Qaeda on the global level because it would not eliminate the remaining al-Qaeda networks elsewhere, whether in the Arabian Peninsula, North Africa (al-Qaeda in the Islamic Maghreb), or Southeast Asia, so this case study is limited to the Afghan theater. Finally, the war is ongoing as of this writing, so this analysis is incomplete until the war actually ends and many academic resources become available for a complete analysis.

What do theories of asymmetric war say about how the war may end in Afghanistan? While none of them explains outcomes of extrasystemic war per se, each can work with sequencing theory to better account for it. First, Arreguin-Toft's theory of strategic interaction would posit that al-Qaeda and the Taliban would win the war if their guerrilla strategy was matched by the American strategy of conventional war or if they adopted a conventional war strategy while the U.S. side used a guerrilla strategy. If both sides use the same strategy (conventional war or guerrilla war strategy), the winner is likely to be the United States. However, this expectation is problematic because, while much of the post-2001 period saw the U.S. use of special operations and CIA forces, the same strategy as al-Qaeda's, the insurgency has yet to win this war (and is not sure to). While the theory may later prove right, at this moment it is difficult to lend strong support to it. However, the theory provides a useful framework. It shows that the U.S. forces have had trouble fighting the insurgents since 2001 during the COIN phase because they initially failed to adapt to the guerrilla environment when the insurgents used guerrilla strategy. It also shows that the reason the initial campaign in the immediate aftermath of the September 11 attacks was generally easy for the U.S. forces was because the U.S. forces used their firepower and airpower well against the Taliban. That does not, however, undermine sequencing theory because the primitive model, one of the theory's components, argues essentially the same thing. That is, the model posits that insurgent efforts to fight guerrilla war against powerful states like the United States would be futile in the end as long as they do not fight on the military front (conventional war) and political front (state building). Insurgent failure is the case in point because the insurgent groups have failed to garner strong support from the population, who continues to defy their orders despite a series of coercive efforts on their part. For this reason, the theory of strategic interaction does not disprove sequencing theory; instead, it works quite well along with it.

The theory of external support seems to explain the insurgency's endurance well.[10] External aid has tremendously helped the insurgency to sustain its operation and presence over the last decade. The insurgency takes in a few million dollars from the drug industry alone, not to mention being robustly supported by the Pakistani Inter-Services Intelligence (ISI). The insurgency enjoys shelter within Pakistani borders, receives military, medical, training, and logistical aid from both Iran and Pakistan, and has an almost endless supply of recruits coming not just from Pashtun communities but also from Uzbeks, Tajiks, and other ethnic minority communities across the region. However, the fact that these factors have yet to translate into results shows the empirical weakness of the theory. The theory nevertheless has some appeal because material resources may after all be a key determinant of the war. The reason why the insurgents are underperforming in Afghanistan may be because the U.S. forces have collaborated with the governments in the region while the insurgents have had meager sources of supply from similar-minded, resource-constrained allies within Afghanistan. Therefore, the theory of external support goes hand in hand with sequencing theory on the material significance of asymmetric war.

Finally, the theory of mechanization would posit that as long as the United States puts mechanized war before manpower and intelligence, it is likely to lose the war. From this perspective, U.S. operations in Afghanistan were initially hindered by their focus on the use of mechanized divisions in COIN missions. A shift away from mechanized divisions, however, allowed the forces to collect intelligence from local populations, to discriminate combatants from noncombatants, and to selectively apply rewards and punishments to the populations. This change proved to be critical in the guerrilla phase. If the United States is to win the war, it is because U.S. forces began to do better once they cut their focus on mechanized divisions and moved toward the use of manpower and intelligence. The theory works better with sequencing theory because the focus on manpower in COIN operations, particularly in the postsurge period, is a key ingredient for success in the guerrilla war phase. The theory identifies one of the most important necessary conditions for the successful execution of guerrilla war—popular support—by means of using manpower to protect civilians. But how exactly does sequencing theory work in Afghanistan?

Sequencing Theory in Afghanistan

Sequencing theory boosts our analysis of the war by deploying the primitive model to capture the way the war has progressed. In 2001, the United States sent special operations forces to carry out Operation Enduring Freedom, quickly destroying large chunks of Taliban forces in Afghanistan where key members of al-Qaeda were believed to be hiding. The fall of the Taliban nevertheless allowed many local tribal leaders to resurge and form new groups. These groups were characterized to be of the "fragmented configuration," while the Afghan government under Karzai controlled a small portion of the country, mostly Kabul and some areas in the west and south.[11] The fragmented nature of these competing groups had much to do with the geographic dispersion of their relatively flexible and fluid units, which made mass movement and transportation difficult but continuation of guerrilla attacks relatively easier. In addition, the primitive nature of the war manifested itself in the discussion on the "Afghan model." That is, indigenous allies in Afghanistan quickly replaced American ground troops as the main combatants in the guerrilla war by exploiting U.S. airpower and small numbers of special operations and CIA forces. The use of local forces, special forces, and airpower epitomized the coalition's interest in avoiding direct engagement with the insurgents and the maximum use of indirect force.[12]

The primitive model best describes the Afghan war. On the military front, the insurgents have yet to garner a modern conventional force that can square off with the U.S./ISAF forces on the same level. Insurgent choice to fight guerrilla warfare is also shaped by the topographic features that favor clandestine movements across high valleys and mountains. Furthermore, the insurgency deliberately avoids fighting U.S. forces directly because it does not want to waste resources in fights it knows it will almost certainly lose. Instead, the insurgency is establishing alternative means of governance and is taking a wide variety of steps to weaken the Afghan government, such as propaganda campaigns, direct attacks, and subversion. On the political front, neither side of the war has succeeded in its efforts to build a stable government. The United States has failed to build strong governance around the Karzai administration for years, and U.S. attempts to develop the Afghan National Army have faced numerous obstacles, such as resource shortage, corruption, weak leadership, and lack of unity across tribal regions in the country.[13] The insurgent side has deliberately avoided nation-building efforts and instead sought the absence of governance. If any, their efforts to create some kind of order in the region have encountered problems unifying their movement across different tribal loyalties, which has long hampered their effort to generate a collective political voice to determine who ultimately is in charge of the insurgency, especially since the death of Osama bin Laden.

Of course, the primitive model does not capture details of all the events that have happened since 2001. For example, it does not incorporate changes in intrainsurgency dynamics (e.g., if the Taliban, Haqqani, or al-Qaeda dominates the Afghan theater at a given time) and the entry of new insurgent groups into the conflict (e.g., Lashkar-e-Taiba) as part of its explanation. For this reason, there may be ways other than the primitive model to explain how the war may end in Afghanistan. After all, insurgencies have won many times without ever making evolutions. In the realm of sequencing theory, however, the primitive model is best suited to address the key dynamics that determine the relationship between the United States and al-Qaeda and the Taliban because the model captures a set of key factors that most strongly shape the strategic landscape. As such, the war in Afghanistan has never turned into a form other than insurgency-COIN interactions. If anything, the 2010 killing of Osama bin Laden reinforced the already disorganized nature of guerrilla insurgency and the institutional linkage with other terrorist groups like the Haqqani network. Inflows of guerrilla members continue, partly associated with the accidental matching of al-Qaeda members to local environments. This also stems from America's tendency to conflate these trends, blurring the distinction between local and global struggles.[14] In this background, it is reasonable to expect that the war is going into the direction close to a primitive model.

The U.S. campaign in Afghanistan in the immediate aftermath of the September 11 incidents went quite successfully, causing the Taliban to fall in a matter of weeks. A few months after the initial breakthrough, however, the U.S. forces began to lose control of Afghanistan, which allowed the Taliban to resurrect in less attended areas and made the country less stable for the next few years. One major reason for this was the fact that the war in Iraq took away resources from Afghanistan, resources to be used not only in military operations but also for the reconstruction of Afghanistan. Of course, resource problems hindered the insurgent side as well. As Sinno points out, the Iraq conflict drew on transnational Islamist militants and proved more attractive to Arab militants than did Afghanistan "because of its greater proximity, ethnic and linguistic ties, and because the Iraqi resistance has been much more vibrant than the Afghan ones."[15] The U.S. forces have since been regaining the momentum, in part because the war in Iraq is over and resources meant for Iraq have been freed up for use in South Asia. Yet the ongoing strategic rebalancing toward Asia and the Pacific, the financial crisis, and defense budget reductions are likely to mean more challenges for operations in Afghanistan in the short run.

The most important *theoretical* implication of the primitive model in Afghanistan is the eventuality of insurgent defeat. My data show that, historically speaking, insurgents have won the primitive types of extrasystemic war only 16 percent of the time. The model predicts that the United States will come out more or less victorious in the end because the insurgent side lacks the capability to wage conventional war effectively and resources to build a powerful state to challenge the U.S.-led authority in Kabul. The United States could win this war on its own terms insofar as it has weakened al-Qaeda and killed Osama bin Laden. This optimism is largely consistent with Martel's argument that the United States has achieved a "limited strategic victory" in Afghanistan through the accomplishment of its principal objective—removal of the Taliban regime. This objective was central to America's success because it was "perceived as the only serious option for destroying the base of operations from which al Qaeda had operated with impunity since the mid-1990s."[16] The strategic victory eroded, of course, around 2006 with the resurgence of the Taliban, which continued through the late 2000s. The turning point came, however, for the Obama administration in 2009 with a comprehensive review of Afghanistan policy followed by the decision to surge U.S. troop levels in Afghanistan as well as the 2011 killing of bin Laden. These two events reflected the newfound U.S. focus on COIN, which eventually made America's success in the war more likely than before. These changes do not, of course, change the fact that the war is primitive; if at all, they are part of America's strategic adjustment to the need to battle insurgency effectively. Thus, we should view these developments as an extension of COIN strategy.

Unlike the conventional wars, key factors that are likely to determine the outcomes of this type of war will not be military factors but a combination of statecraft, which involves efforts to work with locals, build up infrastructure, and train and fund police and intelligence apparatus, as well as the establishment of functional administrations, services, and socioeconomic programs.[17] At its core, however, it is the sustainable level of popular support in Afghanistan that is likely to determine success. That is why the Taliban's strategy is basically to win hearts and minds at the local level, and the key for American success is to build a nation from the bottom up.[18] The war is likely to end as a primitive model because the insurgents have not made much progress in state building or in conventional war efforts. Nor have they succeeded in winning popular support due to internal difficulties that they have encountered within their organization. Fawaz Gerges writes, "Defections, internal cleavages and leadership crises, military setbacks, theological assaults by leading radical clerics, as well as a sharp decline in public support among Muslim, have all sapped al-Qaeda's strength."[19] Al-Qaeda nowadays is a marginal entity, kept alive largely by the public exaggeration of threats it is purported to present and the West's antiterror bureaucracy it helped to spawn. Al-Qaeda is a threat now only because this enormous bureaucracy needs it to remain as such by occasionally exaggerating the threat itself.[20] U.S. victory is achievable if, as Martel sees it, it is associated with "building stability in Afghanistan, protecting the Afghan people while increasing their security, and ending the Taliban insurgency."[21]

An American "victory" will be disputed, however, as some scholars have viewed it differently. John Mearsheimer, for instance, has called on the United States to "accept defeat" and "withdraw its forces from Afghanistan."[22] A former State Department official argues that "we are mortgaging our Nation's economy on a war, which, even with increased commitment, will remain a draw for years to come."[23] By the logic of Henry Kissinger's dictum that "the enemy wins if he does not lose," the Taliban and al-Qaeda actually are winning the war in Afghanistan because they have not been defeated and because the U.S. and coalition forces are leaving, which is one of their objectives of the war. The insurgency could also win insofar as it achieves its principal goal of removing, regardless of on whose terms, U.S. and coalition forces from Afghanistan. This could occur via a series of battlefield victories or by attrition, and the coalition's withdrawal is likely to give the insurgents ammunition to declare victory. These groups are antistate and have little interest in replacing the central government with another such structure. These groups benefit, furthermore, from the fact that once the objective of state failure is achieved, incumbents who wish to see a state structure rebuilt are forced into a hard choice between staying to suffer the protracted conflict and suffering the onerous costs of stateless territory in the form of drug trafficking, terrorism, and human rights abuses. From Omar to Zarqawi at the top all the way down to operational leaders, the Taliban and al-Qaeda have published, both on paper and electronically, a series of statements claiming victory and the weakness of foreign forces. Indeed, because insurgencies can always claim achievement of their objective through the U.S. withdrawal, they could technically win without evolving. For these reasons, I do not argue that the only one way for the Afghan insurgency to win is to defeat U.S. and coalition forces in conventional conflict. While a "draw" is possible in the short run, recent developments in the country indicate slow progress toward it.

A U.S. defeat, if at all, would contradict the prediction of the primitive model. However, even if the final outcome of the war winds up contradicting the mildly optimistic assessment of the primitive model, the case will not invalidate the model or sequencing theory itself because it alone will not challenge the historically grounded patterns of insurgent defeat recognized in this model. While only 16 percent of all primitive insurgents have won their wars, that number may not directly apply to the Taliban or al-Qaeda, two highly complex and unique organizations. These global organizations have the resolve and means to manipulate international public opinion about the war and the way they have fared against the coalition forces. The presence of a large number of actors involved in this conflict also complicates the work of defining victory and defeat. Especially critical in this respect is the interpretation of actors who tend to sympathize with the insurgency, such as Pakistan's ISI and its own Taliban (the Pakistani Taliban), government and paramilitary agencies in Iran, and various tribal and ethnic leaders across Afghanistan and Pakistan, who have great stakes in the stability of the region and are likely to view the insurgency as victorious rather than defeated. In other words, the reality will be more subtle than the theory can describe. What is likely to occur in the next few years is the emergence of competing claims of victory and multiple interpretations of outcome. Because both sides have arguably achieved their respective objectives, the war may be viewed as a draw in essence, a claim that itself will be disputed.

Of course, the war may survive the expected termination of military operations in 2014. This is not only because negotiations on the terms of U.S. withdrawal and the size and composition of remaining forces are still ongoing with the Afghan government and the Taliban, but also because the insurgents have methods of survival based on evasion, topography, and the drug trade, because the departure may embolden the insurgents, and because the Karzai regime may be too weak and corrupt to sustain itself once these strategic changes take place. The last point is particularly important because one of the major reasons why the Taliban resurged after 2001 was the weakness of the central government.[24] What is likely to occur after the termination of the military mission there is the mixture of strategic gains and losses on both sides of the war that makes it difficult for even security specialists to determine the victor. While the primitive model points to the eventual victory for the United States, the path will continue to be rough and difficult at best.

Future Research

Sequencing theory makes broader contributions to the study of international security in two notable ways. First, the theory presents a new analytical framework for a number of security issues. It allows us to prepare a range of methods to deal with multiple circumstances into which a conflict can unfold through intellectually generated but historically consistent patterns. The theory also permits us to determine how insurgent leaders learn from their experiences in the middle of war and innovate. The book deploys a set of ideas in a new package about how extrasystemic war unfolds, ideas that until recently have been understudied in our field. Second, the theory makes a contribution to two key themes in international security studies: the literatures of asymmetric war and war termination. The book advances our knowledge of conflict by applying the theory to the understudied realm of extrasystemic conflict. The book also helps improve the literature of war termination, which has in recent years encompassed a variety of methods—from political psychology to strategic interactions to bargaining and signaling—to examine how actors come to terms to conclude a conflict.[25] This book makes a departure from these general approaches to the study of war by setting war termination in the background of state-insurgency rivalry and pinpointing the cause of extrasystemic war termination in the use of sequences. Doing so not only offers a more concrete way of analyzing extrasystemic wars but also provides new prescriptions for current American security policies.

What does this analysis leave for the future of security studies? I stress three benefits that sequencing theory offers for scholarly research on international security affairs. The first concerns the *transportability* of the theory across issue areas. In this book I have presented the theory for the specific purpose of exploring how insurgents defeat states in war and used it to explain why some insurgents win wars but most fail. Yet the method of sequencing per se is general; it has a wide range of applicability over similar phenomena in international security, including civil war, terrorism, and interstate conflict. The theory can specifically be used to answer such puzzles as these: What sequence of actions might nation-states adopt to fight each other in interstate war? What phases can we use to see terrorist groups wage war against governments? What actions can nation-states take in sequence in order to minimize the threats of insurgents in places like Somalia? This book encourages researchers to employ the theory as a means of answering these questions and broadening the scope of analysis in our field.

Of all types of conflict, particularly relevant is the implication that this analysis has for civil wars, where we mostly see irregular combat between guerrilla revolutionaries and defending regimes in the same country. Civil wars have risen in frequency and significance since the end of the Cold War and in recent months have erupted in various Middle Eastern cities as part of the Arab Spring. The salience of civil war is apparent to anyone aware of its linkage with extrasystemic war. Research shows that civil war becomes internationalized, essentially extrasystemic, when outside states have strong incentives to affect the war through strategies of intervention and externalization.[26] Literature on civil war has advanced to have a number of theories to explain how it ends and why it does not. Works on negotiated settlements in civil war, for instance, have proliferated over the years, but a recent study shows that these types of settlements have proven largely ineffective: Civil wars ended by negotiated settlement are more likely to recur than those ending in victory by one side or the other. Monica Duffy Toft argues that rebel victories are more likely to secure the peace than are negotiated settlements.[27] Charles Call argues that the inclusion of former opponents in postwar governance plays a decisive role in sustained peace.[28] David Cunningham argues that civil wars are longer and more difficult to stop when they involve more veto players. Civil wars may end if peace processes effectively address barriers unique to multiparty conflicts.[29] These are different approaches than sequencing theory, although each of them takes account of various challenges inherent in the process of war termination.

What sequencing theory stresses here, however, is that rebel groups must evolve from underorganized bands into a modern army and build a legitimate state institution to replace the regime. It takes into account stages of civil war and the process of how rebel forces evolve over time to undermine the government. While requirements for rebels are likely different in civil wars than in extrasystemic wars, the key to success lies in whether the rebels evolve into a legitimate government with a modern army, quite a challenge for most such groups. This challenge was salient during recent wars in

Libya and Syria as well, although with different degrees of relevance for sequencing theory. On the one hand, in Libya rebels brought down the Gaddafi regime *without evolution* in 2011 because they received a significant amount of Western support and materials to achieve their ends. In other words, the primitive model would be appropriate to explain the rebels' victory in Libya. On the other hand, in Syria antigovernment rebels have so far failed to displace the Assad regime because they have yet to evolve, with no organized army and no legitimate leader to unite the movement and represent it on the international stage. Here, too, the primitive model would be the most appropriate to explain how guerrilla operations continue to dominate the conflict between rebels and Assad loyalists, although the final result is yet to be seen.

The second benefit relates to the power of the theory to *extend* our scope of analysis beyond the current six patterns of sequence. We gain a measure of analytical breadth by remaking the theory to address more complex security issues. There are two ways to do so. First, we can start by increasing the number of phases in the presented models in order to provide an extensive coverage of analysis. The second way is to combine the models. For example, one can put together the second phase of the degenerative model with the first phase of the progressive model in order to make a four-phased sequence. The degenerative-progressive combination would proceed as follows: conventional war shifts to guerrilla war first, before the guerrilla war shifts to an intensive phase of mutual state-building efforts. In the final phase, the state-building process converts into another intensive conventional war phase. The projected outcome of the war would be insurgent victory accomplished through the four phases. Of course, there are limitations to how well the combined models would be able to explain reality without compromising too much fact. Yet by adding phases to models, one can follow wars in great detail and generate a more precise expectation for their outcomes.

Finally, sequencing theory brings *depth* to our understanding of security issues on the levels of analysis: grand strategic, strategic, operational, and tactical. In this book, sequencing theory addresses strategic and grand strategic levels of activity because it deals mostly with political, economic, and military strategies of both sides as the main determinant of war outcomes. Such is the aim of this book, but the utility of sequencing theory is not limited to those levels. Indeed, it can be used to address operational and tactical dynamics and generate expectations for conflict outcomes on those levels.

Operational factors include those that affect combat in theaters of war, and tactical ones include those that affect combat in battlefields. The concept of evolution is pervasive across these other levels of analysis and is not limited to the strategic and political changes. New weapons, tactics, doctrines, and operational concepts have been introduced. State and insurgents alike have improved their training methods and issued a number of tactical and operational manuals over the years.

Two examples of tactical innovation in Iraq reinforce this point. First, Ahmed Hashim shows that Sunni insurgents operated in small teams of five to ten men to move fast and reduce complexity in command and control. However, operating in small teams reduced the amount of firepower that could be brought to bear against coalition forces. Initially, insurgents would engage their adversaries in set-piece firefights, in which they would usually lose. Gradually, however, insurgent groups have developed the capacity to launch larger, more complex attacks involving as many as 150 fighters. In 2005, they launched two large attacks against Abu Ghraib prison and a Marine base in Husayba. In the first attack, two separate groups of insurgents assaulted the prison following an accurate bombardment of coalition positions using mortars and two IED-laden vehicles. In the second case, a group of one hundred insurgents launched a well-coordinated attack on the Marine outpost.[30] In other words, one of the recent insurgent innovations was characterized by operational enlargement. The other example comes from instances of adaptations of insurgents' tactics and techniques in target selection. Realizing that their devices were not strong enough to cause significant damage to soldiers, they switched to targeting the infrastructure, which caused a huge outcry from the population. Coupled with two other attacks on pipelines in the north, Peter Mansoor argues, these explosions indicated that the insurgents shifted infrastructure targets.[31] Thus the other example of insurgent innovations was the change in target selection in ways that provided them greater efficiency in increasing the cost of operation. These examples demonstrate sustained attempts of insurgents at tactical innovation in a hostile environment.

In sum, this book contributes to our field of analysis with the proposal of a new theoretical framework to address a variety of international security issues. The theory underscores the importance of analysis that incorporates the role of strategic innovation in the outcome of war. The utility of the theory centers on the examination of the evolving nature of war. *Adapting to Win* presents empirical support for the central argument that insurgent groups are likely to win when they evolve. They evolve by progressing through key phases of war in proper sequence and picking up strength from the process. A successful sequence comes to manifest the evolution across the phases of state building, guerrilla war, and conventional war. After all, variation in the combination of the sequences—and the inherent difficulty in matching proper sequences at the time of war—explains why some insurgent groups win extrasystemic wars but most lose.

APPENDIX A

List of Extrasystemic Wars (1816–2010)

#	COW #	Name of War	Years Fought	Duration (Years)	Model	Warriors	Winner
1	301	Ottoman-Wahhabi	1816–1818	2	Conventional	Ottoman Empire and Wahhabis	State
2	302	Chile	1817–1818	1	Conventional	Spain vs. Chile	Insurgent
3	303	First Bolivar Expedition	1817–1819	2	Conventional	Spain vs. New Granada revolutionaries	State
4	304	Mexican Independence	1817–1818	1	Conventional	Spain vs. Mina expedition	Insurgent
5	306	Pindari (Maratha)	1817–1818	1	Conventional	Britain vs. Maratha (India)	State
6	305	Kandyan	1817–1818	1	Conventional	Britain vs. Kandyans (India)	State
7	307	Ottoman Conquest of Sudan	1820–1821	1	Conventional	Ottoman Empire vs. Sudan	State
8	308	Second Bolivar Expedition	1821–1822	1	Conventional	Spain vs. New Granada	Insurgent
9	New	Haiti-Santo Domingo	1821–1822	1	Conventional	Haiti vs. Santo Domingo	State
10	310	First Burmese	1823–1826	3	Conventional	Britain vs. Burma	State
11	311	First Ashanti	1824–1831	7	Conventional	Britain vs. Ashanti (Ghana)	State
12	312	Peru	1824–1825	1	Conventional	Spain vs. Peru	Insurgent
13	313	Java	1825–1830	5	Primitive	Holland vs. Javanese (Indonesia)	State
14	314	Bharatpuran	1825–1826	1	Conventional	Britain vs. Bharatpure (India)	State
15	315	Brazil-Argentine	1826–1828	2	Conventional	Brazil vs. Argentina	Insurgent
16	316	Russo-Persian	1826–1828	2	Conventional	Russia vs. Persia	State
17	317	Mexico	1829	0	Conventional	Spain vs. Mexico	Insurgent
18	320	Ottoman-Bilmez-Asiri	1832–1837	5	Conventional	Ottoman Empire vs. Bīmez and Asiris	State
19	321	First Zulu	1838–1839	1	Conventional	Britain vs. Zulu (South Africa)	State
20	322	First Afghan	1838–1842	4	Conventional	Britain vs. Afghanistan	Insurgent
21	324	First Algerian	1839–1847	8	Primitive	France vs. Algeria	State
22	ES-314	First Khivan	1839–1842	3	Conventional	Russia vs. Khiva	State
23	325	Peru-Bolivian	1841	0	Conventional	Peru vs. Bolivia	Insurgent
24	326	Baluchi	1843	0	Conventional	Britain vs. Baluchi (Pakistan)	State

(*continued*)

#	COW #	Name of War	Years Fought	Duration (Years)	Model	Warriors	Winner
25	330	Moroccan	1844	0	Conventional	France vs. Morocco	State
26	331	First Sikh	1845–1846	1	Conventional	Britain vs. Sikh (India)	State
27	333	Seventh Kaffir (Xhosa)	1846–1847	1	Conventional	Britain vs. Kaffir (South Africa)	State
28	332	Cracow Revolt	1846	0	Primitive	Austria vs. Cracow (Poland)	State
29	334	First Bali	1848–1849	1	Conventional	Netherlands vs. Bali	State
30	335	Second Sikh	1848–1849	1	Conventional	Britain vs. Sikh	State
31	337	Eighth Kaffir	1850–1852	2	Conventional	Britain vs. Kaffir	State
32	338	Ottoman-Yam	1851	0	Conventional	Ottoman-Empire vs. Yam tribe	State
33	339	Second Burmese	1852	0	Conventional	Britain vs. Burma	State
34	341	Santal Insurrection	1855	0	Primitive	Britain vs. Santal (India)	State
35	340	First Senegalese (Tukulor)	1854–1865	11	Conventional	France vs. Senegal	State
36	342	Hodeida Siege	1856	0	Conventional	Ottoman Empire vs. Beni Aseer tribe	State
37	345	Kabylia	1856–1857	1	Conventional	France vs. Kabylia	State
38	347	Indian Mutiny	1857–1858	1	Primitive	Britain vs. India	State
39	349	First Vietnamese	1858–1862	4	Conventional	France vs. Vietnam	State
40	350	Bone	1859–1860	1	Conventional	Netherlands vs. Bone	State
41	351	Argentine-Buenos Aires	1859	0	Conventional	Argentine vs. Buenos Aires	State
42	355	Maori	1860–1870	10	Degenerative	Britain vs. Maori (New Zealand)	State
43	357	Umbeyla	1863	0	Primitive	Britain vs. Pathan tribe	State
44	353	Spanish-Santo Dominican	1863–1865	2	Primitive	Spain vs. Santo Domingo	Insurgent
45	359	Kokand	1864–1865	1	Conventional	Russia vs. Kokand	State
46	360	Duar (Bhutanese)	1865	0	Conventional	Britain vs. Bhutan	State
47	361	Bukharan	1866	0	Conventional	Russia vs. Bukhara	State
48	362	Ethiopian	1867–1868	1	Conventional	Britain vs. Ethiopians	State
49	363	Ten Years	1868–1878	10	Conventional	Spain vs. Cuba	State

#	COW #	Name of War	Years Fought	Duration (Years)	Model	Warriors	Winner
50	364	Bahr el-Ghazal	1869–1870	1	Conventional	Egypt vs. Bahr el-Ghazal	Insurgent
51	365	Ottoman Conquest of Arabia	1870–1872	2	Conventional	Ottoman Empire vs. Yemen, Asir, Hasa	State
52	366	Second Algeria	1871–1872	1	Primitive	France vs. Algeria	State
53	New	Second Khiva	1873	0	Conventional	Russia vs. Khiva	State
54	367	Second Ashanti	1873–1874	1	Conventional	Britain vs. Ashanti	State
55	369	Second Vietnamese	1873–1874	1	Conventional	France vs. Vietnam	Insurgent
56	370	Aceh	1873–1913	40	Degenerative	Holland vs. Acehnese	State
57	371	Kokand Rebellion	1875–1876	1	Conventional	Russia vs. Kokand	State
58	372	First Egypt-Ethiopia	1875–1876	1	Conventional	Egypt vs. Ethiopia	Insurgent
59	374	Ninth Kaffir (Third Xhosa)	1877–1878	1	Conventional	Britain vs. Kaffir	State
60	375	Egypt-Sudanese	1878–1879	1	Conventional	Egypt vs. Sudanese slavers	State
61	376	Russo-Turkoman	1878–1881	3	Conventional	Russia vs. Turkomans	State
62	377	Bosnia	1878	0	Degenerative	Austria vs. Bosnia	State
63	379	Second Afghan	1878–1880	2	Conventional	Britain vs. Afghanistan	State
64	380	Second Zulu	1879	0	Conventional	Britain vs. Zulu	State
65	New	Little	1879–1880	1	Primitive	Spain vs. Cuba	State
66	381	Gun	1880–1881	1	Conventional	Britain vs. Basuto	Insurgent
67	382	First Boer	1880–1881	1	Conventional	Britain vs. Transvaal	Insurgent
68	383	Tunisian	1881–1882	1	Conventional	France vs. Tunisia	State
69	384	First Mahdist	1882–1885	3	Conventional	Britain vs. Sudan	Insurgent
70	389	Mandingo	1882–1889	7	Conventional	France vs. Mandingo (Mali)	State
71	386	First Madagascan	1883–1885	2	Primitive	France vs. Madagascar	State
72	387	Third Burmese	1885–1889	4	Conventional	Britain vs. Burma	State
73	390	Russo-Afghan	1885	0	Conventional	Russia vs. Afghanistan	State

(continued)

#	COW #	Name of War	Years Fought	Duration (Years)	Model	Warriors	Winner
74	391	Serbian-Bulgarian	1885	0	Conventional	Serbia vs. Bulgaria	Insurgent
75	New	Second Ethiopia-Egypt	1885	0	Conventional	Ethiopia vs. Egypt	State
76	392	First Ethiopia-Italy	1887	0	Conventional	Italy vs. Ethiopia	State
77	ES-375	Second Senegalese	1887–1890	3	Degenerative	France vs. Senegal	State
78	394	Dahomey	1890–1894	4	Conventional	France vs. Dahomey (Benin)	State
79	395	Franco-Jolof	1890–1891	1	Conventional	France vs. Jolof-Tukulor	State
80	397	Belgian-Tib (Congolese)	1892–1894	2	Conventional	Belgium vs. Tib Empire	State
81	New	First Matabele	1893	0	Conventional	Britain vs. Matabele (South Africa)	State
82	399	Melilla (Rif)	1893–1894	1	Conventional	Spain vs. Rif tribes	State
83	400	Mahdist-Italian	1893–1894	1	Conventional	Italy vs. Mahdists	State
84	401	Second Madagascan	1894–1895	1	Conventional	France vs. Madagascar	State
85	ES-379	Lombok (Second Balinese)	1894	0	Conventional	Holland vs. Bali (Indonesia)	State
86	398	Third Ashanti	1895–1896	1	Conventional	Britain vs. Ashanti	State
87	403	Portuguese-Gaza Empire	1895	0	Conventional	Portugal vs. Gaza Empire (Mozambique)	State
88	407	Second Ethiopia-Italy	1895–1896	1	Conventional	Italy vs. Ethiopia	Insurgent
89	404	Spanish-Cuban	1895–1898	3	Maoist	Spanish vs. Cuba	Insurgent
90	405	Japanese-Taiwanese	1895	0	Conventional	Japan vs. Taiwan	State
91	406	Mazrui Rebellion	1895–1896	1	Primitive	Britain vs. Mazrui clan of Kenya	State
92	New	Second Matabele	1896–1897	1	Conventional	Britain vs. Matabele	State
93	409	Second Mahdist	1896–1899	3	Degenerative	Britain vs. Sudan	State
94	411	British-South Nigerian	1897	0	Conventional	Britain vs. South Nigeria	State
95	412	Mohmand (British-Pathan)	1897	1	Conventional	Britain vs. Pashtun	State
96	413	Hut Tax	1898	0	Primitive	Britain vs. Sierra Leone	State
97	414	Philippine	1899–1902	3	Degenerative	United States vs. Philippines	State
98	415	Chad	1899–1900	1	Conventional	France vs. Rabih az-Zubayr	State
99	416	Second Boer	1899–1902	3	Degenerative	Britain vs. Boer	State

#	COW #	Name of War	Years Fought	Duration (Years)	Model	Warriors	Winner
100	419	Anglo-Somali	1899–1920	21	Degenerative	Britain vs. Somalis	State
101	417	Last Ashanti	1900	0	Primitive	Britain vs. Ashanti	State
102	420	Bailundu Revolt	1902–1903	1	Conventional	Portugal vs. Bailundu (Angola)	State
103	421	Kuanhama Rebellion	1902–1904	2	Primitive	Portugal vs. Kuanhama (Angola)	State
104	422	Kano & Sokoto	1903	0	Conventional	Britain vs. Nigeria	State
105	New	British-Tibetan	1903–1904	1	Conventional	Britain vs. Tibet	State
106	423	South West African Revolt	1904–1907	3	Degenerative	Germany vs. Namibia	State
107	426	Maji-Maji Revolt	1905	0	Degenerative	Germany vs. Tanzania	State
108	429	Third Zulu	1906–1907	1	Conventional	Britain vs. Zulu	State
109	430	Dembos	1907–1910	3	Primitive	Portugal vs. Dembos (Angola)	State
110	431	Anti-Foreign Revolt	1907	0	Primitive	France vs. Sheika Ma Al-Ainine	State
111	432	Japanese-Korean	1907–1910	3	Primitive	Japan vs. Korea	State
112	433	French Conquest of Wadai	1909–1911	2	Conventional	France vs. Wadai Sultanate (Chad)	State
113	434	French-Berber	1912	0	Primitive	France vs. Berbers	State
114	435	First Sino-Tibetan	1912	0	Conventional	China vs. Tibet	Insurgent
115	437	Moro Rebellion	1913	0	Primitive	United States vs. Moro	State
116	440	Second Sino-Tibetan	1918	0	Conventional	China vs. Tibet	State
117	441	Caco Revolt	1918–1920	2	Primitive	United States vs. Cuba	State
118	442	Third Afghan	1919	0	Conventional	Britain vs. Afghanistan	Insurgent
119	443	First Waziristan	1919–1920	1	Primitive	Britain vs. Waziristan tribes	State
120	444	Syrian	1920	0	Premature	France vs. Syria	State
121	445	Iraqi-British	1920–1921	1	Primitive	Britain vs. Iraq	State
122	446	Mongolia	1920–1921	1	Conventional	China, Russia vs. Baron von Ungern-Sternberg	State
123	447	Italo-Libyan (Saunusi)	1920–1931	11	Premature	Italy vs. Libya	State

(continued)

#	COW #	Name of War	Years Fought	Duration (Years)	Model	Warriors	Winner
124	449	Rif	1920–1926	6	Conventional	Spain vs. Morocco	State
125	450	Moplah Rebellion	1921	0	Degenerative	Britain vs. India	State
126	451	Druze Revolt	1925–1927	2	Conventional	France vs. Syria	State
127	452	Yen Bai Uprising	1930–1931	1	Primitive	France vs. Vietnam	State
128	453	Saya San's Rebellion	1930–1932	2	Degenerative	Britain vs. Burma	State
129	454	British-Palestinian	1936–1939	3	Primitive	Britain vs. Palestine	State
130	455	Second Waziristan	1936–1938	2	Conventional	Britain vs. Pathan tribes	State
131	456	Indonesian	1945–1950	5	Maoist	Holland vs. Indonesia	Insurgent
132	457	Indochina	1946–1954	8	Progressive	France vs. Vietnam	Insurgent
133	459	Third Madagascan	1947–1948	1	Primitive	France vs. Madagascar	State
134	460	Malayan Emergency	1948–1960	12	Primitive	Britain vs. Malay Chinese	State
135	461	Indo-Hyderabad	1948	0	Conventional	India vs. Hyderabad	State
136	463	Tunisia	1952–1956	4	Maoist	France vs. Tunisia	Insurgent
137	464	Mau Mau Rebellion	1952–1960	8	Maoist	Britain vs. Kenya	State
138	465	Morocco	1953–1956	3	Primitive	France vs. Morocco	Insurgent
139	466	Algeria	1954–1962	8	Primitive	France vs. Algeria	Insurgent

#	COW #	Name of War	Years Fought	Duration (Years)	Model	Warriors	Winner
140	467	Cameroon	1955–1960	5	Primitive	France vs. Cameroon	State
141	469	Angola	1961–1975	14	Progressive	Portugal vs. Angola	Insurgent
142	ES-436	Guinea-Bissau	1963–1974	11	Maoist	Portugal vs. Guinea	Insurgent
143	471	Mozambique	1964–1975	11	Progressive	Portugal vs. Mozambique	Insurgent
144	472	East Timor	1975–1978	3	Primitive	Indonesia vs. East Timor	State
145	473	Namibia	1975–1988	13	Maoist	South Africa vs. Namibia	Insurgent
146	476	Soviet-Afghan	1980–1989	9	Primitive	Soviet Union vs. Mujahideen	Insurgent
147	New	Somalia	1992–1995	3	Primitive	United States vs. Somalia	Insurgent
148	New	Iraq	2003–2011	8	Degenerative	United States vs. Sunni insurgency	State

Notes: The dataset was compiled using Gleditsch, "Revised List of Wars Between and Within Independent States"; Sarkees and Wayman, *Resort to War*; Dupuy and Dupuy, *Encyclopedia of Military History*; and Farwell, *Encyclopedia of Nineteenth-Century Land Warfare*. In Appendix B additional references are noted where applicable. ES, for extrasystemic, is followed by a number that corresponds to the designation of the Correlated of War2 version 3.0. "New" entries are those that were found to be extrasystemic war and were inserted. The following counting rule was applied: Zero years were counted when the war began and ended the same year. One year was counted when the war ended the following year; for instance, the Pindari War was counted as one year because it lasted from 1817 to 1818. Two years were counted when the war ended in two years, and so on.

APPENDIX B

Description of 148 Wars and Sequences

Coding Rules

Each war is designated as one of six sequencing models. When a war was dominated by conventional operations throughout, it is coded as conventional. When a war was dominated by guerrilla operations, it is coded as primitive. If it began with conventional operations but shifted to guerrilla operations, it is coded as degenerative. If it began with guerrilla operations but shifted to conventional operations, it is coded as premature. If it began with state building, shifted to guerrilla war, and ended in conventional war, it is coded as Maoist. If it began with guerrilla war, shifted to state building, and ended in conventional war, it is coded as progressive.

1. OTTOMAN-WAHHABI: CONVENTIONAL MODEL

In 1816, part of the Ottoman Empire, Egypt, sent its army consisting of Egyptian soldiers, Ottoman recruits, and Moroccans to suppress a series of Wahhabi revolts in the Holy Places of the Arabian Peninsula. Even though Wahhabis were nonstate and considered to be an insurgent movement, they acquired weapons and organized their forces to meet the Ottoman/Egyptian forces because of a history of prior conflict and because taking advantage of modern weapons and proper unit formation would increase the chances of Wahhabi victory. At the end of the war, however, the state side overwhelmed the insurgent side. As a result of the war, Egyptian forces regained control of Mecca and Medina.

2. CHILE: CONVENTIONAL MODEL

Led by Bernard O'Higgins, Chilean independence rebels fought the Spanish army in Chile. After O'Higgins declared independence in 1818, the Spanish royalists defeated the revolutionaries, but their ally in Peru, General Jose de San Martin, rallied the revolutionaries and defeated the Spanish troops, which led to Chile's independence. This represented one of the many wars fought in Latin America in conventional fashion and formed one of the major orthodox wars held between foreign and local armies.

3. FIRST BOLIVAR EXPEDITION: CONVENTIONAL MODEL

Under Simon Bolivar and General Manuel Carlos Piar, independence-seeking revolutionaries recruited British veterans from the Napoleonic Wars and waged a series of battles against Spain. Piar won a major victory at the Battle of San Felix, before both sides picked up a few victories in subsequent battles. In 1819 Bolivar attacked weakened Spanish forces in Colombia, and his forces also won the Battle of Queseras del Medio, followed by the decisive victory at the Battle of Boyaca. The insurgent side won the war in 1819.

4. MEXICAN INDEPENDENCE: CONVENTIONAL MODEL

Under Francisco Javier Mina, rebels formed an international army consisting of European recruits from the continental wars. Landing in Mexico in 1817, they captured Soto la Marina and joined Mexican soldiers. But the insurgents were defeated after Mina was executed despite several battlefield victories. Mexico was quickly settled by Spanish royalists.

5. PINDARI: CONVENTIONAL MODEL

The Pindari War began in 1817 when a large number of Pindari tribes incited violence in central and southern India against British forces—the Army of the Deccan led by Sir Thomas Hyslop and the Grand Army under Lord General Francis Rawdon-Hastings, consisting of nearly twenty thousand troops. They fought two hundred thousand Marathas armed with guns and crushed them at the Battle of Mahidput, before the Marathas surrendered in 1818. The Pindaris made no effort to fight a guerilla war and seek support from the locals; instead, their attitudes to the population were extremely harsh, as William Lee-Warner writes: "The Pindaris ... were united neither by social nor religious ties. They were a community of human jackals, who herded together attracted by the love of plunder and murder."

Additional references: William Lee-Warner, *The Native States of India* (London: Macmillan, 1910), p. 106; George Alfred Henty, *At the Point of the Bayonet: A Tale of the Mahratta War* (London: Blackie and Son, 1902).

6. KANDYAN: CONVENTIONAL MODEL

Britain deployed its forces to suppress rebel groups in Udarata. The use of force prompted widespread guerrilla reactions, which turned into powerful resistance movements and insurgency. It was so recognized by Tennakoon Vimalananda, who writes that Governor Robert Brownrigg, in charge of the British Imperial Army Kandy, reckoned that destruction of villages "would bring the people to their senses." Choosing terrorism over reconciliation, Brownrigg adopted brutal methods against the peasants to suppress them in a short time and localize the rebellion. Despite the brutality, the Kandyan aristocratic leaders failed to unite the population as they split over the issue of race. While British forces continued to raid villages, Sinhalese guerrillas remained largely powerless and stood hopelessly, before their rebellion was extinguished within a year.

Additional reference: Tennakoon Vimalananda, *The Great Rebellion of 1818: The Story of the First War of Independence and Betrayal of the Nation* (Colombo: M. D. Gunasena, 1970).

7. OTTOMAN CONQUEST OF SUDAN: CONVENTIONAL MODEL

Still part of the Ottoman Empire, Egypt deployed four thousand troops to the Sudan to conquer its provinces. The Sudanese tribes, including the Mamluks, were largely powerless against the Ottoman-supported Egyptian army and were defeated by Egypt's second expedition in 1821.

8. SECOND BOLIVAR EXPEDITION: CONVENTIONAL MODEL

In 1821, the city of Maracaibo revolted against the Spanish, who subsequently resumed military actions against the independence-minded revolutionaries. The Spanish were defeated by a six-thousand-strong revolutionary force at the Battle of Carabobo and again in naval battles along the Venezuelan coast. Simon Bolivar, now in New Granada, sent reinforcements for the revolutionaries, who suffered defeat at Guachi. One of the rebel leaders, General Sucre, raised an army of Argentines, Colombians, and others to defeat the adversary at Quito in 1822, winning the war and independence for Ecuador.

9. HAITI-SANTO DOMINGO: CONVENTIONAL MODEL

Haitian troops invaded Santo Domingo, the capital of today's Dominican Republic, in November 1821, then under Spanish control. The president of Haiti, Jean Pierre Boyer, mobilized his army, composed of one column commanded by himself and another by Guy-Joseph Bonnet, and drove out the Spanish over nine weeks to reunite the island under one government. In the meantime, a Creole group led by Jose Caceres overcame Spanish resistance in a different part of the island and created the independent state of Spanish Haiti as part of Simon Bolivar's Federation of Gran Columbia, but Boyer's army handily crushed the separatist movement.

Additional references: Robert Rotberg, *Haiti: The Politics of Squalor* (Boston: Houghton Mifflin, 1971), pp. 65–66; Robert Heinl and Nancy Heinl, *Written in Blood: The Story of the Haitian People 1492–1995* (Lanham, Md.: University Press of America, 2005), pp. 152–156.

10. FIRST BURMESE: CONVENTIONAL MODEL

Insurgents in Burma, under British control since 1757, faced the powerful British forces in their search to annex Bengal in 1824. Rangoon quickly fell to the British, but sixty thousand Burmese insurgents attacked the city until they were driven back. The next year Anglo-Indian troops moved into the interior of Burma before they were pushed back at Donabew. Burma's king ended the war by signing the Treaty of Yandabo and ceded his territories including Assam to Britain.

11. FIRST ASHANTI: CONVENTIONAL MODEL

The war began in 1824 when the Ashantis, led by Osei Bunsu, decided to drive out the British. They were conventionally organized, with many detachments led by military officers. With reinforcements, British forces defeated the main Ashanti army commanded by Osei Yaw at the Battle of Katamanso. The British were outnumbered but took advantage of the Ashanti vulnerability to artillery, their problems with fighting over open terrain, and a previously unknown weapon, the British Congreve rocket. The Ashantis grew divided about the direction of the war and failed to manage large-scale defections by provincial rulers. Peace was established in 1831 with the Ashanti surrendering their suzerainty over the coastal region.

Additional references: Ivor Wilks, *Asante in the Nineteenth Century: The Structure and Evolution of a Political Order* (London: Cambridge University Press, 1975), pp. 180–189; J. K. Fynn, "Ghana-Asante (Ashanti)," in Crowder, *West African Resistance*, pp. 19–79.

12. PERU: CONVENTIONAL MODEL

Simon Bolivar led four revolutionary armies, from Peru, Chile, Colombia, and Rio de la Plata, against the Spanish for independence of Peru. After losing Lima to the Spanish royalists, Bolivar's army won the Battle of Junin and recaptured Lima in 1824. General Sucre in the meantime won the Battle of Ayacucho, before he and Bolivar's armies defeated the royalists in Upper Peru, winning the war and liberation for Peru.

13. JAVA: PRIMITIVE MODEL

Frustrated by harsh taxes, droughts, and harvest failures, Javanese leader Diponegoro rebelled against Hamengkubuwono V and his Dutch backers in 1825. His aim was to prevent the collection of rural income. So armed rebels attacked the residences and workplaces of Europeans and Chinese and cut the capital off from revenue and supply and pressured farmers to refuse to sell their produce. The war was a series of provincial uprisings loosely coordinated and controlled by Diponegoro. But because Chinese suppliers refused to sell ammunition to the rebels after Diponegoro's followers massacred their families, his men began to lose battles. The Dutch responded by building forts across the countryside and sending out mobile columns of troops to patrol villages, which proved to be effective. Diponegoro was captured in 1830, when the war ended.

Additional reference: Jean Gelman Taylor, *Indonesia: Peoples and Histories* (New Haven, Conn.: Yale University Press, 2003), pp. 232–234.

14. BHARATPURAN: CONVENTIONAL MODEL

This war was a second siege after the first in 1805. Bharatpure was a walled fortress in India that threatened the expansionist interests of the East India Company. Bharatpure's stability was undermined by British authorities who sought to set up a pro-British regime under Raja Baldeo Singh. In 1825, British Lord Combermere, with a cavalry, two infantry divisions, and a large artillery force, invaded Bharatpure to settle a disputed succession. The city's strong defenses were brought down in January 1826. British casualties numbered one thousand, while India losses were estimated at eight thousand.

Additional reference: H. H. Dodwell, ed., *The Cambridge History of the British Empire, vol. 4: British India 1497–1858* (Cambridge: Cambridge University Press, 1929), p. 577.

15. BRAZIL-ARGENTINE: CONVENTIONAL MODEL

Rebels in the Banda Oriental, a buffer zone between Argentina and Brazil, took those in Argentina to fight this conventional war against Brazil in 1826, on land, on the sea, and along the Uruguay River. The war proved to be painful particularly for the Brazilian side, which signed a treaty allowing Uruguay to become independent and resulting in victory for the Argentine forces.

16. RUSSO-PERSIAN: CONVENTIONAL MODEL

Persia, not a member of the international system until 1855, fought Russia to recover the territory of Georgia. These two powers had built up forces to fight in the open terrain and considered this type of war to be effective in achieving their great power ambitions. Led by Fath Ali, the Persian forces lost the Battle of Ganja and withdrew inside their borders. They let the Russian forces take the Battles of Eriva, Tabriz, and Tehran. Russia won this conventional war in 1828.

17. MEXICO: CONVENTIONAL MODEL

This war occurred as an extension of Spain's imperial ambitions in the Western Hemisphere in the early nineteenth century. Insurgent forces gathered around General Antonio Lopez de Santa Anna, governor of Vera Cruz in Mexico, to face the Spanish fleet sent from Cuba in their search for independence. Taking advantage of mobility across their territories, Santa Anna's forces attacked the Spanish near Tampico and in the north of the territory, forcing the Spanish to surrender and winning the war.

18. OTTOMAN-BILMEZ-ASIRI: CONVENTIONAL MODEL

Led by Turkce Bilmez, a few thousand rebels formed an army with some combat ships in the Hejaz region of the Arabian Peninsula in 1832 to occupy various Yemeni cities. They were joined by the Asiri Wahhabis and faced expedition forces from the Ottoman Empire. They garrisoned the cities of Kuhayya and Hodeida against fierce Egyptian attacks. Yet a series of reinforcements strengthened the Egyptian forces, who downed Hodeida and Mocha, before the war ended with the Ottomans' victory.

19. FIRST ZULU: CONVENTIONAL MODEL

The first Zulu War was between British settlers in Zulu known as Voortrekkers and led by Andrius Pretorius on the one hand and the Zulus in South Africa, led by chief Dingane kaSenzangakhona on the other. It broke out in 1838 in an era of the so-called Great Trek, when Boer immigrants moved northeastward of South Africa. British and local Boer allies were conventionally armed and organized; the Zulus, too, took proper formations that were then divided into regiments. Although several Zulu battlefields victories and the murder of Piet Retief as a Voortrekker leader put the Zulus at an advantage, British and Boer firepower proved overwhelming, leaving a few thousand Zulus dead. The fighting culminated in the Battle of Blood River, where British forces annihilated the Zulus. After this war most of the Voortrekkers settled in Natal and Zulu.

Additional references: Leonard Thompson, "Cooperation and Conflict: The Zulu Kingdom and Natal," in Monica Wilson and Leonard Thompson, eds., *The Oxford History of South Africa*, vol. 1 (Oxford: Oxford University Press, 1969), pp. 359–365; C. F. J. Muller, ed., *Five Hundred Years: A History of South Africa* (Pretoria, South Africa: Academica, 1969), pp. 162–167.

20. FIRST AFGHAN: CONVENTIONAL MODEL

Britain invaded Afghanistan, led by Dost Mohammad, as part of the Battle of Ghazni, which prompted Afghan tribes to gather quickly to support Dost Mohammad's son, Mohammad Akbar Khan. Following a brief occupation of territory, in 1842 Britain suffered the massacre of General William Elphinstone's army of 16,500 men. Then Lord Ellenborough ordered forces at Kandahar and Jalalabad to leave Afghanistan after securing the release of prisoners. General Nott advanced from Kandahar and seized Ghazni. Meanwhile, General Pollock, advancing through the Khyber Pass, defeated Akbar Khan. The British forces took Kabul, but soon withdrew from Afghanistan through the Khyber Pass. Dost Muhammad was released and restored to power in Kabul. Britain stopped Russian expansion for the time being, but failed to restore control in Afghanistan.

Additional references: J. A. Norris, *The First Afghan War, 1838–1842* (Cambridge: Cambridge University Press, 1967); Archibald Forbes, *The Afghan Wars, 1839–42 and 1878–80* (London: Seeley and Co., 1982), pp. 1–157.

21. FIRST ALGERIAN: PRIMITIVE MODEL

The war began when Algeria rebels led by Abd el-Kader engaged the French in small skirmishes. The insurgents initially benefited from Moroccan support, but France bombarded Tangiers, stopped the support network, and further attacked Algerian rebels. Low-intensity conflict continued for a few years, with the Algerians fighting mobile warfare and the French attacking civilians supporting the rebels and destroying the countryside. The war ended with Abd el-Kader's surrender to the French.

22. FIRST KHIVAN: CONVENTIONAL MODEL

The Correlates of War (COW) project shows that the Russo-Khivan war was fought between 1839 and 1847 and won by the weaker side, Khiva, but it actually ended in 1842 and was won by Russia, which, although losing the war militarily, forced its treaty upon Khiva. In 1839, to extend Russian boundaries southward into Turkestan and open up trade, Russia's Nicholas ordered an army of four thousand men under General Basil Perovsky to move into the Khanate of Khiva, in what are now Kazakhstan, Uzbekistan, and Turkmenistan. The expedition ended in failure for the Russians, who suffered the effects of challenging terrain and snowstorms before they withdrew. Khiva had a stable political institution, with a tax system and mature conventional force, but it succumbed to Russia's pressure to accept an unfavorable treaty.

Additional reference: Hugo Stumm, *Russia in Central Asia: Historical Sketch of Russia's Progress in the East up to 1873, and of the Incidents Which Led to the Campaign Against Khiva* (London: Harrison and Sons, 1865), pp. 25–45.

23. PERU-BOLIVIAN: CONVENTIONAL MODEL

In 1841, Peru, then a nonstate entity, sent its army to invade Bolivia, which had been part of the Spanish viceroyalty during colonial times, to clash in a conventional war. Bolivian forces won the Battle of Ingavi, which led to the death of the Peruvian leadership and terminated the war. Battles were decided largely by the capture of towns and destruction of enemy forces. The gap in military power was so enormously in favor of Bolivia that the war lasted just a month.

24. BALUCHI: CONVENTIONAL MODEL

British forces under Sir Major-General Charles Napier confronted Baluchi forces led by Talpur amirs who attempted to liberate the Pakistani region of Sindh. Both sides were conventionally armed and had cavalry forces, with the Baluchis armed with muskets and wearing colored uniforms. The British easily defeated the Baluchi entrenchments led by Mir Sher Muhammad Talpur in the Battles of Miani and Dubbo. Napier wrote, "The Beloochies were clustered on both banks, and covered the plain beyond. Guarding their heads with dark shields, they shook their sharp swords, gleaming in the sun, and their shouts rolled like peals of thunder as with frantic gestures they dashed against the front of the 22nd regiment. But... British forces met them with the queen of weapons, and laid their foremost warriors wallowing in blood" (Farwell, p. 88).

Additional references: Byron Farwell, *Eminent Victorian Soldiers: Seekers of Glory* (New York: Norton, 1988); George Anderson and M. Subedar, *The Expansion of British India (1818–1858)* (London: G. Bell and Sons, 1918), pp. 27–31; Sir W. Napier, *The Conquest of Scinde*, vol. 2 (T. & W. Boone), pp. 309–318.

25. MOROCCAN: CONVENTIONAL MODEL

Morocco's assistance of Algeria in its war against France put Morocco into a war with France in 1844. The French fleet bombarded Tangier and Mogador while the French army, well armed with artillery and with eight thousand infantry and cavalry soldiers, fought and defeated the Moroccan rebels numbering forty thousand at the Battle of Isly, which ended the war with a French victory.

26. FIRST SIKH: CONVENTIONAL MODEL

The Sikhs had developed a regular army in the Punjab area of India. In 1845, their twenty-thousand-man army invaded a British Indian territory where Sir Hugh Gough faced them in the Battle of Mudki. Soon, in the Battle of Ferozeshah, the British defeated the army of Lal Singh, fifty thousand strong. The British guns ran out and British officers ordered the cavalry to retire to Ferozeshah. Yet at this point a new Sikh army under Tej Singh reached the battlefield to retake the camp. The following year saw the Battle of Aliwal, where British and Indian forces decimated the Sikhs numbering twenty thousand under Runjar Singh before the Battle of Sobranon, where Gough's forces defeated a Sikh army of fifty thousand. As a result, the Punjab became a British protectorate.

Additional references: Byron Farwell, *Queen Victoria's Little Wars* (New York: Harper & Row, 1972), pp. 37–50; Charles Allen, *Soldier Sahibs: The Men Who Made the North-West Frontier* (London: John Murray, 2000), pp. 58–87.

27. SEVENTH KAFFIR: CONVENTIONAL MODEL

In the Seventh Kaffir War, Britain's development of the Cape Colony in South Africa was accompanied by tribal outbreaks. The seventh war broke out in 1847 when Britain sent in infantry and cavalry divisions. British troops were conventionally armed and organized, and the Kaffirs were equipped with modern weapons, as the country was an ideal place for a cavalry charge. Britain overwhelmed the Kaffirs, particularly at the Battle of Gwanga. The Kaffirs failed to build institutions to strengthen armies, nor did they engage in guerrilla operations. The Kaffirs were not decisively defeated in the field but they disbanded after the year.

Additional reference: A. J. Smithers, *The Kaffir Wars 1779–1877* (London: Leo Cooper, 1973), pp. 192–215.

28. CRACOW REVOLT: PRIMITIVE MODEL

The Cracow Revolt for Polish independence from Austria was led by Edward Dembowski. The revolt took place in 1846 when Jan Tyssowski organized and led an insurrection in Galicia. Teofil Wisniowski then led the Uprising Tribunal in the eastern part of the country and battled with the Austrian army in Narajow. The uprising had weak local support because the citizens were more concerned about their own freedom and did not risk their lives in this insurrection. Thus it was overwhelmed and suppressed by Austrian troops, although the war witnessed a major case of orthodox combat at the Battle of Gdow. Consequently, Cracow was absorbed into the Austrian Empire two years later.

29. FIRST BALI: CONVENTIONAL MODEL

The Dutch began to bombard a coalition of rajas who formed kingdoms in Bali in 1848. Four thousand Dutch soldiers landed to find themselves confronted by thousands of Balinese warriors and were forced to retreat from their fortification in Jagaraga in what may appear to have been a Dutch defeat. In early 1849, the Dutch came back with an army and navy of thirteen thousand men aided by four thousand warriors sent from Lombok, against the united raja forces of thirty-three thousand warriors. This time, the Dutch did better in grabbing the fortress at Jagaraga, attacking Karangasem, and forcing the Balinese rajas to sign an armistice and a treaty in which the Balinese accepted Dutch sovereignty.

30. SECOND SIKH: CONVENTIONAL MODEL

After the destruction of institutions in the First Sikh War, there was little organized movement left in the Punjab, an area that was controlled by the British. The murder of two British subjects in 1848 prompted Sir Hugh Gough's British army, consisting of about fifteen thousand men, to invade the region. In the Battles of Ramnagar and Chilianwala, his cavalry was nevertheless repulsed, costing him his command. With reinforcements, however, British forces crushed the outnumbered Sikhs in the Battle of Gujrat in 1849. Preceded by heavy artillery bombing, British ground forces marched forward, with cavalry charges leading both sides, bringing about the destruction of Sikh towns and their quick surrender afterward.

Additional reference: Farwell, *Queen Victoria's Little Wars*, pp. 51–60.

31. EIGHTH KAFFIR: CONVENTIONAL MODEL

War-related institutions of the Kaffirs had been destroyed in the seventh war, but remaining forces challenged British authority again. The eighth war was provoked in 1850 when Sandile led a movement against British seizure of territory. Sandile mobilized his volunteer forces drawn from Boer and local settlers, whose number grew to fourteen thousand. The Africans this time again were organized, as a British report called them "bodies of regularly marshaled assailants, moving in columns, and protected by clouds of skirmishers." Furthermore, Basil Davidson writes, "Sandile's units, attacking the little British forces in Kaffraria, had to face shelling as well as small-arms fire. Only disciplined and well-conducted fighters withstand such bombardment, and continue to advance. Sandile's men did both."

Additional references: Basil Davidson, *The People's Cause: A History of Guerrillas in Africa* (Essex, UK: Longman, 1981), p. 26; Smithers, *Kaffir Wars*, pp. 226–246.

32. OTTOMAN-YAM: CONVENTIONAL MODEL

In 1851, the independence-minded Yam tribes, some of the five most powerful tribes in northern Yemen, and their followers of seven thousand men led by Sharif Hasan attacked and briefly occupied the Ottoman fort at Luhayyah. Soon after, the Ottoman troops crushed the Yam tribesmen at the Battle of Zaydidah. As predicated in the conventional model, the gap in military power played an important role in the war, which ended with Ottoman victory the same year.

33. SECOND BURMESE: CONVENTIONAL MODEL

The war began in 1852 when the British fleet sailed up the Irawadi River and seized the fort of Martaban and then destroyed the capital of Rangoon. Countering British attempts at colonization, Burmese rebels fought back at Martaban but were repulsed. The British capture of the Pegu region allowed them to capture other towns, culminating in a major battle that they won decisively. The war ended with British victory.

34. SANTAL INSURRECTION: PRIMITIVE MODEL

The Santal Insurrection occurred in 1855 when some six to seven thousand Santal tribes rose in violent protest against the British occupation of Damin-i-koh in India. As the rebels swept forward, they murdered innocent foreigners in villages and initially overwhelmed the British by catching them by surprise. However, they soon found themselves counterattacked by concerted operations of better-armed British forces. British forces instituted martial law and built a de facto garrison manned by some fourteen thousand soldiers, before, in 1855, they successfully suppressed the insurrection.

Additional reference: Edward Duyker, *Tribal Guerrillas: The Santals of West Bengal and the Naxalite Movement* (Delhi: Oxford University Press, 1987), pp. 33–35.

35. FIRST SENEGALESE: CONVENTIONAL MODEL

Between 1854 and 1860, there were French operations against Tukulor tribesmen in Senegambia. Army General Louis Faidherbe led operations to quell the Moorish Trarzas in 1955 and overcome Walo. Then Faidherbe slowly pushed Hajj Omar, the Tukulor ruler, to upper Niger. Among his major operations was the siege of Medine between 1857 and 1860. French strategy and formations were largely conventional, as were those of the Senegalese. In 1957 Faidherbe organized the Senegalese Riflemen and created two battalions drawing from local tribes. Ultimately, Omar failed to oust the French from the Senegambia.

Additional reference: J. F. A. Ajayi and Michael Crowder, eds., *The History of West Africa* (New York: Columbia University Press, 1973), p. 363; Kevin Shillington, ed., *Encyclopedia of African History*, vol. 3 (New York: Fitzroy Dearborn, 2005), pp. 1331–1332.

36. HODEIDA SIEGE: CONVENTIONAL MODEL

After a series of confrontations in the previous few decades, rebels from the Beni Aseer tribe on the Arabian Peninsula attacked enemy forces along the Red Sea coast. The sixty thousand warriors besieged Ottoman forces at Hodeida. A cholera epidemic weakened the rebels after a month of siege, forcing them to withdraw without achieving their objective.

37. KABYLIA: CONVENTIONAL MODEL

France sent nearly forty-five thousand troops to subdue unrest in Algeria and confronted the outnumbered Kabyle rebel groups there because France was concerned that a liberated Kabyle would threaten stability among the Arabs in Algeria. So France sent General Jacques Loius Random to lead a mission to pacify the area, while they adopted a strategy of "total warfare" on villages. As in most conventional wars, the state side overwhelmed the insurgent side, with the French outnumbering and outgunning the rebels, capturing Lalla Fadhma. The French succeeded in setting up military installations there to end the war in 1857.

38. INDIAN MUTINY: PRIMITIVE MODEL

The Indian Uprising of 1857, also casually known as the Sepoy Munity, began as a mutiny in a small portion of the Bengal Army and spread across the neighboring areas as a series of peasant rebellions, involving a number of Indian civilians. A majority of people were strongly motivated to join the insurrection, but the rebellion lacked coherent leadership or unified strategic plans. While the British committed acts of civilian brutality, such acts were also committed by the Sepoys. Denis Judd writes that "bands of mutineers roamed the countryside looting, and as a consequent antagonized the peasantry to whom they might otherwise have looked for support." The revolt was quashed by the British army in a year.

Additional references: J. A. B. Palmer, *The Mutiny Outbreak at Meerut in 1857* (Cambridge: Cambridge University Press, 1966); Denis Judd, *The Lion and the Tiger: The Rise and Fall of the British Raj, 1600–1947* (Oxford: Oxford University Press, 2004), pp. 70–90, 78.

39. FIRST VIETNAMESE: CONVENTIONAL MODEL

Supported by Spain, French forces attacked warriors belonging to the Annam, one of the three political entities in Vietnam, at Da Nang in 1858 to establish a fort there. Siege and disease, however, forced their withdrawal in 1860. Backed by the French fleet attacking Saigon beginning in 1859, the French forces swept through the region to occupy the key upper provinces of Cochin China before, in 1861, they defeated the Annamite forces at the intense Battle of Chi Hoa and achieved a decisive victory there. This forced the Annam to recognize French control of three additional provinces, ending the war with French victory.

40. BONE: CONVENTIONAL MODEL

The Bone, an indigenous monarchy on the island of Sulawesi, Indonesia, faced a Dutch military expedition in 1859 to conquer the territory. The Dutch forces and ships proved to be powerful enough to overcome strong Bone resistance, move inward on the island and capture the capital. A second Dutch expedition grabbed the Boni capital of Watampone, which ended the war, forcing the Bone to feudal status.

41. ARGENTINE-BUENOS AIRES: CONVENTIONAL MODEL

The war between the Argentine confederation and the province of Buenos Aires was provoked in part by a congressional decision to delegate greater authority to General Urquiza to place the province under its control. Then the capital of the state of Buenos Aires, Buenos Aires, dispatched two steamers to deny the Argentine provinces the power to carry out river operations, but those aboard defected against the province. The two sides met at the Battle of Cepeda in 1859, where Urquiza's force of ten thousand men defeated Mitre's forces of eight thousand. This defeat forced the governor to resign, and Buenos Aires became part of the Argentine Confederation.

Additional reference: Robert Scheina, *Latin America's Wars: The Age of the Caudillo, 1791–1899* (Washington, D.C.: Brassey's, 2003), pp. 123–124.

42. MAORI: DEGENERATIVE MODEL

The Maori War was fought between Britain and Maoris in today's New Zealand. The weaker side, the Maori, overwhelmingly relied on guerrilla war, in which they used *pa*, a form of defensive settlement, to protect villages. Although the British colonial forces began the war with conventional weapons and strategies of fighting in open terrain, soon they converted to irregular operations. It took them a decade to defeat a weaker opponent. The choice of guerrilla war helped the Maoris prolong the war, although village support for their operations continued to decline. The war had four major components over the 10 years of fighting: (1) First Taranaki (1860–1861), (2) Second Taranaki (1863–1866), (3) East Cape (1865–1868), and (4) Titokowalu (1868–1870). Characteristics included unorthodox combat, such as hit-and-run, large-scale involvement of civilian populace, and protracted jungle war.

Additional references: James Belich, *The New Zealand Wars and the Victorian Interpretation of Racial Conflict* (Auckland: Penguin, 1986); Farwell, *Queen Victoria's Little Wars*, pp. 176–178.

43. UMBEYLA: PRIMITIVE MODEL

With Indian soldiers, Britain's forces carried out the Umbeyla campaign in 1863 to subdue the Pathan tribe rebels in the area of the North West Frontier province of India. The geographic nature of the war encouraged both sides to use guerrilla operations more than conventional attacks. British and Indian forces were forced to make long treks and suffered from surprise and raid attacks in areas such as Crag Picket and Eagle's Nest. Six thousand rebel forces further attacked the opponent at Crag Picket, now known as the "Place of Slaughter," but they proved powerless in front of the British forces at Malka and subdued the raids for the time being.

44. SPANISH-SANTO DOMINICAN: PRIMITIVE MODEL

Then under Spanish control, rebel forces in Santo Domingo on the Caribbean island of Hispaniola waged guerrilla operations for independence. In 1863 the Spanish forces won initial victories against Dominican revolutionary forces. The guerrilla forces overcame a series of Spanish attacks and reinforcements as well as yellow fever and forced the Spanish forces to withdraw in 1865.

45. KOKAND: CONVENTIONAL MODEL

Russian forces numbering four thousand regular men attacked Kokand, the capital city of a then relatively autonomous Khanate kingdom, in 1864. The Kokand side had twelve thousand regular soldiers and twenty-eight thousand irregular men. The Russian troops defeated the Kokand forces at Chimkent, short of seizing the city, but later defeated the six-thousand-man army commanded by Alim Kul. Another Russian attack against Tashkent ended the war with Russian victory.

46. DUAR: CONVENTIONAL MODEL

Most combat of this war took place in an orthodox manner between the British and Bhutanese armies. In 1864, Britain invaded Bhutan to occupy the Bengal Duars there. Led by Brigadier Generals Mulcaster and Dunsford, British forces had more weapons but soon found themselves overwhelmed by the enemy's effective use of "jingal bullets." The Bhutanese had a well-formed military organization, rich in supplies and well-trained snipers. With reinforcements from India, however, British forces reversed the strategic tide and occupied the city of Dewangiri, before forcing Bhutan to sign the so-called Ten-Article Treaty of Rawa Rani, which awarded Britain the Assam and Bengal areas and returned British captives home.

Additional references: Surgeon Rennie, *Bhotan and the Story of the Dooar War* (London: John Murray, 1866); Bikrama Jit Hasrat, *History of Bhutan: Land of the Peaceful Dragon* (Thimphu: Education Department, Royal Government of Bhutan, 1980), pp. 109–117.

47. BUKHARAN: CONVENTIONAL MODEL

The leader of Bukhara, a city in Central Asia, Muzaffar ad-Din sought greater control over his territory against Russian interests. Russia sent expeditionary forces in 1866 after Bukhara seized its diplomatic mission there. Russian forces failed to occupy a key fort at Dzhizak but defeated Bukharan forces later at the Battle of Irdzhar and later grabbed more Bukharan territory, forcing Bukhara to sign a peace treaty.

48. ETHIOPIAN: CONVENTIONAL MODEL

The war began in 1867 when Ethiopia's Emperor Tewodros seized British officials and caused an uproar in the British Empire. Britain responded by sending thirteen thousand soldiers who landed at Zula. They defeated the Ethiopian army at the Battle of Arogi in 1868 and further attacked Tewodros's fortress, where the emperor committed suicide and handed victory to the British forces.

49. TEN YEARS: CONVENTIONAL MODEL

The war began in 1868 when rebel leader Carlos Céspedes declared independence from Spain. Partly for war purposes, Cuban elites had built executive and legislative branches with an eye toward eventual self-rule, a move that was nevertheless widely unpopular among the people. The unstable political system was also due to its shaky presidency and its weak armed forces. After years of stalemate, the presidency collapsed, before Céspedes was overthrown and General Ignacio Agramonte was killed at the Battle of Jimaguayu, leaving the so-called Mambi warriors fatally vulnerable to the Spanish.

Additional reference: Louis Perez, Jr., *Cuba: Between Reform and Revolution*, 3rd ed. (New York: Oxford University Press, 2006).

50. BAHR EL-GHAZAL: CONVENTIONAL MODEL

This war broke out when Egypt targeted the Bahr el-Ghazal region of southern Sudan and sent Samuel Baker to lead an expedition there. Baker initially succeeded in the suppression of slave trade there, but Zubayr Pasha defeated his forces decisively, ending the war with his victory.

51. OTTOMAN CONQUEST OF ARABIA: CONVENTIONAL MODEL

Partly in response to the 1869 uprising of the Asir tribe in Yemen, Ottoman forces advanced into central Arabia. The two thousand Ottoman soldiers pushed back Asiri attacks at Hodeida and another six thousand Ottoman troops repulsed the Asiris and consolidated control over the coastal Tihama region. They occupied several tribal areas as they moved into Yemen, including the fortress of Manakha. Moreover, they attacked al-Hasa, besieged it, and, after a few months of starvation, conquered it to end the war with Ottoman victory in 1872.

52. SECOND ALGERIA: PRIMITIVE MODEL

Taking advantage of the French defeat by the Prussians in 1870, Muhammad al-Muqrani organized a rebellion against France in Algiers and Constantine a year later. He raised 25,000 troops and 100,000 followers. Leader of the Rahmaniyya Sufi order, Sheik al-Haddad, joined the Kabyle rebellion with 120,000 troops and pushed the revolt eastward. Those of the Hodna tribe, and the Beni Menacer rose up as well to make an 800,000-man rebellion. The size of the movement made internal coordination difficult, leaving warriors with minimal arms against the powerful French troops. Harsh colonial repression ensued, forcing al-Haddad and his followers to surrender in 1871, killing al-Muqrani, and suppressing his uprising in 1872.

Additional references: Mahfoud Bennoune, *The Making of Contemporary Algeria, 1830–1987: Colonial Upheavals and Post-Independence Development* (Cambridge: Cambridge University Press, 1988), p. 40; Charles-Robert Ageron, *Modern Algeria: A History from 1830 to the Present* (Trenton, N.J.: Africa World Press, 1990), pp. 51–52.

53. SECOND KHIVA: CONVENTIONAL MODEL

The Second Khiva War broke out in 1873 when General Kauffmann led the Russian effort to organize an army of fourteen thousand men. Subsequently he succeeded in offensive operations from the corners of Caspian, running through Orenburg, Perovski, and Tashkent and overrunning the Khanate. The Khan was unable to defend the capital and soon forced to surrender, allowing Russia to render the Khanate become Russia's protectorate the same year.

Additional references: Seymour Becker, *Russia's Protectorates in Central Asia: Bukhara and Khiva, 1865–1924* (London: Routledge, 2004), pp. 65–80; Francis Skrine and Edward Ross, *The Heart of Asia: A History of Russian Turkestan and the Central Asian Khanates from the Earliest Times* (Methuen, 1899), pp. 258–259.

54. SECOND ASHANTI: CONVENTIONAL MODEL

Following years of small Ashanti raids between 1863 and 1872, British colonial authorities expanded on the Gold Coast, which prompted a major Ashanti offensive and became the Second Ashanti War. The British again managed to overwhelm the Gold Coast tribes, this time under Governor Garnet Wolseley, who had received numerous reinforcements supported by powerful cannon and modern rifles. Overall, Ashanti forces were decisively defeated in most combat opportunities. With regulars and native levies the colonial forces fought to and destroyed Kumasi, the Ashanti capital, in 1874. One of the causes of Ashanti defeat during the period of the war was that they invested little in developing their war institutions and rather focused on combat, which resulted in poor performance.

Additional reference: Robert Edgerton, *Africa's Armies: From Honor to Infamy* (Boulder, Colo.: Westview, 2002), pp. 54–58.

55. SECOND VIETNAMESE: CONVENTIONAL MODEL

France sent troops in 1873 to solve a trading issue in Vietnam and confronted a Vietnamese-Chinese force called the Black Flag. The Black Flag fought the French and killed the French commander Francis Gamier. It also captured French battleships in 1874 and destroyed Western towns in the territory, forcing the French forces to withdraw from the north, ending the war with Vietnamese victory that year.

56. ACEH: DEGENERATIVE MODEL

The war broke out in 1873 when the Dutch East India Army invaded forts in northern Sumatra. The war took a shift from conventional war to guerrilla war in the late 1870s. J. I. Bakker writes that "during the first phase of the war, the armed struggled changed from outright warfare to guerrilla warfare." From the mid-1890s to the mid-1900s, there were no large-scale expeditions but a series of skirmishes against locals, including the Gayo Expedition of 1900, leading to the murder of up to three thousand. Between the mid-1900s and 1913, Dutch forces continued to attack local religious reformers using guerrilla methods. In 1913, the rebellion was finally suppressed. The duration of the war is disputed among historians; some say it lasted until the late 1930s.

Additional references: J. I. Bakker, "The Ache War and the Creation of the Netherlands East Indies State," in Hamish Ion and Elizabeth Jane Errington, eds., *Great Powers and Little Wars: The Limits of Power* (Westport, Conn.: Greenwood, 1993), p. 57; Elizabeth Drexler, *Aceh, Indonesia: Securing the Insecure State* (Philadelphia: University of Pennsylvania Press, 2008), pp. 70–71.

57. KOKAND REBELLION: CONVENTIONAL MODEL

Rebels seeking more autonomy rose up in Kokand, whose leader declared a holy war against Russia in 1875. The rebel force retreated to the fortress of Makhram, which allowed Russian forces to move into their territory and defeat the Kokand troops in several battles including one at Namangan. The rebel force lost another battle at Andizhan, losing the war against Russia when the remaining revolutionary leaders lost the fortress at Uch-Kurgan.

58. FIRST EGYPT-ETHIOPIA: CONVENTIONAL MODEL

In a war between Egypt and Abyssinia, today's Ethiopia, Abyssinia was the nonstate side that defeated the Egyptians. In 1875, Egyptian eastward expansion, mainly directed by the Ottoman Empire, threatened to cut off Abyssinia's access to the ocean. When the Egyptians occupied Harar and adjacent ports, King John of Abysiania declared war. Having raised a functional army by the early 1870s, he had his troops move fast and nearly destroyed Khedive Isma'il's forces at Gundet. Khedive Isma'il sent two expeditions to Ethiopia, only to experience disasters, which confirmed its bankruptcy and surrendered the port of Massawa to the Italians. A second Egyptian expedition faced another defeat at Gura in 1876 when the war was over. Then Menilek sought to expand territory in the north, but returned home because two rebellions erupted in Shoa.

Additional references: Lewis Gann and Peter Duignan, eds., *Colonialism in Africa, 1870–1960*, vol. 1 (Cambridge: Cambridge University Press, 1975), pp. 421–422; Kevin Shillington, ed., *Encyclopedia of African History*, vol. 1 (New York: Fitzroy Dearborn, 2005), p. 458.

59. NINTH KAFFIR: CONVENTIONAL MODEL

In the Ninth Kaffir War, the Kaffirs, mainly the Xosa tribes in today's South Africa, rose up in violent protest in 1877. The British forces were organized in an orthodox manner under General Bartle Frere and were supported by only a few irregulars. The rebels, too, were organized in modern fashion, as A. J. Smithers writes that they engaged in "throwing to the winds of the old methods of warfare and . . . trusting to numbers to swamp the [British] by mass attacks." Yet as they failed to match British firepower, the rebellion was soon put down, and Britain annexed all of Kaffraria in 1878.

Additional reference: Smithers, *Kaffir Wars*, pp. 265–275, 269.

60. EGYPT-SUDANESE: CONVENTIONAL MODEL

Egypt's Ismail Pasha appointed British Colonel Charles Gordon to suppress the slave trade in Sudan. The Sudanese rebels led by Zebehr Pasha and his son Suleiman reacted violently by attacking Bahr el-Ghazal in the southern part of Sudan and taking over Dem Idris. In response an Egyptian army of nearly eight thousand troops defeated the Sudanese forces and liberated the slaves, achieving its objective.

61. RUSSO-TURKOMAN: CONVENTIONAL MODEL

In 1878 Russia sent twenty thousand troops into Turkestan but failed to beat the Turkomans at Geok Tepe. Russia's second offensive was led by General Mikhail Skobelev and targeted Geok Tepe, the fort that fell as a result to Russian assaults. After the Turkoman defeat, Russians proceeded to kill thousands of civilians fleeing the city.

62. BOSNIA: DEGENERATIVE MODEL

A year after the 1877 Treaty of Berlin awarded control over Bosnia to Austria, the Hapsburg monarchy proceeded to occupy it. Its four imperial divisions of seventy-two thousand troops overran the territory quickly. Bosnian Muslims had built up an army of forty-one Ottoman battalions. The army was a common institution in Bosnia, as Robert Donia and John Fine report: "The Bosnians attacked Austro-Hungarian convoys from the hills that rose above the primitive roads leading into the province's interior. Snipers harassed the monarchy's troops and forced them to deploy large flanking units to provide security for the main columns." But combat turned into guerilla warfare soon. By the fall of that year, the resistance forces were defeated.

Additional references: Robert Donia and John Fine, *Bosnia and Hercegovina: A Tradition Betrayed* (New York: Columbia University Press, 1995), pp. 93–94, 94; Peter Sugar, *Industrialization of Bosnia-Herzegovina, 1875–1918* (Seattle: University of Washington Press, 1963), pp. 23–26.

63. SECOND AFGHAN: CONVENTIONAL MODEL

The British opened the war when Afghanistan's Sher Ali refused to accept their diplomatic mission in 1878. Britain decisively won the Battles of Ali Musjid and Peiwar Kotal. Major General Sir Frederick Roberts led the Kabul Field Force over the Shutargardan Pass into central Afghanistan, defeating the Afghan army and occupying Kabul. Ayub Khan, governor of Herat, defeated a British detachment at the Battle of Maiwand. Roberts then led the main British force from Kabul and decisively defeated Ayub Khan at the Battle of Kandahar, ending the rebellion. Unlike the first war, Britain overwhelmed Afghan forces largely by military power.

Additional references: Indian Army, *The Second Afghan War, 1870–80, Abridged Official Account* (London: John Murray, 1908); Forbes, *Afghan Wars*, pp. 161–327.

64. SECOND ZULU: CONVENTIONAL MODEL

Britain invaded Zululand in South Africa in 1879 with a force of about 5,000 British and 8,200 native troops. Cetewayo, a Zulu leader, confronted the British with 40,000 well-organized warriors. The Zulu battle formation was a crescent; that is, while the center forces engaged the enemy directly, with detachments on both sides for a double envelopment, designed to encircle enemies. Their attack operations involved the use of spears. They had trouble at the Battles of Isandhlwana and Kambula and the siege of Eshowe until reinforcements arrived to allow them to defend Rorke's Drift. In the Battle of Ulundi, British rifle fire and bayonets broke up brave Zulu attacks and a small detachment charged into the collapsing enemy formations, pushing Cetewayo to flee. The war forced the Zulus to essentially become a British protectorate.

Additional references: David Clammer, *The Zulu War* (Newton Abbot, UK: David & Charles, 1973); Ian Knight, *The Anatomy of the Zulu Army: From Shaka to Cetshwayo, 1818–1879* (London: Greenhill Books, 1995).

65. LITTLE: PRIMITIVE MODEL

The rebellion was initiated in 1879 by forces composed mostly of slaves and led by Calixto Garcia, who had established the Cuban Revolutionary Committee for the sake of independence. Hiding in the mountains and forest and looking for weaknesses in the Spanish troops, the insurgent movement frightened the Spanish authorities in Cuba, so Madrid sent two thousand reinforcements to suppress it. Garcia had to confront severe problems besides the war, however, such as securing more resources and support from outside and local citizens now that they had been fighting since the earlier Ten Years' War. The movement was crushed in two years, however.

Additional reference: Willis Johnson, *The History of Cuba*, vol. 3 (New York: B. F. Bucks and Company, 1920), p. 308.

66. GUN: CONVENTIONAL MODEL

The Basuto tribe revolted against Cape Colony forces controlled by Britain in Basutoland, Lesotho, adjacent to South Africa. The Cape forces were sent to put down the rebellion but instead found themselves suffering serious casualties in a single battle, as the Basotho had been empowered with firearms obtained from the neighboring Orange Free State and enjoyed a strong defensive position in the mountains. They successfully ambushed an experienced cavalry unit at Qalabani, which discouraged colonial authorities and led in part to a peace treaty signed in 1881, in which Britain was forced to agree to Basuto demands that the territory remain with the local tribes and that they have good access to weapons.

Additional references: Gann and Duignan, *Colonialism in Africa*; Jack Halpern, *South Africa's Hostages: Basutoland, Bechuanaland and Swaziland* (Harmondsworth: Penguin, 1965), pp. 79–80.

67. FIRST BOER: CONVENTIONAL MODEL

In 1880, the Boers in the Transvaal Republic organized three commando groups and fought the British at Potchefstroom, seeking independence after an incident related to their refusal to pay taxes. They continued their offensive, occupying Laing's Nek in British Natal, defeating the British at Ingogo, and storming Majuba Hill to kill half the British soldiers there.

68. TUNISIAN: CONVENTIONAL MODEL

The Royal Bey of Tunis had a certain structure of political institutions, such as constitution and ministries, for various functions. In 1881, French naval forces occupied Bizerte while land forces crossed the border. The French force of about thirty thousand men met only weak resistance and occupied the territory in just three weeks. Yet a rebellion broke up once the French withdrew, spreading quickly and forcing another deployment, although it ended in 1882. John Munholland writes that the "French operations were more than punitive," suggesting their reliance on orthodox power. The nonstate side accepted a protectorate status in the Treaty of Bardo.

Additional references: John Munholland, *The Emergence of the Colonial Military in France, 1880–1905* (PhD diss., Princeton University, 1964), pp. 63–67; A. Pellegrin, "A Century of Tunisian History," in *Tunisia 54* (New York: Negro Universities Press, 1954), pp. 9–12.

69. FIRST MAHDIST: CONVENTIONAL MODEL

The First Mahdist War was fought by Britain and its colonial Egypt against Sudanese Mahdists. In 1882, armed with spears and swords, the Mahdi overwhelmed a seven-thousand-man Egyptian force and later besieged El Obeid, to which Egyptian forces responded via the so-called Kordofan expedition, only to surrender. At El Teb and

Sheikan, furthermore, Osman Digna's forces defeated Egyptian forces. The war culminated at Khartoum, where British forces led by General Gordon, responsible for evacuating the capital, were besieged and killed by the Muhammad Ahmad. So the Mahdists as the nonstate actor managed to defeat the powerful British in conventional war without building a mature political system to support their military operations.

Additional references: Winston Churchill, *The River War: An Account of the Reconquest of the Sudan* (Mineola, N.Y.: Dover, 2006); Anders Bjorkelo, *Prelude to the Mahdiyya: Peasants and Traders in the Shendi Region, 1821–1885* (Cambridge: Cambridge University Press, 1989).

70. MANDINGO: CONVENTIONAL MODEL

The Mandingo war was fought between France and the Wassoulou Empire of Mandingo people led by Samori Toure. In 1882, a French expedition attacked Samori's armies at Keniera where Samori drove the French off. At the siege of Bamako in 1883, Samori sought to expand the front line via diplomatic means. The British sold Samori modern weapons. When an 1885 French expedition attempted to seize the Buré gold fields, the better armed Samori forces counterattacked dividing his army into three columns. The war ended there, with Samori forced to accept the Niger as a new frontier. He signed the Treaty of Kenyeba-Kura in 1885 to accept a protectorate status.

Additional references: John Hargreaves, "West African States and the European Conquest," in Gann and Duignan, *Colonialism in Africa*; M. Legassick, "Firearms, Horses and Samorian Army Organization, 1870–1898," *Journal of African History* 7 (1966).

71. FIRST MADAGASCAN: PRIMITIVE MODEL

Independence-seeking rebels in Madagascar revolted against the French, who sent military forces in 1883 to suppress the movement. The French fleet occupied Majunga and twenty-three thousand French troops took over the town of Tamatave after bombarding it. For the duration of the war the combat was low in intensity.

72. THIRD BURMESE: CONVENTIONAL MODEL

The war broke out in 1885 when the Burmese objected to Britain's assertion of control. British forces bombarded Burmese fortifications and defeated the minimal resistance from the Burmese army at the Battle of Minhla. British forces continued to seize more territory and advanced to the capital, where the Burmese king surrendered, ending the war with British victory.

73. RUSSO-AFGHAN: CONVENTIONAL MODEL

In 1885, after a few border clashes, Russian troops attacked Afghan rebels out of the Penjdeh region and at Ak-Teppe. With British intervention, Russia signed a treaty with the Afghans to gain the Penjdeh region.

74. SERBIAN-BULGARIAN: CONVENTIONAL MODEL

Upon the declaration of union between Roumelia and Bulgaria, then not a state system member, Serbia invaded Bulgaria with King Milan leading twenty-eight thousand troops in 1885. The Bulgarian leader Prince Alexander confronted the Serbia forces with fifteen thousand soldiers at Slivnitza and secured a victory. The Battle of Pirot ensued, which the Bulgarians won again. A threatened intervention by Austria ended the war, and the union of Roumelia and Bulgaria was recognized with the rebels' victory in this war.

75. SECOND ETHIOPIA-EGYPT: CONVENTIONAL MODEL

In 1884, King Johannes of Abyssinia, in today's Ethiopia, was asked by Britain to rescue Egyptian garrisons placed under siege by Mahdist forces from Sudan near the Abyssinian-Egyptian border. In 1885, this led to the successful operation by the Abyssinian governor of Tigre to liberate some garrisons. He went on to defeat Osman Digna, forcing Egypt to withdraw from Massawa.

Additional reference: Shillington, *Encyclopedia of African History*, vol. 1, p. 507.

76. FIRST ETHIOPIA-ITALY: CONVENTIONAL MODEL

As emperor of Ethiopia, Yohannis had built a large army, anticipating an advancing Italian force. Once in war in 1887, his forces surrounded an Italian garrison in Ethiopia. Italy's Lieutenant-Colonel Tommaso de Cristoforis received reinforcements but found himself outnumbered and overwhelmed by the Ethiopians under General Alula's command and suffered a massacre at Dogali. But the Italians came back stronger this time with an African expeditionary force led by General Tancredi Saletta, forcing Yohannis to the negotiation table with British mediation and soon into retreat, exploiting his vulnerability to the Mahdists near Sudan and essentially ending the war. The battle also led the Italians to inquire whether Yohannis's rival, Menelik II, then ruler of Shewa, would ally with them.

Additional references: Haggai Erlich, *Ras Alula and the Scramble for Africa* (Lawrenceville, N.J.: Red Sea Press, 1996); Paul Henze, *Layers of Time: A History of Ethiopia* (New York: St. Martin's, 2000), pp. 156–159; Gann and Duignan, *Colonialism in Africa*, pp. 425–427.

77. SECOND SENEGALESE: DEGENERATIVE MODEL

The COW inaccurately claims that the Second Senegalese War was fought only in 1890. Instead, it began in 1887 with the eruption of a Muslim reform movement. A Soninke cleric, al-Hajj Mamadu Lamine Drame, launched his attack on pro-French rulers in the Upper Senegal Valley, then invading Bondou, murdering its ruler, and continuing on to attack French garrisons at Senedoubou. French forces then intervened to help their allies and chased after Lamine, killing him in 1887, which allowed them to assert their control over the valley. By 1890, a French column led by General Dodds placed the territory of modern-day Senegal under firm French control.

Additional references: Michael Crowder, *West Africa Under Colonial Rule* (Evanston, Ill.: Northwestern University Press, 1968), pp. 74–81; B. Olatunji Oloruntimehin, "Senegambia—Mahmadou Lamine," in Crowder, *West African Resistance*, pp. 80–110.

78. DAHOMEY: CONVENTIONAL MODEL

Refer to my case study for details.

79. FRANCO-JOLOF: CONVENTIONAL MODEL

France invaded two regions—the Jolof Empire and Futa Toro—in 1890 to attack Tukulor rebels led by Albury Njay. Albury Njay soon withdrew his rebel forces into what is now Mali and Mauritania. French forces attacked other Tukular forces under Ahmadu Seku and seized the religious center of Segu the same year before they picked up additional victories at Koniakary, Nioro, and Diena. These conventional wins added up to a French victory in the war.

80. BELGIAN-TIB: CONVENTIONAL MODEL

Belgium's King Leopold II sent his nearly ten thousand Congolese troops to conquer the Tib Empire in what is now the Congo in 1892. They attacked the local forces at a fort on the Lomami River and subsequently took over nearby cities. Reinforcements strengthened the Belgian forces to attack the capital city of Kasongo before the Tib Empire was destroyed in 1894.

Additional references: L. H. Gann and Peter Duignan, *The Rulers of Belgian Africa 1884–1914* (Princeton, N.J.: Princeton University Press, 1979), p. 57; Gann and Duignan, eds., *Colonialism in Africa*, pp. 261–277.

81. FIRST MATABELE: CONVENTIONAL MODEL

The war began when Lobengula, king of the Ndebele, planned to impose a tributary system upon the Mashona tribe in Fort Victoria, causing the British South Africa Company to intervene. Led by Leander Starr Jameson, the British forces had fewer than 750 police members, some volunteers, and 700 Tswana allies, while Lobengula had 100,000 men. But they used the Maxim gun effectively to overwhelm the Matabeles and conquered Bulawayo. After Lobengula's death in 1894, the British restored control over Matabeleland.

Additional references: Stafford Glass, *The Matabele War* (London: Longmans, 1968); Robert Blake, *A History of Rhodesia* (New York: Knopf, 1978), pp. 107–121; Roland Oliver and G. N. Sanderson, eds., *Cambridge History of Africa*, vol. 6, from 1870 to 1905 (Cambridge: Cambridge University Press, 1985).

82. MELILLA (RIF): CONVENTIONAL MODEL

In 1893, six thousand Rif soldiers attacked a Spanish garrison at Melilla. The Rif warriors increased in number after the Spanish bombed a mosque there, which helped them repulse a series of attacks by the Spanish who nevertheless had weapons of better quality. Reinforcements strengthened the Spanish further, but the Rifs besieged Spanish fortifications. Yet Spanish firepower after all played a decisive role, forcing the Rifs to sign a treaty that allowed the Spanish to complete fortifications at Melilla, receive an indemnity, and suppress the Rifs.

83. MAHDIST-ITALIAN: CONVENTIONAL MODEL

In a struggle to get rid of foreign influence in the Sudan, the Mahdi led an army of eleven thousand to attack the Italian forces in Eritrea in 1893. Mahdi troops clashed with the Italian army at the Battle of Agordat but lost the battle and retreated. They met more offensives by the Italian forces who occupied Kasala to end the war with Italian victory the next year.

84. SECOND MADAGASCAN: CONVENTIONAL MODEL

In 1894, the Hova rebels resisted a French attempt to control their territory in Madagascar. But they found it increasingly difficult to deal with French bombardment and reinforcements, as twenty-three thousand French troops landed on the island. The Madagascan army sought to prevent French attacks at Tsarasoatra but was defeated there and later at Andriba. The Hova tribes surrendered after another major bombardment attack by the French, giving up the entire island to the French.

85. LOMBOK (SECOND BALINESE): CONVENTIONAL MODEL

The Lombok War (originally designated as "Dutch-Balian" in the COW dataset) began in 1894 when the Dutch sent an expedition to Lombok in Bali, Indonesia, to colonize it. Well-armed Balinese forces were initially successful, pushing the Dutch soldiers back to the shore, at one point taking an entire regiment hostage, where they built a fortification to discourage further invasion. Numerous reinforcements sent from Java strengthened the Dutch, however, allowing them to be on the offensive, using artillery fire to suppress the resistance and killing the crown prince. Following the war Lombok merged into the Dutch East Indies.

Additional references: Miguel Covarrubias, *Island of Bali* (Read Books, 2006), pp. 30–32; Keat Gin Ooi, *Southeast Asia: A Historical Encyclopedia, from Angkor Wat to East Timor* (Oxford: ABC-CLIO, 2004), pp. 790–791.

86. THIRD ASHANTI: CONVENTIONAL MODEL

After a series of defeats by the British colonial forces, the Ashanti tribes chose a new ruler, Agheman Prempeh, who carried out a series of violent acts in his territory starting in 1895. A British punitive expedition was soon organized under Colonel Sir Francis Scott, who successfully besieged Mafeking. In 1896, the British entered Kumasi, the Ashanti capital, without any opposition, forcing the Ashantis to accept a British protectorate status. The COW dataset describes the war as beginning in 1893 and lasting until 1894, but that is historically inaccurate.

87. PORTUGUESE-GAZA EMPIRE: CONVENTIONAL MODEL

In the autonomous entity in southern Mozambique, rebels from the Gaza Empire sought to prevent Portuguese expansionism in 1895. The king Gungunyane's forces besieged the Portuguese port of Lourenco Marques but faced a growing number of Portuguese forces with reinforcements. The Gazan army subsequently lost the Battle of Marracuene and later Coolela. Rebel leader Gungunyane was then captured, which ended the war with Portuguese victory.

88. SECOND ETHIOPIA-ITALY: CONVENTIONAL MODEL

A few years after suffering defeat in the first war, the Ethiopians restarted the war with a victory at Amba Alagi. Led by Menilek II, they followed the retreating Italians and besieged them. The Italian army numbered about 20,000 men armed with rifles, artillery, and machine guns. The opposing Ethiopian forces were numerically superior, composed of nearly 110,000 riflemen, cavalry, artillery, and machine gun soldiers. The Ethiopians use their manpower advantage to envelop and overwhelm the Italian army. One of the key battles was held at Adowa, where the Ethiopians won a decisive victory, forcing the Italians to sign the Treaty of Addis Ababa, which granted Ethiopia independence.

Additional reference: Erlich, *Ras Alula and the Scramble for Africa*; Henze, *Layers of Time*, pp. 167–171; Gann and Duignan, *Colonialism in Africa*, pp. 435–438.

89. SPANISH-CUBAN: MAOIST MODEL

After two abortive wars of independence, the Cuban Revolutionary Party and the Cuba Libre movement confronted Spain's puppet regime of Valeriano Weyler. The Cubans had strong popular support, including, as Perez writes, "displaced professionals, impoverished planters, an expatriate proletariat, former slaves, a dispossessed peasantry, poor blacks and whites ... responded to the summons to arms" and occupied the countryside. Cuba's Army chief Maximo Gomez proclaimed that the Ten Years' War was "the top down, that is why it failed; [but] this one surges from the bottom up, that is why it will triumph." The Cubans used American aid to organize and execute artillery attacks, forcing the Spanish army to cease fighting. Although they were to be occupied by the United States afterward, the Cubans achieved the objective of independence from Spain.

Reference: Louis Perez, Jr., *Cuba: Between Reform and Revolution*, 3rd ed. (New York: Oxford University Press, 2006), pp. 120–121, 128–135, 121.

90. JAPANESE-TAIWANESE: CONVENTIONAL MODEL

Japan's victory over China in 1895 brought Taiwan under its rule through the Treaty of Shimonoseki, which provoked a short-lived organized Taiwanese resistance against Japanese occupying forces. Troops from the Republic of Formosa took up leftover weapons and fought mostly a conventional war in a series of skirmishes against mixed brigades of the Imperial Japanese Body Guard, led by Prince Kitashirakawa, tasked to pacify the islands. The rebellion was severely put down within a year.

Additional references: S. C. M. Paine, *The Sino-Japanese War of 1894–1895: Perceptions, Power, and Primacy* (Cambridge: Cambridge University Press, 2003); James Davidson, *The Island of Formosa: Past and Present* (Oxford: Oxford University Press, 1988), pp. 290–370.

91. MAZRUI REBELLION: PRIMITIVE MODEL

In 1895, Mbaruk bin Rashid led a Mazrui rebellion against the British in eastern Africa after the British East Africa Company threatened to undermine local autonomy. In an intense guerrilla war lasting nine months, the rebels gradually ceded territory and fled to Mwele, where they built a fort. But they soon lost the fort to the overwhelming British, although they did gain some victories along the way, prompting other rebel groups to join in. Yet British reinforcements from India played a decisive role in suppressing the revolt.

92. SECOND MATABELE: CONVENTIONAL MODEL

The Second Matabele War broke out again between the British and Matabele. This time, the Matabele forces were led by Mlimo but turned out to be vulnerable to British forces and so-called Cape Boys soldiers who received reinforcements for earlier combat with the Transvaal Republic. They had Bulawayo besieged this time again, and the British occupied Babyan's stronghold in the Matopos and the Mashona frontier before they murdered Mlimo. In the meantime, a separate conflict emerged in Mazowe after Nehanda Nyakasikana killed Mazowe Native Commissioner Pollard. As the war in Matabeleland neared its end, the British concentrated forces on Mashonaland and forced the rebels into retreat.

Additional reference: Oliver and Sanderson, eds., *Cambridge History of Africa*, vol. 6, from 1870 to 1905.

93. SECOND MAHDIST: DEGENERATIVE MODEL

The Second Mahdist War began in 1896 when the British returned to the Sudan to reoccupy it. In the early phase, the Mahdist Dervishes, this time led by Abdallahi ibn Muhammad, were annihilated by the British and Egyptian forces in 1899. Not surprisingly, the Mahdists found themselves less organized and fought insurgency wars, unlike the last time, when they were overwhelmed at the Battle of Atbara in 1898. The British forces were well armed and led by General Horatio Kitchener and were composed of over twenty-five thousand soldiers. Kitchener conquered Sudan with steamers and boats and an army of artillery and Maxim gun batteries. The decisive battle was held at Omdurman, where the British dominated, leaving eleven thousand dead among the forty thousand Dervishes. In this war Winston Churchill was an army reporter and described the key factor in British victory as the overwhelming British firepower.

Additional references: Churchill, *River War*, pp. 274–300; Daniel Headrick, *The Tools of Imperialism: Technology and European Imperialism in the Nineteenth Century* (Oxford: Oxford University Press, 1981), pp. 117–119.

94. BRITISH–SOUTH NIGERIAN: CONVENTIONAL MODEL

In 1897, local men led by British officers made an expedition to the Niger and Benue Rivers to fight the Battle of Bida, where they faced the Nupe army of Emir Abubakr, numbering ten thousand troops, and defeated them. Moving to Ilorin they also defeated Emir Sulaymanu's army there. As a result of this Britain increased colonial territory by incorporating Ilorin.

95. MOHMAND (BRITISH–PATHAN): CONVENTIONAL MODEL

The Mohmand War was held between Britain and Mohmand Pashtun tribes in the northwest frontiers of India and near the Afghan border. Britain deployed a main combat division led by Major General Sir Bindon Blood. One of his detachments suffered a small defeat at camp Inayat Killa, but soon they recovered. General Jeffreys then resumed his offensive in the Mamund valley, destroying a number of villages and ending the short-lived war.

Additional references: Lionel James, *The Indian Frontier War: Being an Account of the Mohmund and Tirah Expeditions, 1897* (London: Heinemann, 1898); Hugh Chisholm, *The Encyclopædia Britannica* (Cambridge: Cambridge University Press, 1911), p. 649.

96. HUT TAX: PRIMITIVE MODEL

The war broke out in 1898 when Mende tribes in Sierra Leone mounted a determined resistance to British colonialism there. Two years after Britain had established a protectorate status and introduced a harsh tax system called Hut Tax, the tribesmen revolted and killed a large number of merchants and their families, only to be suppressed by the British army in a year. While Britain had established elaborate tax and court systems there, the Mendes did not engage in counterefforts to build institutions, instead fighting like guerrillas.

Additional reference: Martin Kilson, *Political Change in a West African State: A Study of the Modernization Process in Sierra Leone* (Cambridge, Mass.: Harvard University Press, 1966), pp. 14–17, 256–257.

97. PHILIPPINE: DEGENERATIVE MODEL

In the aftermath of the U.S.-Spanish War, a Filipino nationalist leader, Emilio Aguinaldo, organized a local force to confront the U.S. forces in the Philippines in 1899. The apparent gap in military power and resources nevertheless quickly convinced Aguinaldo to switch to guerrilla strategy. This sudden switch, however, did not serve the insurgent rebels well because now the insurgency was more disorganized and became less capable of organizing the movement, let alone making efforts to build an independent state. So after three years of guerrilla war, the insurgency succumbed to the American leadership in the country.

98. CHAD: CONVENTIONAL MODEL

In 1899 French troops moved in the north, west, and south of Chad to attack rebel forces of Rabih az-Zubayr in a small Battle of Niellim. The French sent reinforcements from Fort Archambault to continue the expedition against the rebels at Kouno to defeat them and the so-called Dazingers, or African slave-soldiers. The French combined troops from Congo, Algeria, and Niger to attack the rebels at Kousseri the following year to kill the insurgent leader in the battle known as the Logone and Lakhta, to win the war and control over Chad.

99. SECOND BOER: DEGENERATIVE MODEL

In 1899 the Boer insurgents attacked British forces in Natal, defeated them at Laing's Nek, and besieged them at Ladysmith. After the famous Battle of Mafeking, the Boer War became less conventional, and Boer guerrilla tactics led to British defeats and casualties at Colenso and the Modder River. Reinforcements nevertheless saved British forces, who then took to the offensive, captured the Orange Free State capital and Pretoria, and defeated the Boers at Valkfontein. But the guerrilla war continued for two more years before the British forced its sovereignty to be recognized by the Boers in 1902.

100. ANGLO-SOMALI: CONVENTIONAL MODEL

Refer to my case study for details.

101. LAST ASHANTI: PRIMITIVE MODEL

In the final battle of insurgency between the Ashantis and British, the rebels, led by Queen Yaa Asantewa, launched a rebellion and besieged the British forces at Kumasi. Reinforcements helped the British to launch another expedition toward Aboasa, where they defeated the insurgents before they annexed the Ashanti kingdom and won the war.

102. BAILUNDU REVOLT: CONVENTIONAL MODEL

In 1902 the king of the Bailundu in what is now Angola led a rebellion of a forty-thousand-man army of mixed ethnic groups against the Portuguese forces there, who in turn deployed three expeditions. A small Portuguese force defeated the rebels near Huambo, allowing them to move into Mixoco, and defeated the Bailundu army.

103. KUANHAMA REBELLION: PRIMITIVE MODEL

Rebels in Kuanhama had confronted Portuguese control in the late nineteenth century. In 1902, at the same time as the Bailundu revolt, a major revolt broke out against the Portuguese, who suffered a series of losses to the Kuanhama near the Kunene and the border area with the Southwest African colony. But the rebellion quickly declined before it was suppressed in a year.

104. KANO & SOKOTO: CONVENTIONAL MODEL

This war took place when the British invaded Nigerian towns and garrisons in 1903. After an initial loss at Babeji, British forces quickly overwhelmed local troops and killed the town king. All operations in both Kano and Sokoto ended in just two months. The primary cause of Nigerian defeat was the unfavorable military balance against Britain. Sokoto had originally developed as a caliphate and fought neighboring regions before the conflict, but during the war it suffered from a lack of institutions beyond the caliphate and absence of effective territorial boundaries, which reduced its military capability. Adeleye writes that "the area within its perimeter did not wholly come under its jurisdiction."

Additional references: R. A. Adeleye, *Power and Diplomacy in Northern Nigeria, 1804–1906* (New York: Humanities Press, 1971), p. 52; William Geary, *Nigeria Under British Rule* (London: Routledge, 1965), pp. 217–220.

105. BRITISH-TIBETAN: CONVENTIONAL MODEL

Britain invaded Tibet in 1903 to face militia groups. British forces were led by Colonel Francis Younghusband and consisted of Gurkha and Sikh soldiers armed with machine guns and artillery and were better trained than those of Tibet. The conventionally organized Tibetan army at one point launched a successful surprise attack on Gyantse Fort, but their shortage of resources and negligence of guerrilla operations led to their defeat. Tibetan forces also lacked good training and used old, locally made muzzle loaders as their primary weapon, which proved ineffective. As army combat went badly, the thirteenth Dalai Lama fled to Mongolia. Of course, geography was an issue for both sides. British forces suffered from problems of transporting supplies from India all the way to Tibet. Yet the war ended when Britain forced Tibet to sign a trade pact known as the Anglo-Tibetan Agreement.

Additional references: John Powers, *History as Propaganda: Tibetan Exiles Versus the People's Republic of China* (Oxford: Oxford University Press, 2004), pp. 79–82; Charles Bell, *Tibet: Past and Present* (New Delhi: Munshiram Manoharlal, 1990), pp. 66–69.

106. SOUTH WEST AFRICAN REVOLT: DEGENERATIVE MODEL

The revolt occurred in Namibia in 1904 when Samuel Maharero led a rebellion to stop the Germans from collecting debts and moving the Herero people into a settlement as part of an occupation. The war was initially conventional; Bridgman writes that "because of their tribal discipline and warlike traditions, Hereros proved to be man for man equivalent to the German soldiers." The eight thousand Hereros under Maharero's command executed a series of concerted attacks on German garrisons. But with reinforcements the German army recuperated and pushed the locals back. Soon the war turned into a guerrilla war. In what is known as the Herero Genocide, which was provoked by Maharero's killing of Europeans, the Germans sought to eliminate all Africans and destroy settlements regardless of their participation in the revolt. But because Hereros attacked villages and looted them, they lost popular support and, later, the war itself.

Additional references: Jon Bridgman, *The Revolt of the Hereros* (Berkeley: University of California Press, 1981), p. 68; Gann and Duignan, *Colonialism in Africa*.

107. MAJI-MAJI REVOLT: DEGENERATIVE MODEL

The rebellion began in Tanzania when Matumbi villagers rose up to resist German policy to impose cotton-growing labor. Led by an independence-minded religious leader, Kinjikitile Ngwale, the violence spread quickly from coastal areas inland, involving neighboring tribes in the conflict. The rebellion spread further to Mahenge and Kilosa, but it soon broke up into a small guerrilla force. Ngwale's death left his brother to lead the movement before it collapsed soon. Overall, four hundred Germans were killed, and two hundred thousand Maji Maji warriors may have been killed.

Additional references: A. Bdu Boahen, *General History of Africa: Africa Under Colonial Domination 1880–1935*, vol. 2 (Berkeley: University of California Press, 1985), pp. 166–168; John Iliffe, "The Organization of the Maji Maji Rebellion," *Journal of African History* 8, no. 3 (1967).

108. THIRD ZULU: CONVENTIONAL MODEL

The Third British-Zulu War was led by Bambatha and took place against British rule and taxation policy in 1906 in Natal Colony in South Africa. Both sides were organized. Redding writes that "Bambatha proceeded to raise a small army of his own and then based the army in the forested Nkandla mountains." At the same time, Bambatha's forces were not strongly supported by the majority of Zulus, and there are questions about the level of commitment to combat among his forces.

Additional reference: Sean Redding, *Sorcery and Sovereignty: Taxation, Power, and Rebellion in South Africa, 1880–1963* (Athens: Ohio University Press, 2006), pp. 92–93, 92.

109. DEMBOS: PRIMITIVE MODEL

The Dembos (Ndembu) are part of the Mbundu people, the second largest ethnic group in Angola. Dembos rebels rose up in 1907 to fight Portuguese forces who wanted to control the Dembos territory. Low-intensity guerrilla operations ensued for much of the duration, in which Captain Joao de Almeida led a one-thousand-man Portuguese force to win numerous skirmishes, before they completed their occupation of the territory.

110. ANTI-FOREIGN REVOLT: PRIMITIVE MODEL

French forces moved from Senegal to Morocco in 1907 to suppress anti-European movements, provoking violent resistance against the French. Calling a holy war, the Moroccan sultan sent aids to the rebels in Mauretania, but the French conduct of pacification operations led Adrar to be occupied and most other ethnic tribal groups to be subjugated. The French forces defeated six thousand rebel soldiers at Tadla, ending the war with French victory.

111. JAPAN-KOREA: PRIMITIVE MODEL

Japan's victory over China generated confusion in which up to seventy thousand Korean peasant guerrillas along with some Korean soldiers fought the occupying Japanese army of two divisions starting in 1907. The insurgents took advantage of the terrain of the Korean Peninsula, both urban and mountainous areas. The guerrilla mission lasted for three years, during which Japanese resident general was assassinated in 1909 until Japan went on to annex Korea.

112. FRENCH CONQUEST OF WADAI: CONVENTIONAL MODEL

Located along the border of Sudan and east of Lake Chad, Wadai had organized an armed force to fight the French forces moving against Abeche in 1909. Abeche was taken quickly, with the French soon building a puppet sultan, which provoked the existing sultan to launch a revolt for two years.

113. FRENCH-BERBER: PRIMITIVE MODEL

In 1912 French forces found themselves fighting the Berbers in Morocco when the Moroccan soldiers revolted and joined the fifteen thousand rebels seeking independence. Reinforcements helped them achieve that goal despite the arms imported for the rebels from neighbors in Egypt. Berber rebel leader El Hiba seized Marrakesh, but soon they were defeated by five thousand French soldiers. The Battle of Sidi Ben Othman ended the war with French victory in the same year.

114. FIRST SINO-TIBETAN: CONVENTIONAL MODEL

Tibet took advantage of the 1912 rebellion that overthrew the Manchu dynasty in China as a chance to seek autonomy by attacking Chinese troops. With the Dalai Lama returning from exile, Tibetans overwhelmed the Chinese troops, forced three thousand of them to surrender and disarm, and ejected them through India. Tibetans then expelled Chung Ying, the Amban who took over Tibet briefly after the rebellion. The war forced China's Yuan Shikai to issue an apology and the Dalai Lama to declare independence. Tibet's political control during the war was challenged because he spent long time in exile after the British-Tibetan war of 1903 and because elements of the Gelugpa monastic segments sought to challenge his authority. Tibet developed a conventional force made up of three thousand men in 1913.

Additional reference: Josef Kolmas, *Tibet and Imperial China: A Survey of Sino-Tibetan Relations up to the End of the Manchu Dynasty in 1912* (Canberra: Australian National University, 1967), pp. 65–67.

115. MORO: PRIMITIVE MODEL

In 1913 U.S. General John Pershing returned to the Philippines with an aim to disarm some of the remaining Moro rebel guerrillas. Numbering more than five thousand, the Moros retreated into the rural and mountainous areas to evade direct attacks by the American forces. Yet they soon found their fortifications destroyed by American firepower before they were defeated in guerrilla war.

116. SECOND SINO-TIBETAN: CONVENTIONAL MODEL

A war broke out between China and autonomy-seeking Tibetan rebels in 1918 when Chinese General Peng Jih-sheng's forces began a full-scale attack. The Tibetan forces pushed the adversary back and forced the general to surrender his fort at Chamdo. The Tibetan offensive nevertheless failed to materialize a favorable treaty between the two sides, in which the rebels acquiesced to give up Inner Tibet and retreat partially from the territorial gains they made in the war; thus I code the war as a Chinese victory.

117. CACO REVOLT: PRIMITIVE MODEL

In the Caco Revolt of 1918, Haitian nationalist leader Charlemagne Péralte, the self-proclaimed "Commander-in-Chief of the Forces Operating against the Americans," led a large group of approximately forty thousand former *caco* peasants against the United States, which occupied the island. The uprising and resultant guerrilla war overwhelmed the Creole-speaking American gendarmerie troops, but marine reinforcements helped the United States suppress the revolt after an estimated two thousand Haitians died. The death of Benoît Batraville, who had taken over guerrilla leadership, ended the revolt in 1920. I consider the war to be a primitive model; Mary Renda and Hans Schmidt, in their respective works, treat the war as mostly insurgency.

Additional reference: Mary Renda, *Taking Haiti: Military Occupation and the Culture of U.S. Imperialism, 1915–1940* (Chapel Hill: University of North Carolina Press, 2001).

118. THIRD AFGHAN: CONVENTIONAL MODEL

In the Third Afghan War, the Afghans faced the British with an army of fifty thousand men in cavalry and infantry units backed up by eighty thousand local forces. The British were well organized while the Afghan units were ill trained and weak. After they crossed the Khyber and occupied the village of Bagh, British and Indian troops took western Khyber with no opposition. Fighting in the war of independence, the Afghans came successfully out of the negotiations in Rawalpindi to gain control of foreign affairs. Arreguin-Toft designates this war as a victory by the strong side, but I code it as a victory by the nonstate side since the weaker Afghans achieved independence in 1919 as a result of the war.

Additional reference: Leon Poullada, *Reform and Rebellion in Afghanistan, 1919–1929* (Ithaca, N.Y.: Cornell University Press, 1973), pp. 46–47.

119. FIRST WAZIRISTAN: PRIMITIVE MODEL

In 1919, militia groups in Waziristan in what is now Afghanistan revolted against British control at Miranshah and Wana. Geography allowed the insurgents to operate like guerrilla forces and take advantage of stealth and terrain to hit and run away from enemy troops. Thus the tribes staged more than one hundred raids against the British in the first few weeks of the war. The British forces nevertheless planned for an offensive to subdue the rebels, with forty thousand soldiers and another forty thousand irregular forces heading toward Kaniguram. Both sides suffered great casualties, but the British forces overwhelmed in the end.

120. SYRIAN: PREMATURE MODEL

The war began with a French invasion on disorganized guerrilla movements in Syria's countryside, specifically at Alawite Mountain and Aleppo. Local resistance forces were weak, however; Hananu forces in Aleppo had only some cannons and small arms. The war moved from countryside to towns, where conventional forces began to play central roles. That is where French forces met with forces of Syria's King Faysal, whose army and rebel bands were composed of tribesmen and irregulars. France entered Syria through Aleppo with a large army under General Goybet and won a decisive Battle of Khan Maysalun. Damascus fell to France in a matter of three weeks, before French General Gouraud marched on Damascus to place Syria under French protectorate.

Additional reference: Philip Khoury, *Syria and the French Mandate: The Politics of Arab Nationalism, 1920–1945* (Princeton, N.J.: Princeton University Press, 1987), pp. 97–118.

121. IRAQI-BRITISH: PRIMITIVE MODEL

Arab nationalists in Iraq rose against the British, who devised ways to satisfy local demands and aspirations for the independence of Mesopotamia. The long-term

British presence in the country caused the outbreak of tribal revolts. In their war for independence, Iraqi forces were led by trained officers, although their movements and overall strategy were largely uncoordinated. The Iraqi forces fought a guerrilla war in a third of their territory, mostly in rural areas, causing little disorder in cities. They were considered to have had ineffective strategy, no cohesion, and little leadership structure. British forces capitalized on these enemy problems and gradually recovered areas to control. As the war came to a close, the British placed Faisal, the former king of Syria, as king of Iraq to assert their will and control the territory.

Additional reference: Stephen Longrigg, *Iraq, 1900 to 1950: A Political, Social, and Economic History* (London: Oxford University Press, 1953), pp. 122–126.

122. MONGOLIA: CONVENTIONAL MODEL

In 1920, Baron Roman von Ungern-Sternberg led an army against Chinese and Russian interests to restore monarchies in Russia, Mongolia, and Manchuria. Open terrain encouraged the strategy of conventional war, rather than a guerrilla strategy. The Chinese forces defeated his forces at the Battle of Huree. Ungern-Sternberg's five-thousand-man army nevertheless occupied Urga and beat the four thousand Chinese soldiers in early 1921. The entry of Russian Bolsheviks in the war on the Chinese side nonetheless changed the direction of the war, by routing the baron's forces back into Mongolia and capturing Huree later before capturing the baron himself, ending the war with Russian victory.

123. ITALO-LIBYAN: PREMATURE MODEL

The war took place as a local revolt against Italian control in the early 1920s. During this time, Italian forces led by Pietro Badoglio and Rodolfo Graziani carried out pacification campaigns. Starting in 1924, some tribes engaged in insurgency using the help of Awlad Sulayman tribes. Yet in the Green Mountain region, Italian Governor Mombelli built a counterguerrilla force that defeated the rebels in 1925. Rebel leader Omar Mukhtar then switched to surprise attacks and began to coordinate with other Libyan forces, setting the stage for confrontation with Graziani. Having failed in a massive offensive against Omar's forces, Graziani looked at counterinsurgency ideas and led Italian efforts to transfer villagers into resettlement projects near Egypt and to curtail external aid to the rebels. In 1930, Italian use of air strikes and chemical weapons finally overcame the resistance, forcing the tribes to flee. Large-scale resistance disappeared after the execution of rebel leader Mukhtar in 1931.

Additional reference: Ali Abdullatif Ahmida, *The Making of Modern Libya: State Formation, Colonization, and Resistance, 1830–1932* (Albany: State University of New York Press, 1994).

124. RIF: CONVENTIONAL MODEL

The war involved Spanish forces and Rif and J'bala tribes in today's Morocco and lasted from 1920 to 1926. Led by Abd el-Krim, Rifian tribesmen comprised elite regular units that excelled in encirclement maneuvers and surprise attacks. The other Rifian forces were mostly tribal militia who made up a large part of the eighty-thousand-man Rifian Army. Spanish regular forces suffered a defeat at the Battle of Annual in 1921. With French support, however, Spanish forces regrouped and pushed back, finally dissolving the Republic of Rif in 1926. El-Krim ran the Rifian regime though some political institutions, but they crumbled as his regional leaders distanced themselves from him.

Additional references: David Woolman, *Rebels in the Rif, Abd el Krim and the Rif Rebellion* (Stanford, Calif.: Stanford University Press, 1968); Pessah Shinar, "Abd al Qadir and Abd al Krim: Religious Influences on Their Thought and Action," *Asian and African Studies* 1 (1965).

125. MOPLAH REBELLION: DEGENERATIVE MODEL

The rebellion began in 1921 by Moplah Muslims against British control in India. Led by Ali Musaliar and Variamkunnath Kunhammad Haji, the rebellion was not well organized. They clashed with the British army and lost Pukkottur and later mounted an unsuccessful resistance at Tirurangadi. Thus the early part of the rebellion was conventional; as Wood writes, the rebellion "at least in its earliest stage, did not present the appearance of a communal rampage in which Hindus were slaughtered wholesale merely as Hindus." Immediately after the losses, however, the rebels switched their strategy to evade direct confrontation. The British countered by using a special force called the Malabar Special Police. Allied with a loosely organized group called the Khilafat noncooperation movement, the insurrection placed its initial focus on the government, which carried out repressive measures. The rebellion was suppressed in a few months.

Additional reference: K. N. Panikkar, *Against Lord and State: Religion and Peasant Uprisings in Malabar, 1836–1921* (Delhi: Oxford University Press, 1989).

126. DRUZE REVOLT: CONVENTIONAL MODEL

Druze leader Sultan al-Atrash led an insurrection against the French regime in Syria in 1925. As the Druzes built a provisional government and supporting institutions, they mobilized peasants into a revolutionary army. The war saw mostly combat operations in cities. Rebels moved fast and rather freely within Syrian territory, in contrast to the French strategy focused more on holding the territory in major cities like Damascus. As Khoury writes, French military strategy was to "crush the revolt by the maximum use of every mechanical contrivance available but with the minimum use of French soldiers." French reinforcements and bombings in Damascus proved to overwhelm the underarmed insurgents, suppressing the rebellion in two years.

Additional references: Khoury, *Syria and the French Mandate*; Michael Provence, *The Great Syrian Revolt and the Rise of Arab Nationalism* (Austin: University of Texas Press, 2005), p. 192.

127. YEN BAI UPRISING: PRIMITIVE MODEL

In 1930, nationalist rebels rose up against the French colonial government in Vietnam. The Vietnamese Nationalist Party, VNQDD, embodied the insurgents' state-building efforts, so essentially the war was not completely a primitive case. Yet the uprising never turned into a conventional conflict, and much of the conflict war a series of terror, assassination attempts, and general uprisings by a mutiny of Vietnamese soldiers against their French officers at Yen Bai in the north. The French forces added reinforcements to police in the affected area and in a year managed to suppress the uprising.

128. SAYA SAN'S REBELLION: DEGENERATIVE MODEL

Saya San led a nationalist peasant movement against British colonial rule in Burma in 1930. The rebel group started the war using the Galong army, targeting enemy police stations and village people, but they soon abandoned open confrontation to adopt guerrilla strategy and moved to jungle areas. British reinforcements came from India, and by 1931, the British governor sent in over eight thousand troops to fight the small, underarmed, and geographically dispersed Saya San group. The rebel group did not have a way to go around Burma's multiethnic society, had few institutions to support the war effort, and as a result suffered from a lack of popular support. By 1932, the Saya San group was put down.

Additional reference: Mary Callahan, *Making Enemies: War and State Building in Burma* (Ithaca, N.Y.: Cornell University Press, 2003), pp. 30, 36, 38, 114.

129. BRITISH-PALESTINIAN: PRIMITIVE MODEL

The war was an Arab rebellion in Palestine against mass Jewish communities. Britain responded using support from the Haganah, a paramilitary organization, to succeed in suppressing the rebellion. The insurgent side did not manage the war well. The Qassamite peasant bands operated independently of the Higher Arab Committee, and commander Fawzi al-Qawuqji failed to unite groups. The rebels split up because of differences in political, family, and regional perspectives and British misinformation strategy. The guerrilla bands had a political structure, but because they lacked resources to modernize their forces and Syria refused to support them, the war was fought more in rural areas and highlands than in cities. As the war culminated in the late 1930s, more problems surfaced when the rebels alienated the population through public terror.

Additional reference: Yehoshua Porath, *The Palestinian Arab National Movement: From Riots to Rebellion*, vol. 2 (London: Frank Cass, 1977).

130. SECOND WAZIRISTAN: CONVENTIONAL MODEL

The war began when two British brigades were attacked by local rebels in Waziristan in 1936. The attack prompted the dispatch of British troops to encounter more raids by an army led by Haji Mirza Ali Khan. Britain's deployment of thirty-seven thousand troops seized Arsal Kot, which forced the enemy leader to flee and cede military operations in two years, although he was able to continue his attacks in the meantime.

131. INDONESIAN: MAOIST MODEL

Known as the National Revolution, the war took place after the Japanese occupation ended in 1945. Nationalist leaders like Sukarno and Mohammad Hatta seized the opportunity to form the Committee for Preparatory Work for Indonesian Independence (BPUPKI) and adopted a constitution in 1945. They then created the Central Indonesian National Committee to build support bases. Indonesians had developed national institutions in the early phase of the revolution. When the Dutch intervened, the war took on a guerrilla style. Yet it became increasingly conventional, as Dutch forces launched Operation Crow in 1948 and the Indonesian National Army, by then a professional armed force, responded vigorously, eventually culminating in the "March 1 Public Attack." Indonesia forced the Netherlands to accept its independence, which it had claimed at the beginning of the war.

Additional references: Benedict Anderson, *Java in a Time of Revolution: Occupation and Resistance, 1944–1946* (Ithaca, N.Y.: Cornell University Press, 1972); Bernhard Dahm, *Sukarno and the Struggle for Indonesian Independence* (Ithaca, N.Y.: Cornell University Press, 1969), pp. 276–350.

132. INDOCHINA: PROGRESSIVE MODEL

Refer to my case study for details.

133. THIRD MADAGASCAN: PRIMITIVE MODEL

In 1947, the nationalist group MDRM in Madagascar revolted against France for independence. The war proved to be a challenge initially; the insurgent group did not have a national-level support base. Yet they succeeded in seizing one-third of the island before French forces received reinforcements and crushed the rebellion in six weeks. France adopted the so-called oil-spot method, designed to counter insurgents by gradually spreading areas of pacification. The violent movement was suppressed in 1948, and Madagascar remained under French protectorate until it gained independence in 1960.

Additional reference: Robert Aldrich, *Greater France: A History of French Overseas Expansion* (London: Macmillan, 1996), pp. 284–285.

134. MALAYAN EMERGENCY: PRIMITIVE MODEL

Refer to my case study for details.

135. INDO-HYDERABAD: CONVENTIONAL MODEL

Capitalizing on unrest following India's independence in 1947, Hyderabad fought India in 1948 for self-rule. A group of communists started engaging in violent acts in Telaugana. The Indian army, made up of twenty thousand troops and over eight thousand irregulars, suppressed the movement in a few days and India successfully annexed Hyderabad. The wars were both conventional and guerrilla nature, but I weigh the conventional side more. Popular support, too, was divided, as the movement was a popular upsurge while the guerrillas engaged in attacking civilians and coerced them. I code the war outcome as a victory by India, as Gantzel and Schwinghammer do.

Additional references: S. N. Prasad, *Operation Polo: The Police Action in Hyderabad, 1948* (New Delhi: Ministry of Defence, 1972); Klaus Jürgen Gantzel and Torsten Schwinghammer, *Warfare Since the Second World War* (New Brunswick, N.J.: Transaction, 2000), p. 60.

136. TUNISIA: MAOIST MODEL

Beginning in the late 1940s, nationalist leader Habib Bourguiba led the Neo-Destour Party as a government in Tunisia. The government benefited from support from a labor union called L'Union Generale Tunisienne du Travail (UGTT) and student organizations. The French began several reform projects in order to undermine the independence cause, to no avail. This initial failure permitted bloody riots to break out, responded to by the French via "Operation Mars." French measures intensified in the early 1950s and culminated in sweep operations of Cape Bon. Bourguiba countered by mobilizing freedom fighters called *fellaghas* in order to fight better. Bourguiba had studied theories of revolution advanced by Mao, Lenin, and Trotsky.

Additional reference: Dwight Ling, *Tunisia: From Protectorate to Republic* (Bloomington: Indiana University Press, 1967), pp. 131–185.

137. MAU MAU REBELLION: MAOIST MODEL

The rebellion began in Kenya in 1952 when the Kikuyu Central Association appealed for independence. The Kenya African Union (KAU) began to demand greater autonomy before Mau Mau leaders mobilized their forces and turned violent. Led by Jomo Kenyatta, KAU activities were limited to the countryside. When Britain announced the state of emergency in 1952, the war became largely a guerrilla war and combat moved into cities like Nairobi. Later the Kenyans built the Land and Freedom Armies, but they were poorly equipped and proved powerless. The last rebel leader, Dedan Kimathi, was captured in 1956 and executed the following year. Kenya became independent in 1963, but the rebellion itself fell and ended in 1960.

Additional reference: Prosser Gifford and Roger Louis, eds., *Decolonization and African Independence: The Transfers of Power, 1960–1980* (New Haven, Conn.: Yale University Press, 1988), pp. 401–426.

138. MOROCCO: PRIMITIVE MODEL

The war of independence broke out in Morocco in 1953 when France and the royal family split, resulting in the sultan's ouster, although a weak nationalist movement had been there for years. Led by the resistance movement called Armee de Liberation, the guerrilla war became so intense that it generated a large number of French desertions. Student movements also joined the insurrection, later followed by the colonial military. In 1955, the movement faced daily attacks on their posts, but in 1956 France permitted the establishment of the Moroccan Royal Armed Forces, which incorporated members of the liberation movement. Paris also accepted Moroccan independence, bringing Sultan Mohammed V back from exile to take over sovereignty.

Additional reference: Moshe Gershovich, *French Military Rule in Morocco: Colonialism and Its Consequences* (London: Frank Cass, 2000), pp. 207–216.

139. ALGERIA: PRIMITIVE MODEL

The Algerian war broke out between the National Liberation Front (FLN) and France. The French army was not defeated militarily, but the Algerians achieved their goal of independence through violent resistance involving terrorist acts. The FLN also took advanced of a long-term crisis in France that was provoked by the protracted and costly imperialist wars in places like Indochina. The Algerian organization grew its popular control in the latter part of the war, but it was decentralized, so I code the war as primitive. The Algerians also made efforts to modernize their forces during the years to counter the French army, but the combat mostly took the form of guerrilla operations.

Additional reference: Alistair Horne, *A Savage War of Peace: Algeria, 1954–1962* (New York: New York Review of Books, 2006)

140. CAMEROON: PRIMITIVE MODEL

Arreguin-Toft codes Cameroon as a case of victory by nonstate actors, but this is inaccurate. I posit that the state side—France—won the war, as the Union of the Peoples of Cameroon (UPC) failed to achieve its war goal of becoming the majority party even though Cameroon became independent in 1960. From its inception the UPC was a minority party in Cameroon, always battered by the more unified Ahidjo government and French forces. Cameroon became independent because Ahidjo won a French concession, not the UPC. Thus I code it as a victory by the state, as do Gantzel and Schwinghammer. The independence movement built political institutions that managed to surround the UPC in the late 1940s and carried out a violent rebellion in 1955, after which it gradually slid into guerrilla war.

Additional references: Victor LeVine, *The Cameroons from Mandate to Independence* (Berkeley: University of California Press, 1964), pp. 141–192; James Coleman and Carl Rosberg, Jr., eds., *Political Parties and National Integration in Tropical Africa* (Berkeley: University of California Press, 1964), pp. 132–184; Gantzel and Schwinghammer, *Warfare Since the Second World War*, p. 45.

141. ANGOLA: PROGRESSIVE MODEL

In this war, the Popular Movement for the Liberation of Angola (MPLA) fought guerrilla style aiming at the willpower of the Portuguese colonial and domestic forces. It also achieved its end of independence by fending off challenges from other nationalist groups, such as the National Union for the Total Independence of Angola (UNITA) and National Front for the Liberation of Angola (FNLA) in order to unify the nationalist movement. Its guerrilla strategy helped draw support from peasants as well as exiled members overseas in building a nationwide political movement. The development of independence-minded political authority helped boost the MPLA's military capability. Facilitated by the nature of war between Portugal and Guinea Bissau, the war turned positively for the Angolans when Portugal collapsed in 1975.

142. PORTUGUESE GUINEA: MAOIST MODEL

Refer to my case study for details.

143. MOZAMBIQUE: PROGRESSIVE MODEL

The nationalist movement Liberation Front of Mozambique (FRELIMO) included a number of members coming from other colonies to wage guerrilla war against Portugal. Taking advantage of the concurrent Portuguese wars elsewhere, FRELIMO was a centralized organization that gained popular support. Once political foundations became firm, FRELIMO grew militarily powerful as well, using support from states like China, which helped it build a modern army. The 1975 Carnation Revolution brought down Portugal and caused its army to withdraw, and FRELIMO completed a successful three-stage evolution.

Additional references: Gifford and Louis, *Decolonization and African Independence*, pp. 427–444; David Birmingham, *Frontline Nationalism in Angola and Mozambique* (Trenton, N.J.: Africa World Press, 1993).

144. EAST TIMOR: PRIMITIVE MODEL

Indonesia invaded East Timor under "Operation Komodo" in 1975 when Portugal's Carnation Revolution left Timor free. Timor's disorganized militia group, Falintil, fought guerrilla warfare using its ten thousand members but was outgunned and repulsed by Indonesia. The guerrilla campaign involved murdering civilians and destroying bases. Thus Falintil abandoned bases in favor of mobile columns, but it never grew large enough to become a powerful organization, nor did it fight like an army, which the Indonesian forces were. In the face of Indonesia's Operation Total Encirclement and Annihilation, the Falintil lost the war. It was only after the war was over that the Falintil developed a National Council of Maubere Resistance as the highest organization in the resistance.

Additional references: Taylor, *Indonesia*, pp. 380–381; Peter Carey and G. Carter Bentley, eds., *East Timor at the Crossroads: The Forging of a Nation* (Honolulu: University of Hawaii Press, 1995), pp. 59–72.

145. NAMIBIA: MAOIST MODEL

Namibia was nominally under UN mandate but practically controlled by South Africa, which recruited mercenaries to serve in Namibia and built bases there and along the border with Angola. It also conscripted Namibians for the Citizen Force units. Namibians reacted by forming the South West Africa People's Organization (SWAPO) as their party machine once the war began. They led guerrilla war with the South African army under the leadership of its military wing, the People's Liberation Army of Namibia (PLAN). The war became more organized, with South Africa building the SWA/Namibia Territory Force (SWATF). The PLAN launched heavy artillery offensives on enemy bases at Omahenene Angola. According to the United Nations and the South African Council of Churches, the SWAPO received wide Namibian popular support. The war ended in 1988, two years before Namibia became independent.

Additional reference: Leonard Thompson, *A History of South Africa* (New Haven, Conn.: Yale University Press, 1996), p. 239.

146. SOVIET-AFGHAN: PRIMITIVE MODEL

The 1979 deployment of Soviet forces led to a ten-year conflict in Afghanistan where we saw much of combat in guerrilla forms between the Soviet forces and the Mujahideen. Geographic dispersion of the insurgent forces helped them wage guerrilla warfare effectively, while the Soviet forces struggled to attack them in proper formation, which helped the war to reach a stalemate for years. The Soviet forces armed themselves with modern weapons, but they proved to be largely ineffective on the strategic level as they found themselves bogged down in insurgent operations. The war ended with a Soviet loss in 1989 when the Soviet Union pulled its forces from Afghanistan.

147. SOMALIA: PRIMITIVE MODEL

I designate the war in Somalia to be extrasystemic, as it was fought mainly between the United States and local insurgents. UN activities during the period of 1992 to 1995 were mostly humanitarian relief operations. Yet violent operations took place between the two sides with the United States seeking to capture the Somali clan leaders during Operation Restore Hope. The conflict had heavy involvement of the civilian populace in Somalia, whom the rebel soldiers used as human shields. In 1993, the war culminated at the Battle of Mogadishu in which Somali rebels defeated American forces via the "Black Hawk Down" incident, forcing the U.S. forces to withdraw. The state of anarchy continues throughout Somalia today, but the conflict between the two sides ended in 1995.

148. IRAQ: DEGENERATIVE MODEL

Refer to my case study for details.

NOTES

Chapter 1

1. Kenneth Waltz, "The Politics of Peace," *International Studies Quarterly* 11, no. 3 (September 1967): 205.

2. Shankar Vedantam, "Don't Send a Lion to Catch a Mouse," *Washington Post*, March 5, 2007.

3. T. V. Paul, *Asymmetric Conflicts: War Initiation by Weaker Powers* (Cambridge: Cambridge University Press, 1994).

4. Andrew Mack, "Why Big Nations Lose Small Wars: Politics of Asymmetric Conflict," *World Politics* 27, no. 2 (January 1975); Ivan Arreguin-Toft, *How the Weak Win Wars: A Theory of Asymmetric Conflict* (Cambridge: Cambridge University Press, 2005); Gil Merom, *How Democracies Lose Small Wars: State, Society, and Failures of France in Algeria, Israel in Lebanon, and the United States in Vietnam* (Cambridge: Cambridge University Press, 2003); Jason Lyall and Isaiah Wilson, "Rage Against the Machines: Explaining Outcomes in Counterinsurgency Wars," *International Organization* 63 (Winter 2009).

5. David Singer and Melvin Small, *Resort to Arms: International and Civil War, 1816–1980* (Beverly Hills, Calif.: Sage, 1982). Also see Meredith Reid Sarkees, "The Correlates of War Data on War: An Update to 1997," *Conflict Management and Peace Science* 18, no. 1 (2000).

6. Jason Lyall, "Do Democracies Make Inferior Counterinsurgents? Reassessing Democracy's Impact on War Outcomes and Duration," *International Organization* 64, no. 1 (2010): 175.

7. Minority Rights Group International, *People Under Threat 2012* (London: Minority Rights Group International, 2012).

8. Lawrence Keeley, *War Before Civilization: The Myth of the Peaceful Savage* (New York: Oxford University Press, 1997), p. 32.

9. John Mueller, *The Remnants of War* (Ithaca, N.Y.: Cornell University Press, 2004).

10. Michael Horowitz, *The Diffusion of Military Power: Causes and Consequences for International Politics* (Princeton, N.J.: Princeton University Press, 2010).

11. Kristian Skrede Gleditsch, Idean Salehyan, and Kenneth Schultz, "Fighting at Home, Fighting Abroad: How Civil Wars Lead to International Disputes," *Journal of Conflict Resolution* 52, no. 4 (2008).

12. Charles Caldwell, *Small Wars: Their Principles and Practice* (London: Harrison and Sons, 1896).

13. Mao Zedong, *On Guerrilla Warfare* (Mineola, N.Y.: Dover, 2005); Herbert Wulf, "Dependent Militarism in the Periphery and Possible Alternative Concepts," in Stephanie Neuman and Robert Harkavy, eds., *Arms Transfers in the Modern World* (New York, 1979).

14. Avery Goldstein, *Deterrence and Security in the 21st Century: China, Britain, France, and the Enduring Legacy of the Nuclear Revolution* (Stanford, Calif.: Stanford University Press, 2000), pp. 26–31.

15. Rupert Smith, *The Utility of Force: The Art of War in the Modern World* (London: Knopf, 2007), p. 16.

16. Russell Weigley, *The American Way of War: A History of United States Military Strategy and Policy* (New York: Macmillan, 1973); Thomas Mahnken, *Technology and the American Way of War Since 1945* (New York: Columbia University Press, 2008).

17. Fred Weyand and Harry Summers, "Vietnam Myths and American Realities," in *Commanders Call* (Fort Leavenworth, Kans.: U.S. Army Command and General Staff College, July–August 1976), p. 3.

18. Samuel Huntington, "Guerrilla Warfare in Theory and Practice," in Franklin Mark Osanka, ed., *Modern Guerrilla Warfare: Fighting Communist Guerrilla Movements, 1941–1961* (New York: Free Press, 1962), p. xvi.

19. Keeley, *War Before Civilization*, pp. 59–69.

20. The literature on guerrilla war is large. See, for instance, Robert Taber, *War of the Flea: The Classic Study of Guerrilla Warfare* (Washington, D.C.: Brassey's, 2002); Andrew Krepinevich, *The Army and Vietnam* (Baltimore: Johns Hopkins University Press, 1981); David Galula, *Counterinsurgency Warfare: Theory and Practice* (Westport, Conn.: Greenwood, 2006); David Kilcullen, *The Accidental Guerrilla: Fighting Small Wars in the Midst of a Big One* (Oxford: Oxford University Press, 2009).

21. William Martel, *Victory in War: Foundations of Modern Strategy*, 2nd ed. (Cambridge: Cambridge University Press, 2011).

22. Dominic Johnson and Dominic Tierney, *Failing to Win: Perceptions of Victory and Defeat in International Politics* (Cambridge, Mass.: Harvard University Press, 2006).

23. Mueller, *Remnants of War*, pp. 21–22.

24. Janice Thomson, *Mercenaries, Pirates, and Sovereigns: State-Building and Extraterritorial Violence in Early Modern Europe* (Princeton, N.J.: Princeton University Press, 1996).

25. Max Abrahms, "Why Terrorism Does Not Work," *International Security* 31, no. 2 (Fall 2006).

26. Arreguin-Toft, *How the Weak Win Wars*.

27. Lyall and Wilson, "Rage Against the Machines."

28. Galula, *Counterinsurgency Warfare*, p. 68.

29. However, Jessica Weeks argues that democracies should have no audience cost advantage over autocracies because elites in democratic states are more able to solve coordination problems and the possibility of coordination is observable to foreign decision makers. Jessica Weeks, "Strongmen and Straw Men: Authoritarian Regimes and the Initiation of International Conflict," *American Political Science Review* 106, no. 2 (May 2012).

30. Scott Bennett and Allan Stam, "The Declining Advantages of Democracy: A Combined Model of War Outcomes and Duration," *Journal of Conflict Resolution* 42, no. 3 (June 1998).

31. Elizabeth Stanley and John Sawyer, "The Equifinality of War Termination: Multiple Paths to Ending War," *Journal of Conflict Resolution* 53, no. 5 (October 2009).

32. Hilde Ralvo, Nils Gleditsch, and Han Dorussen, "Colonial War and the Democratic Peace," *Journal of Conflict Resolution* 47, no. 4 (August 2003).

33. Dan Reiter and Curtis Meek, "Determinants of Military Strategy, 1903–1994: A Quantitative Empirical Test," *International Studies Quarterly* 43, no. 2 (June 1999).

34. Todd Sechser and Elizabeth Saunders, "The Army You Have: The Determinants of Military Mechanization, 1979–2001," *International Studies Quarterly* 54, no. 2 (June 2010).

35. Mack, "Why Big Nations Lose Small Wars."

36. Arreguin-Toft, *How the Weak Win Wars*.

37. Merom, *How Democracies Lose Small Wars*.

38. Lyall, "Do Democracies Make Inferior Counterinsurgents?"; Max Abrahms, "Why Democracies Make Superior Counterterrorists," *Security Studies* 16, no. 2 (2007); Alexander Downes, "How Smart and Tough Are Democracies? Reassessing Theories of Democratic Victory in War," *International Security* 33, no. 4 (2009); Dan Reiter and Allan Stam, *Democracies at War* (Princeton, N.J.: Princeton University Press, 2002); Kenneth Schultz, *Democracy and Coercive Diplomacy* (New York: Cambridge University Press, 2001).

39. Jeffrey Record, *Beating Goliath: Why Insurgencies Win* (Washington, D.C.: Potomac Books, 2007).

40. Paul Staniland, "Organizing Insurgency: Networks, Resources, and Rebellion in South Asia," *International Security* 37, no. 1 (Summer 2012).

41. Daniel Byman, *Deadly Connections: States That Sponsor Terrorism* (New York: Cambridge University Press, 2005).

42. David Carter, "A Blessing or a Curse? State Support for Terrorist Groups," *International Organization* 66, no. 1 (January 2012).

43. Lyall and Wilson, "Rage Against the Machines."

44. Patricia Sullivan, *Who Wins? Predicting Strategic Success and Failure in Armed Conflict* (Oxford: Oxford University Press, 2012).

45. Kristian Skrede Gleditsch, "A Revised List of Wars Between and Within Independent States, 1816–2002," *International Interactions* 30 (2004); Meredith Reid Sarkees and Frank Wayman, *Resort to War: 1816–2007* (Washington, D.C.: CQ Press, 2010).

46. To the dataset I have added Haiti-Santo Domingo, First Afghan, Second Khivan, Second Afghan, Little, Second Ethiopia-Egypt, First Ethiopia-Italy, First Matabele, Second Matabele, British-Tibetan, and Somalia. I have eliminated from the dataset (1) First Moroccan Crisis, (2) Second Moroccan Crisis, and (3) Khivan War because they were diplomatic crises short of war; (4) Oman-Zanzibar war; and (5) Kabylia uprising and (6) Russo-Turkoman war because reliable data were unavailable.

47. To compile the data I referred to Ernest Dupuy and Trevor Dupuy, *The Encyclopedia of Military History* (New York: HarperCollins, 1993); Byron Farwell, *The Encyclopedia of Nineteenth-Century Land Warfare* (New York: Norton, 2001); Klaus Jürgen Gantzel and Torsten Schwinghammer, *Warfare Since the Second World War* (New Brunswick, N.J.: Transaction, 2000); Lyall, "Do Democracies Make Inferior Counterinsurgents?"; Arreguin-Toft, *How the Weak Win Wars*; David Gompert and John Gordon, *War by Other Means: Building Complete and Balanced Capabilities for Counterinsurgency* (Santa Monica, Calif.: RAND, 2008).

48. Alexander George and Andrew Bennett, *Case Studies and Theory Development in the Social Sciences* (Cambridge, Mass.: MIT Press, 2005).

49. William Reno, *Warfare in Independent Africa* (Cambridge: Cambridge University Press, 2011).

50. George and Bennett, *Case Studies*, p. 156.

51. Erica Chenoweth and Maria Stephan, *Why Civil Resistance Works: The Strategic Logic of Nonviolent Conflict* (New York: Columbia University Press, 2011).

Chapter 2

1. Mao, *On Guerrilla Warfare*, p. 96.

2. Kenneth Waltz, *Theory of International Politics* (New York: McGraw-Hill, 1979).

3. Bruce Hoffman, "Terrorism Trends and Prospects," in Ian Lesser et al., eds., *Countering the New Terrorism* (Santa Monica, Calif.: RAND, 1999), p. 25.

4. Dominic Johnson, "Darwinian Selection in Asymmetric Warfare: The Natural Advantage of Insurgents and Terrorists," *Journal of the Washington Academy of Sciences* 95 (2009): 92.

5. Peter Bowler, *Evolution: The History of an Idea* (Berkeley: University of California Press, 2003), pp. 236–244.

6. Peter Hatemi and Rose McDermott, "A Neurobiological Approach to Foreign Policy Analysis: Identifying Individual Differences in Political Violence," *Foreign Policy Analysis* 8 (2011).

7. Ibid.

8. Rafael Sagarin, "Adapt or Die: What Charles Darwin Can Teach Tom Ridge About Homeland Security," *Foreign Policy*, September–October 2003.

9. Scott Sigmund Gartner, *Strategic Assessment in War* (New Haven, Conn.: Yale University Press, 1999).

10. Vladimir Lenin, *Imperialism: The Highest Stage of Capitalism* (Resistance Books, 1999).

11. James DeNardo, *Power in Numbers: The Political Strategy of Protest and Rebellion* (Princeton, N.J.: Princeton University Press, 1985), especially chaps. 1 and 2.

12. Vladimir Lenin, *Collected Works*, vol. 31, 4th English ed. (Moscow: Progress, 1966), pp. 240–265.

13. Vladimir Lenin, *Selected Works*, vol. 1: *One Step Forward, Two Steps Back* (New York: International, 1967), p. 421.

14. Lowell Dittmer, "The Legacy of Mao Zedong," *Asian Survey* 20, no. 2 (May 1980).

15. Chalmers Johnson, *Autopsy on People's War* (Berkeley: University of California Press, 1973), pp. 10–11.

16. Mao, *On Guerrilla Warfare*.

17. Roger Trinquier, *Modern Warfare: A French View of Counterinsurgency* (New York: Praeger, 1961), p. 19. For analyses of insurgency and COIN in the 1960s, see Taber, *War of the Flea*.

18. Galula, *Counterinsurgency Warfare*, pp. 30–39, 75–94.

19. Neta Crawford, *Argument and Change in World Politics: Ethics, Decolonization, and Humanitarian Intervention* (Cambridge: Cambridge University Press, 2002).

20. Harold Jacobson, "The United Nations and Colonialism: A Tentative Appraisal," *International Organization* 16, no. 1 (Winter 1962): 43–44.

21. Chenoweth and Stephan, *Why Civil Resistance Works*.

22. Marc Sageman, "A Strategy for Fighting International Islamist Terrorists," *Annals of the American Academy of Political and Social Science* 618 (July 2008); Marc Sageman, *Leaderless Jihad: Terror Networks in the Twenty-First Century* (Philadelphia: University of Pennsylvania Press, 2008).

23. Kilcullen, *Accidental Guerrilla*.

24. John Nagl, *Learning to Eat Soup with a Knife: Counterinsurgency Lessons from Malaya and Vietnam* (Chicago: University of Chicago Press, 2005).

25. David Ucko, *The New Counterinsurgency Era: Transforming the U.S. Military for Modern Wars* (Washington, D.C.: Georgetown University Press, 2009).

26. Janine Davidson, *Lifting the Fog of Peace: How Americans Learned to Fight Modern War* (Ann Arbor: University of Michigan Press, 2010).

27. Chad Serena, *A Revolution in Military Adaptation: The US Army in the Iraq War* (Washington, D.C.: Georgetown University Press, 2011).

28. Dan Reiter, *Crucible of Beliefs: Learning, Alliances, and World Wars* (Ithaca, N.Y.: Cornell University Press, 1996); Robert Jervis, *Perception and Misperception in International Politics* (Princeton, N.J.: Princeton University Press, 1976); Yuen Foong

Khong, *Korea, Munich, Dien Bien Phu, and the Vietnam Decisions of 1965* (Princeton, N.J.: Princeton University Press, 1992).

29. Che Guevara, Rolando Bonachea, and Nelson Valdes, eds., *Che: Selected Works of Ernesto Guevara* (Cambridge, Mass.: MIT Press, 1969), p. 368.

30. Ibid., p. 149.

31. Jack Woddis, *New Theories of Revolution* (New York: International, 1972), p. 267.

32. Ibid., pp. 238–239.

33. Guevara, Bonachea, and Valdes, *Che*, pp. 96–97, 100–101, 152–153; Regis Debray, *Revolution in the Revolution? Armed Struggle and Political Struggle in Latin America* (New York: Monthly Review Press, 1967); Woddis, *New Theories of Revolution*, pp. 195–196, 259–260.

34. Amilcar Cabral, *Return to the Source* (New York: Monthly Review Press, 1974), p. 87.

35. Christian Potholm, *The Theory and Practice of African Politics* (Englewood Cliffs, N.J.: Prentice Hall, 1979), p. 90.

36. Kwame Nkrumah, *Neo-colonialism: The Last Stage of Imperialism* (New York: International, 1965).

37. Frantz Fanon, *The Wretched of the Earth* (New York: Grove Press, 1963), pp. 131–138, 222–223.

Chapter 3

1. Huntington, "Guerrilla Warfare in Theory and Practice," p. xvi.

2. Chalmers Johnson, *Peasant Nationalism and Communist Power: The Emergency of Revolutionary China 1937–1945* (Stanford, Calif.: Stanford University Press, 1962), p. 186.

3. Krepinevich, *Army and Vietnam*; Joel Migdal, *Peasant, Politics, and Revolution: Pressures Towards Political and Social Change in the Third World* (Princeton, N.J.: Princeton University Press, 1974).

4. Mao Zedong, "On Protracted War," in *Selected Works*, vol. 2 (May 1958), pp. 143–144.

5. Galula, *Counterinsurgency Warfare*, p. 88.

6. Samuel Huntington, *Political Order in Changing Societies* (New Haven, Conn.: Yale University Press, 1968), p. 266.

7. Byman, *Deadly Connections*.

8. Idean Salehyan, *Rebels Without Borders: Transnational Insurgencies in World Politics* (Ithaca, N.Y.: Cornell University Press, 2009).

9. Carter, "A Blessing or a Curse?"

10. Timothy Crawford and Alan Kuperman, eds., *Gambling on Humanitarian Intervention: Moral Hazard, Rebellion, and Civil War* (London: Routledge, 2006), especially Crawford's chapter.

11. Clifford Bob, *The Marketing of Rebellion: Insurgents, Media, and International Activism* (Cambridge: Cambridge University Press, 2005).

12. Daniel Headrick, *Power over Peoples: Technology, Environments, and Western Imperialism, 1400 to the Present* (Princeton, N.J.: Princeton University Press, 2010).

13. Richard Bean, "War and the Birth of the Nation State," *Journal of Economic History* 33, no. 1 (1973).

14. Michael Howard, *War in European History* (Oxford: Oxford University Press, 1976), pp. 87–106; Joshua Goldstein, *Long Cycles: Prosperity and War in the Modern Age* (New Haven, Conn.: Yale University Press, 1988), pp. 332–333.

15. Daniel Headrick, *The Tools of Imperialism: Technology and European Imperialism in the Nineteenth Century* (Oxford: Oxford University Press, 1981).

16. Theodore Ropp, *War in the Modern World* (Baltimore: Johns Hopkins University Press, 2000).

17. Martin Van Creveld, *Supplying War: Logistics from Wallenstein to Patton* (Cambridge: Cambridge University Press, 1977), pp. 5–70, 244–252.

18. Douglass North, *Institutions, Institutional Change, and Economic Performance* (New York: Cambridge University Press, 1990).

19. Tilly's analysis is so grounded in modern European history that it has only limited application to the periphery, but the fundamental notion of political competition for coercive means in times of conflict can be extended to extrasystemic wars.

20. Michael Doyle, *Empires* (Ithaca, N.Y.: Cornell University Press, 1986), p. 371.

21. Jeremy Weinstein, *Inside Rebellion: The Politics of Insurgent Violence* (Cambridge: Cambridge University Press, 2007).

22. The concept of state building is similar to that of political development, whose literature is vast. See Joel Migdal, *Strong Societies and Weak States: State-Society Relations and State Capabilities in the Third World* (Princeton, N.J.: Princeton University Press, 1988); Francis Fukuyama, *State-Building: Governance and World Order in the 21st Century* (Ithaca, N.Y.: Cornell University Press, 2004).

23. David Edelstein, *Occupational Hazards: Success and Failure in Military Occupation* (Ithaca, N.Y.: Cornell University Press, 2008); David Edelstein, "Foreign Militaries, Sustainable Institutions, and Postwar Statebuilding," in Roland Paris and Timothy Sisk, eds., *The Dilemmas of Statebuilding: Confronting the Contradictions of Postwar Peace Operations* (London: Routledge, 2009), pp. 81–103.

24. Charles Tilly, *From Mobilization to Revolution* (New York: McGraw-Hill, 1978).

25. Alastair Smith, "Fighting Battles, Winning Wars," *Journal of Conflict Resolution* 42, no. 3 (June 1998).

26. Amílcar Cabral, *Unity and Structure: Speeches and Writings* (New York: Monthly Review Press, 1979), p. 207.

27. Keeley, *War Before Civilization*, p. 79.

28. A version of this argument is made in Nori Katagiri, "Suicidal Armies: Why Do Rebels Fight Like an Army and Keep Losing?," *Comparative Strategy* 32, no. 4 (2013): 354–377.

29. Alexander Wendt and Michael Barnett, "Dependent State Formation and Third World Militarization," *Review of International Studies* 19, no. 4 (1993).

30. John Meyer, John Boli, George Thomas, and Francisco Ramirez, "World Society and the Nation-State," *American Journal of Sociology* 103 (1997): p. 164.

31. Michael Adas, *Machines as a Measure of Men: Science, Technology, and Ideologies of Western Dominance* (Ithaca, N.Y.: Cornell University Press, 1989), p. 160.

32. Harry Turney-High, *Primitive War: Its Practice and Concepts* (Columbia: University of South Carolina Press, 1971).

33. Quincy Wright, *A Study of War*, vol. 1 (Chicago: University of Chicago Press, 1942), pp. 80–85; Turney-High, *Primitive War*, pp. 21–137.

34. Harry Turney-High, *The Military: The Theory of Land Warfare as Behavioral Science* (West Hanover, Mass.: Christopher Publishing House, 1981).

35. Bowyer Bell, *The Myth of the Guerrilla: Revolutionary Theory and Malpractice* (New York: Knopf, 1971).

36. Vladimir Lenin, "Guerrilla Warfare," *Proletary* 3, no. 5 (September 30, 1906), in Lenin, *Collected Works* (Moscow: Progress, 1965), pp. 216–222.

37. Johnson, *Autopsy on People's War*, p. 27.

38. Ibid., p. 15.

39. Mao Zedong, *On Coalition Government*, 3rd ed. (Beijing: Foreign Languages Press, 1965).

40. Dominic Johnson and Joshua Madin, "Population Models and Counterinsurgency Strategies," in Raphael Sagarin and Terence Taylor, eds., *Natural Security: A Darwinian Approach to a Dangerous World* (Berkeley: University of California Press, 2008), p. 162.

41. Fanon argued that "the native's violence unifies the people" and "violence is a cleansing force." Fanon, *Wretched of the Earth*, p. 94.

42. This logic applies to revolutions as well. Revolutions are likely when state control is weak. To quote Jacques Ellul, "Each successful revolution has left the state enlarged, better organized, more potent, and with wider areas of influence; that has been the pattern even when revolution has assaulted and attempted to diminish the state." For Ellul, "This relationship between revolution and the growth of the state leads to a definition: revolution is finally the crisis of the development of the state." Jacques Ellul, *Autopsy of Revolution* (New York: Knopf, 1971), pp. 160, 162–163; Theda Skocpol, *States and Social Revolutions: A Comparative Analysis of France, Russia, and China* (New York: Cambridge University Press, 1979).

Chapter 4

1. Eugene Golob, *The Meline Tariff: French Agriculture and Nationalist Economic Policy* (New York: Columbia University Press, 1944), p. 147.

2. A. S. Kanya-Forstner, *The Conquest of the Western Sudan: A Study in French Military Imperialism* (Cambridge: Cambridge University Press, 1969), pp. 239–249.

3. Stuart Michael Persell, *The French Colonial Lobby, 1889–1938* (Stanford, Calif.: Hoover Institution Press, 1983).

4. James Cooke, *New French Imperialism, 1880–1910: The Third Republic and Colonial Expansion* (Hamden, Conn.: Archon Books, 1973), p. 17.

5. Anthony Clayton, *France, Soldiers, and Africa* (London: Brassey's, 1988), p. 72.

6. Boniface Obichere, *West African States and European Expansion: Dahomey-Niger Hinterland, 1885–1898* (New Haven, Conn.: Yale University Press, 1971), p. 105.

7. Clayton, *France, Soldiers, and Africa*, p. 68.

8. Obichere, *West African States and European Expansion*, p. 74.

9. Archibald Dalzel, *The History of Dahomy: An Inland Kingdom of Africa* (London: Frank Cass, 1967), pp. x, 55.

10. Obichere, *West African States and European Expansion*, p. 252.

11. Ibid., p. 67.

12. For a discussion of how these factors shaped Dahomey's choice to fight conventionally, see Nori Katagiri, "Drawing Strategic Lessons from Dahomey's War," *Air Power and Space Journal—African and Francophonie* 3, no. 3 (October 2012).

13. Jack Goody, *Technology, Tradition, and the State in Africa* (Cambridge: Cambridge University Press, 1980), p. 36.

14. Obichere, *West African States and European Expansion*, p. 75.

15. Stanley Alpern, *Amazons of Black Sparta: The Women Warriors of Dahomey* (New York: New York University Press, 1998), pp. 195, 201.

16. Douglas Porch, "Bugeaud, Gallieni, Lyautey: The Development of French Colonial Warfare," in Peter Paret, Gordon Craig, and Felix Gilbert, eds., *Makers of Modern Strategy: From Machiavelli to the Nuclear Age* (Oxford: Oxford University Press, 1986), pp. 383–384.

17. W. J. Argyle, *The Fon of Dahomey: A History and Ethnography of the Old Kingdom* (Oxford: Oxford University Press, 1966), pp. 81, 89.

18. Ibid., p. 86; David Ross, "Dahomey," in Michael Crowder, ed., *West African Resistance: The Military Response to Colonial Occupation* (London: Hutchinson University Press), p. 151; Porch, "Bugeaud, Gallieni, Lyautey," p. 396.

19. Ross, "Dahomey," p. 160.

20. According to Trevor Dupuy, it takes a while for a country to get familiar with a weapon and make it adequately integrated to use it. The country must experiment with it to see how it works best under which circumstances. Its military doctrine shapes how long it takes it to adopt the weapon. Generally it takes about twenty years of time lag between the adoption of a new weapon and its integration. Trevor Dupuy, *The Evolution of Weapons and Warfare* (Cambridge: Da Capo Press, 1990), pp. 302–305.

21. Persell, *French Colonial Lobby*, p. 80.

22. Porch, "Bugeaud, Gallieni, Lyautey," pp. 396, 399–401.

23. Ibid., pp. 389–391.

24. Ibid., p. 384.

25. Raymond Betts, *Assimilation and Association in French Colonial Theory 1890–1914* (Lincoln: University of Nebraska Press, 2005), pp. 155–156.

26. Jeffrey Herbst, *States and Power in Africa: Comparative Lessons in Authority and Control* (Princeton, N.J.: Princeton University Press, 2000), p. 42.

27. John Iliffe, *Africans: The History of a Continent* (Cambridge: Cambridge University Press, 1995), p. 70.

28. Alpern, *Amazons of Black Sparta*, p. 147.

29. Argyle, *Fon of Dahomey*, p. 63.

30. Ibid., pp. 75, 85.

31. Robert Stephenson, *Population and Political Systems in Tropical Africa* (New York: Columbia University Press, 1968).

32. Igor Kopytoff, "The Internal African Frontier: The Making of African Political Culture," in Igor Kopytoff, ed., *The African Frontier: The Reproduction of Traditional African Societies* (Bloomington: Indiana University Press, 1987), p. 29.

33. Elizabeth Colson, "African Society at the Time of the Scramble," p. 44, and John Hargreaves, "West African States and European Conquest," p. 199, both in Gann and Duignan, *Colonialism in Africa*.

34. Karl Polanyi, *Dahomey and the Slave Trade: An Analysis of an Archaic Economy* (Seattle: University of Washington Press, 1966), p. 36; Ross, "Dahomey," p. 147.

35. Richard Burton, *A Mission to Gelele, King of Dahome*, vol. 1 (London: Tylston and Edwards, 1893), p. 263.

36. John M'Leod, *A Voyage to Africa with Some Account of the Manners and Customs of the Dahomian People: With Some Account of the Manners and Customs of the Dahomian People* (John Murray, 1820), pp. 37–38.

37. Robert Bates, "The Centralization of African Societies," in Robert Bates, *Essays on the Political Economy of Africa* (Berkeley: University of California Press, 1987), p. 42.

38. Obichere, *West African States and European Expansion*, p. 71.

39. Philip Curtin, *Disease and Empire: The Health of European Troops in the Conquest of Africa* (Cambridge: Cambridge University Press, 1998), p. 100.

40. Obichere, *West African States and European Expansion*, pp. 67–68.

41. Mack, "Why Big Nations Lose Small Wars."

42. Ross, "Dahomey," pp. 148, 150.

43. Persell, *French Colonial Lobby*, p. 74.

44. Obichere, *West African States and European Expansion*, pp. 60, 65–66.

45. Ibid., pp. 76–77.

46. Persell, *French Colonial Lobby*, pp. 122–123.

47. Curtin, *Disease and Empire*, pp. 102–105.

48. Charles John Balesi, *From Adversaries to Comrades-in-Arms: West Africans and the French Military, 1885–1918* (Waltham, Mass.: Crossroads Press, 1979), p. 17.

49. Arreguin-Toft, *How the Weak Win Wars*.

Chapter 5

1. See, for instance, Ahmed Hashim, *Insurgency and Counterinsurgency in Iraq* (Ithaca, N.Y.: Cornell University Press, 2006); Nagl, *Learning to Eat Soup.*

2. Richard Stubbs, "Guerrilla Strategies and British Counterinsurgency Strategies of the 1950s and 1960s: Why Was the War Lost?," in C. C. Chin and Karl Hack, eds., *Dialogues with Chin Peng: New Light on the Malayan Communist Party* (Singapore: Singapore University Press, 2004), p. 302.

3. British War Office, "Military Situation in Malaya in August, 1948" (PRO, ref. FO 371/69698).

4. CAB 129/48, C(51)26, "'The Situation in Malaya': Cabinet Memorandum by Mr. Lyttelton," in A. J. Stockwell, ed., *British Documents on the End of Empire: Malaya: Part II* (London: HMSO, 1995), pp. 313–315; Richard Clutterbuck, *The Long, Long War: Counterinsurgency in Malaya and Vietnam* (New York: Praeger, 1966), pp. 42–43.

5. R. W. Komer, *The Malayan Emergency in Retrospect: Organization of a Successful Counterinsurgency Effort* (Santa Monica, Calif.: RAND, 1972), pp. 7–8.

6. Riley Sunderland, *Army Operations in Malaya, 1947–1960* (Santa Monica, Calif.: RAND, 1964), p. 64.

7. Ibid., p. 80.

8. Robert Tilman, "The Non-Lessons of the Malayan Emergency," *Military Review* 46 (December 1966): 417–418. This number does not include the Royal Air Force and elements of the Royal Navy operating out of the Singapore Naval Base.

9. Robert Thompson, *Defeating Communist Insurgency: Experiences from Malaya and Vietnam* (New York: Praeger, 1966).

10. Deborah Avant, *Political Institutions and Military Change: Lessons from Peripheral Wars* (Ithaca, N.Y.: Cornell University Press, 1994).

11. John Coates, *Suppressing Insurgency: An Analysis of the Malayan Emergency, 1948–1954* (Boulder, Colo.: Westview, 1992).

12. Huw Bennett, "'A Very Salutary Effect': The Counter-Terror Strategy in the Early Malayan Emergency, June 1948 to December 1949," *Journal of Strategic Studies* 32, no. 3 (2009): 417.

13. A. J. Stockwell, "'A Widespread and Long-Concocted Plot to Overthrow Government in Malaya?' The Origins of the Malayan Emergency," *Journal of Imperial and Commonwealth History* 21, no. 3 (1993): 85.

14. Bennett, "'Very Salutary Effect'"; Karl Hack, "British Intelligence and Counter-Insurgency in the Era of Decolonisation. The Example of Malaya," *Intelligence and National Security* 14, no. 2 (Summer 1999): 127–128; Riley Sunderland, *Antiguerrilla Intelligence in Malaya* (Santa Monica, Calif.: RAND, 1964), pp. 3–10.

15. Stubbs, "Guerrilla Strategies," p. 302.

16. Sunderland, *Army Operations in Malaya*, p. 127.

17. Nagl, *Learning to Eat Soup*, p. 67.

18. Komer, *Malayan Emergency in Retrospect*, p. 50.
19. Clutterbuck, *Long, Long War*, pp. 55–56.
20. Chin Peng, *My Side of Story* (Singapore: Media Masters, 2003), p. 257.
21. Anthony Short, *The Communist Insurgency in Malaya 1948–1960* (London: Frederick Muller, 1975), p. 211.
22. Riley Sunderland, *Organizing Counterinsurgency in Malaya, 1947–1960* (Santa Monica, Calif.: RAND, 1964).
23. T. N. Harper, *The End of Empire and the Making of Malaya* (Cambridge: Cambridge University Press, 2001), p. 149.
24. Tim Jones, "The British Army, and Counter-Guerrilla Warfare in Transition, 1944–1952," *Small Wars and Insurgencies* 7, no. 3 (Winter 1996): 265–308.
25. PREM 8/1406/2, MAL C(50)12 [May 12, 1950], reprinted in Stockwell, *Documents*, vol. 2, p. 214.
26. CAB 129/48, C(51)59 [December 21, 1951], "Malaya: Cabinet Memorandum by Mr. Lyttelton," in Stockwell, *Documents*, vol. 2, pp. 322–323.
27. Gregorian, "Jungle Bashing," *Small Wars and Insurgencies* 5, no. 3 (Winter 1994): 349–350.
28. John Cloake, *Templer, Tiger of Malaya: The Life of Field Marshal Sir General Templer* (London: Harrap, 1985), p. 242.
29. Nagl, *Learning to Eat Soup*, p. 193.
30. David Ucko, "Countering Insurgents Through Distributed Operations: Insights from Malaya 1948–1960," *Journal of Strategic Studies* 30, no. 1 (February 2007): 47–72.
31. Short, *Insurrection in Malaya*, p. 414.
32. Komer, *Malayan Emergency in Retrospect*, p. 53.
33. Ibid., p. 56.
34. See, for instance, Robert Jackson, *The Malayan Emergency: The Commonwealth's Wars 1948–1966* (London: Routledge, 1991), p. 116.
35. The primary message was that the civilians must be won over. Malayan Communist Party, "Past Errors" (Party Directive of October 1951, pt. 1), *Times* (London), December 1, 1952.
36. Karl Hack, *Defense and Decolonization in Southeast Asia: Britain, Malaya, and Singapore 1941–1968* (London: Curzon, 2001), p. 121.
37. Lucien Pye, *Guerrilla Communism in Malaya* (Princeton, N.J.: Princeton University Press, 1956), pp. 95, 98–102.
38. Richard Stubbs, *Hearts and Minds in Guerrilla Warfare: The Malayan Emergency, 1948–60* (New York: Oxford University Press, 1990), pp. 226–228.
39. Hack, *Defence and Decolonisation in Southeast Asia*, pp. 136, 138.
40. Margaret Shennan, *Out in the Midday Sun: The British in Malaya 1880–1960* (London: John Murray, 2000), p. 241. His policies were criticized, however, for the harshness of collective punishment programs and imposition of curfews and cuts in

food rations, but most notably for his Emergency Regulation 17D, which allowed mass detention and deportation. See, for instance, Cloake, *Templer*, p. 272.

41. Nagl, *Learning to Eat Soup*.

42. Harper, *End of Empire and the Making of Malaya*, p. 7; Stubbs, *Hearts and Minds in Guerrilla Warfare*, pp. 260–264.

43. Riley Sunderland, *Winning the Hearts and Minds of the People: Malaya, 1948–1960* (Santa Monica, Calif.: RAND, 1964), p. 1.

44. Komer, *Malayan Emergency in Retrospect*, p. 62.

45. AIR20/777, Secret, Early History of Emergency, Report on the Emergency in Malaya from April, 1950 to November 51 by Harold Briggs, excerpt from the Malaysian National Archive, August 2009.

46. Federation of Malaya, "Weekly News Summary," July 4, 1953.

47. Komer, *Malayan Emergency in Retrospect*, p. 64.

48. On a more elaborate treatment of this aspect, see Nori Katagiri, "Winning Hearts and Minds to Lose Control: Exploring Various Consequences of Popular Support in Counterinsurgency Missions," *Small Wars and Insurgencies* 22, no. 1 (March 2011).

49. Heng Pek Koon, *Chinese Politics in Malaysia: A History of the Malaysian Chinese Association* (Singapore: Oxford University Press, 1988), p. 136.

50. The Alliance sought to win an "early independence for Malaya." CAB 134/1202, CA (56)3, "Conference on Constitutional Advance in the Federation of Malaya," memorandum by Mr. Lennox-Boyd for Cabinet Colonial Policy Committee, in A. J. Stockwell, ed., *British Documents on the End of Empire: Malaya: Part III* (London: HMSO, 1995), p. 235.

51. Pye, *Guerrilla Communism in Malaya*, pp. 343–363.

52. Harper, *End of Empire and the Making of Malaya*, p. 229.

53. Sunderland, *Organizing Counterinsurgency in Malaya*, pp. 40–41.

54. Cloake, *Templer*, chap. 10; Stubbs, *Hearts and Minds in Guerrilla Warfare*, pp. 156–164.

55. Hack, *Defence and Decolonisation in Southeast Asia*, pp. 119–120.

56. Short, *Insurrection in Malaya*, p. 360n19.

57. Komer, *Malayan Emergency in Retrospect*, p. 7.

58. Ibid., p. 8.

59. For a similar conclusion, see James Ongkili, *Nation-building in Malaysia 1946–1974* (Singapore: Oxford University Press, 1985), p. 80.

60. See Hack, *Defence and Decolonisation in Southeast Asia*, p. 19, chap. 4.

Chapter 6

1. Daniel Byman and Kenneth Pollack, *Things Fall Apart: Containing the Spillover from an Iraqi Civil War* (Washington, D.C.: Brookings Institution Press, 2007).

2. Martel, *Victory in War*, p. 319.

3. Mohammed Hafez, *Suicide Bombers in Iraq: The Strategy and Ideology of Martyrdom* (Washington, D.C.: U.S. Institute for Peace, 2007), pp. 243–249.

4. Hashim, *Insurgency and Counterinsurgency in Iraq*.

5. Martel, *Victory in War*, p. 312.

6. Frank Harvey, *Explaining the Iraq War: Counterfactual Theory, Logic and Evidence* (New York: Cambridge University Press, 2011).

7. Brendan O'Leary, *How to Get Out of Iraq with Integrity* (Philadelphia: University of Pennsylvania Press, 2009), chap. 1.

8. Noah Feldman, *What We Owe Iraq: War and the Ethics of Nation Building* (Princeton, N.J.: Princeton University Press, 2006), pp. 18–20.

9. Bruce Moon, "Long Time Coming: Prospects for Democracy in Iraq," *International Security* 33, no. 4 (Spring 2009).

10. Joseph Stiglitz and Linda Bilmes, *Three Trillion Dollar War: The True Conflict of the Iraq Conflict* (Rochester, N.Y.: Boydell & Brewer, 2008).

11. One of the critics of the war was Jeffrey Record, who argued that "the U.S. war against Iraq in 2003 was not only unnecessary but also damaging to long-term U.S. political interests in the world." Jeffrey Record, *Dark Victory: America's Second War Against Iraq* (Annapolis, Md.: Naval Institute Press, 2004), pp. xiv–xv.

12. Ole Holsti, *American Public Opinion on the Iraq War* (Ann Arbor: University of Michigan Press, 2011), p. 6.

13. Peter Feaver and Christopher Gelpi, *Choosing Your Battles: American Civil-Military Relations and the Use of Force* (Princeton, N.J.: Princeton University Press, 2004).

14. For instance, see Eric Larson, *Casualties and Consensus: The Historical Role of Casualties in Domestic Support for U.S. Military Operations* (Santa Monica, Calif.: RAND, 1996).

15. Christopher Gelpi, Peter Feaver, and Jason Reifler, *Paying the Human Costs of War: American Public Opinion and Casualties in Military Conflicts* (Princeton, N.J.: Princeton University Press, 2009), p. 2.

16. Christopher Gelpi, Peter Feaver, and Jason Reifler, "Success Matters: Casualty Sensitivity and the War in Iraq," *International Security* 30, no. 3 (Winter 2005–2006).

17. Martel, *Victory in War*, pp. 312–340.

18. Nora Bensahel, et al., *After Saddam: Prewar Planning and the Occupation of Iraq* (Santa Monica, Calif.: RAND, 2008), p. xviii.

19. National Security Archive, George Washington University, declassified PowerPoint slides available at http://www.gwu.edu/~nsarchiv/NSAEBB/NSAEBB214/Tab%20F.pdf.

20. Donald Wright and Timothy Reese, *On Point II: Transition to the New Campaign* (Fort Leavenworth, Kans.: U.S. Army Combined Arms Center, 2008), p. 3, emphasis added.

21. Keith Shimko, *The Iraq Wars and America's Military Revolution* (New York: Cambridge University Press, 2010), p. 200.

22. Martel, *Victory in War*, p. 344.

23. For a good comparison between 1991 and 2003 (but not after 2003), see Williamson Murray and Robert Scales, *The Iraq War: A Military History* (Cambridge, Mass.: Harvard University Press, 2003).

24. For details on the conventional phase of the war, see Gregory Fontenot, E. J. Degen, and David Tohn, *On Point: The United States Army in Operation Iraqi Freedom* (Annapolis, Md.: Naval Institute Press, 2005); Micheal Gordon and Bernard Trainor, *Cobra II: The Inside Story of the Invasion and Occupation of Iraq* (New York: Vintage, 2006).

25. Martin Van Creveld, *The Changing Face of War: Lessons of Combat from the Marne to Iraq* (New York: Ballantine Books, 2006), pp. 247–248.

26. Charles Duelfer and Stephen Benedict Dyson, "Chronic Misperception and International Conflict: The U.S.-Iraq Experience," *International Security* 36, no. 1 (Summer 2011).

27. Stephen Hosmer, *Why the Iraqi Resistance to the Coalition Invasion Was So Weak* (Santa Monica, Calif.: RAND, 2007); Kevin Woods, James Lacy, and Williamson Murray, "Saddam's Delusions: The View from the Inside," *Foreign Affairs* 85, no. 3 (May–June 2006).

28. Stephen Biddle, "Speed Kills? Reassessing the Role of Speed, Precision, and Situation Awareness in the Fall of Saddam," *Journal of Strategic Studies* 30, no. 1 (February 2007): 22.

29. Phil Williams, *Criminals, Militias, and Insurgents: Organized Crime in Iraq* (Carlisle, Pa.: Strategies Studies Institute, 2009).

30. Robert Jervis, *Why Intelligence Fails: Lessons from the Iranian Revolution and the Iraq War* (Ithaca, N.Y.: Cornell University Press, 2010).

31. Shimko, *Iraq Wars and America's Military Revolution*, p. 203.

32. Jennifer Morrison Taw, *Mission Revolution: The U.S. Military and Stability Operations* (New York: Columbia University Press, 2012).

33. Bruce Pirnie and Edward O'Connell, *Counterinsurgency in Iraq (2003–2006)* (Santa Monica, Calif.: RAND, 2007).

34. Wright and Reese, *On Point II*, p. 80.

35. Peter Mansour, *Baghdad at Sunrise: A Brigade Commander's War in Iraq* (New Haven, Conn.: Yale University Press, 2007).

36. See, for instance, Michael O'Hanlon, *Defense Strategy for the Post-Saddam Era* (Washington, D.C.: Brookings Institution Press, 2005).

37. Hashim, *Insurgency and Counterinsurgency in Iraq*, p. 339.

38. Pirnie and O'Connell, *Counterinsurgency in Iraq*, p. xviii.

39. Stephen Biddle, "Seeing Baghdad, Thinking Saigon," *Foreign Affairs* 85, no. 2 (March–April 2006).

40. Quoted in Nigel Aylwin-Foster, "Changing the Army for Counterinsurgency Operations," *Military Review* (November–December 2005).

41. Edelstein, "Foreign Militaries."

42. Robert Pape, *Dying to Win: The Strategic Logic of Suicide Terrorism* (New York: Random House, 2005).

43. Edelstein, *Occupational Hazards*, p. 159.

44. Eric Herring and Glen Rangwala, *Iraq in Fragments: The Occupation and Its Legacy* (Ithaca, N.Y.: Cornell University Press, 2006), p. 161.

45. Kimberly Marten, *Warlords: Strong-Arm Brokers in Weak States* (Ithaca, N.Y.: Cornell University Press, 2012).

46. Herring and Rangwala, *Iraq in Fragments*, p. 161.

47. U.S. Army and Marine Corps, *Field Manual 3-24* (Washington, D.C.: U.S. Department of the Army, 2006).

48. Austin Long, "The Anbar Awakening," *Survival* 50, no. 2 (March–April 2008); John McCary, "The Anbar Awakening: An Alliance of Incentives," *Washington Quarterly* 32, no. 1 (January 2009).

49. Ucko, *New Counterinsurgency Era*.

50. Kilcullen, *Accidental Guerrilla*, p. 129.

51. Ibid., p. 116.

52. Jeffrey Friedman, "Manpower and Counterinsurgency: Empirical Foundations for Theory and Doctrine," *Security Studies* 20, no. 4 (2011).

53. Stephen Biddle, Jeffrey Friedman, and Jacob Shapiro. "Testing the Surge: Why Did Violence Decline in Iraq in 2007?," *International Security* 37, no. 1 (Summer 2012).

54. Sarah Kenyon Lischer, "Security and Displacement in Iraq: Responding to the Forced Migration Crisis," *International Security* 33, no. 2 (Fall 2008): 95–119.

55. Lyall and Wilson, "Rage Against the Machines."

56. Colin Kahl, "In the Crossfire or the Crosshairs? Norms, Civilian Casualties, and U.S. Conduct in Iraq," *International Security* 32, no. 1 (Summer 2007): 7–46.

57. U.S. Department of Defense, "Measuring Stability and Security in Iraq" (July 2009), p. 24.

58. Paul Staniland, "Between a Rock and a Hard Place: Insurgent Fratricide, Ethnic Defection, and the Rise of Pro-State Paramilitaries," *Journal of Conflict Resolution* 56, no. 1 (2012).

59. James Russell, *Innovation, Transformation, and War: Counterinsurgency Operations in Anbar and Ninewa Provinces, Iraq, 2005-2007* (Stanford, Calif.: Stanford University Press, 2011).

60. International Crisis Group, *In Their Own Words: Reading the Iraqi Insurgency*, Middle East Report no. 50 (February 15, 2006), pp. 9, 15.

61. See, for instance, Lawrence Wright, "The Rebellion Within: An Al Qaeda Mastermind Questions Terrorism," *New Yorker*, June 2, 2008.

62. Herring and Rangwala, *Iraq in Fragments*, pp. 260–261.

63. Feldman, *What We Owe Iraq*.

64. Herring and Rangwala, *Iraq in Fragments*, pp. 5–6.

65. Audrey Kurth Cronin, "Cyber-Mobilization: The New Levee en Mass," *Parameters* (Summer 2006).

66. Thomas Hegghammer, "The Rise of Muslim Foreign Fighters: Islam and the Globalization of Jihad," *International Security* 35, no. 3 (Winter 2010–2011).

67. Hashim, *Insurgency and Counterinsurgency in Iraq*, pp. 125, 135, 138–151, 200–213.

68. Marten, *Warlords*, p. 3.

69. Ibid., pp. 11–12.

70. Mia Bloom, *Dying to Kill: The Allure of Suicide Terror* (New York: Columbia University Press, 2005).

71. Arreguin-Toft, *How the Weak Win Wars*.

72. Nagl, *Learning to Eat Soup*; Ucko, *New Counterinsurgency Era*; Taw, *Mission Revolution*; Davidson, *Lifting the Fog of Peace*; Serena, *Revolution in Military Adaptation*.

73. Record, *Beating Goliath*.

74. Lyall and Wilson, "Rage Against the Machines."

75. Shimko, *Iraq Wars and America's Military Revolution*; Murray and Scales, *Iraq War*.

76. Mahnken, *Technology and the American Way of War*.

Chapter 7

1. Saadia Touval, *Somali Nationalism: International Politics and the Drive for Unity in the Horn of Africa* (Cambridge, Mass.: Harvard University Press, 1963), p. 3.

2. David Laitin, *Politics, Language, and Thought: The Somali Experience* (Chicago: University of Chicago Press, 1977), p. 19.

3. Robert Hess, *Italian Colonialism in Somalia* (Chicago: University of Chicago Press, 1966).

4. Great Britain, War Office, General Staff, *Official History of the Operations in Somaliland, 1901–1904*, vol. 1 (London: Harrison and Sons, 1907), p. 40.

5. Douglas Jardine, *The Mad Mullah of Somaliland* (New York: Negro Universities Press, 1969), p. 159.

6. Abdisalam Issa-Salwe, *The Collapse of Darwish State: The Impact of the Colonial Legacy* (London: HAAN Associates, 1994), p. 17.

7. Jardine, *Mad Mullah of Somaliland*, p. 59.

8. David Laitin and Said Samatar, *Somalia: Nation in Search of a State* (Boulder, Colo.: Westview, 1987), p. 56.

9. "Notes Exchanged Between the British and Italian Governments Respecting the Italian Agreement of 1905 with Seyid Mahamed-bin-Abdulla. London, 19th March, 1907," in Information Services of the Somali Government, *The Somali Peninsula: A New Light on Imperial Motives* (Mogadishu: Information Services of the Somali Government, 1962), Appendix XVII(b), pp. 121–122.

10. Abdi Sheik-Abdi, *Divine Madness: Mohammed Abdulle Hassan (1856–1920)* (London: Zed Books, 1993), pp. 92–93, 104.

11. Laitin and Samatar, *Somalia*, p. 58.

12. Great Britain, *Official History of the Operations in Somaliland*, vol. 1, p. 306.

13. Ibid., vol. 1, p. 321.

14. Ibid., vol. 2, pp. 347–348.

15. Ibid., vol. 1, pp. 308–309.

16. John Gooch, *The Plans of War: The General Staff and Military Strategy c. 1900–1916* (New York: John Wiley, 1974).

17. Sheik-Abdi, *Divine Madness*, pp. 107–108.

18. Sheik-Abdi, *Divine Madness*, p. 111.

19. Issa-Salwe, *Collapse of Darwish State*.

20. See Sheik-Abdi, *Divine Madness*, p. 155.

21. Jardine, *Mad Mullah of Somaliland*, p. 85.

22. Malcolm McNeill, *In Pursuit of the "Mad" Mullah: Service and Sport in the Somali Protectorate* (London: C. Arthur Pearson Ltd., 1902), pp. 278–281.

23. Great Britain, *Official History of the Operations in Somaliland*, vol. 1, p. 323.

24. Jardine, *Mad Mullah of Somaliland*, pp. 90–94.

25. Great Britain, *Official History of the Operations in Somaliland*, vol. 1, p. 314.

26. Ibid., vol. 1, p. 241. Bracketed material added by the author.

27. I. M. Lewis, *A Modern History of Somalia: Nation and State in the Horn of Africa* (Boulder, Colo.: Westview, 1988), p. 72.

28. Great Britain, *Official History of the Operations in Somaliland*, vol. 2, p. 355.

29. Touval, *Somali Nationalism*, p. 53.

30. Issa-Salwe, *Collapse of Darwish State*.

31. For more details, see Great Britain, *Official History of the Operations in Somaliland*, vol. 1, pp. 324–326.

32. Touval, *Somali Nationalism*, p. 56.

33. Ibid., p. 51.

34. Issa-Salwe, *Collapse of Darwish State*.

35. Said Samatar, *Oral Poetry and Somali Nationalism: The Case of Sayyid Mahammad 'Abdille Hasan* (Cambridge: Cambridge University Press, 1982), p. 128.

36. Touval, *Somali Nationalism*, p. 54.

37. Lewis, *Modern History of Somalia*, p. 78.

38. Sheik-Abdi, *Divine Madness*, p. 118.

39. That summer, a Dervish force clashed with the Somaliland Camel Corps in Dulmadooba and killed Corfield. In 1913, the British army founded a small striking force in the fifteen-man Somali Camel Constabulary.

40. Ahmed Samatar, *Socialist Somalia: Rhetoric and Reality* (London: Zed Books, 1988), p. 34.

41. Lewis, *Modern History of Somalia*, p. 81.

42. Issa-Salwe, *Collapse of Darwish State*.

43. McNeill, *In Pursuit of the "Mad" Mullah*, pp. 148–149.
44. Jardine, *Mad Mullah of Somaliland*, p. 231.
45. B. G. Martin, *Muslim Brotherhoods in Nineteenth Century Africa* (Cambridge: Cambridge University Press, 1976), p. 191, bracketed comment added by Sheik-Abdi, *Divine Madness*, pp. 163–164.
46. Mack, "Why Big Nations Lose Small Wars."
47. Jardine, *Mad Mullah of Somaliland*, p. 315.
48. Ibid., pp. 176–179.
49. Ibid., p. 189.
50. Ibid., p. 204, Appendix 4.
51. John Keegan, *The First World War* (New York: Vintage, 1998), p. 319.
52. See, for instance, Robert Ergang, *Europe Since Waterloo* (London: Heath, 1966), p. 272; John Gooch, *The Boer War: Direction, Experience, and Image* (London: Routledge, 2000), p. 13.
53. Robert Roberts, *The Classic Slum: Salford Life in the First Quarter of the Century* (Harmondsworth: Penguin, 1990), pp. 167, 184–185.
54. David Cannadine, *The Rise and Fall of Class in Britain* (New York: Columbia University Press, 1993), pp. 130–131.
55. Harold James Perkin, *Rise of Professional Society: England Since 1880* (London: Taylor & Francis, 1989), pp. 192, 207.
56. Cannadine, *Rise and Fall of Class in Britain*, p. 131; Ross McKibbin, *Ideologies of Class: Social Relations in Britain, 1880–1950* (Oxford: Oxford University Press, 1990), p. 298.
57. Paul Kennedy, *The Rise and Fall of British Naval Mastery* (New York: Penguin, 2004), p. 273.

Chapter 8

1. Basil Davidson, *A Política da Luta Armada: Libertação Nacional nas Colónias Africanas de Portugal* (Lisbon: Editorial Caminho, 1979), p. 13, English original in Basil Davidson, Joe Slovo, and Anthony Wilkinson, *Southern Africa: The New Politics of Revolution* (Harmondsworth: Penguin, 1976).
2. Mustafah Dhada, "The Liberation War in Guinea-Bissau Reconsidered," *Journal of Military History* 62, no. 3 (July 1998): 571.
3. James Duffy, *Portugal in Africa* (Cambridge, Mass.: Harvard University Press, 1962), pp. 19, 25.
4. Joao Paulo Borges Coelho, "African Troops in the Portuguese Colonial Army, 1961–1974: Angola, Guinea-Bissau and Mozambique," *Portuguese Studies Review* 10, no. 1 (2002): 149.
5. Patrick Chabal, "People's War, State Formation and Revolution in Africa: A Comparative Analysis of Mozambique, Guinea-Bissau, and Angola," *Journal of Commonwealth and Comparative Politics* 21, no..3 (1983): 112–113.

6. Cabral, *Unity and Struggle*, p. 214.

7. Patrick Chabal, *Amílcar Cabral: Revolutionary Leadership and People's War* (Cambridge: Cambridge University Press, 1983), p. 2.

8. Kenneth Maxwell, *The Making of Portuguese Democracy* (Cambridge: Cambridge University Press, 1995), p. 81

9. Dhada, "Liberation War in Guinea-Bissau Reconsidered," p. 593.

10. Lars Rudebeck, *Guinea-Bissau: A Study of Political Mobilization* (Uppsala: Scandinavian Institute of African Studies, 1974), p. 105.

11. Ronald Chilcote, *Amílcar Cabral's Revolutionary Theory and Practice: A Critical Guide* (Boulder, Colo.: Lynne Rienner, 1991), p. 66.

12. *No Pintcha*, nos. 39–41 (March–May 1972), emphasis added.

13. Chabal, "People's War," pp. 105–106.

14. Basil Davidson, *No Fist Is Big Enough to Hide the Sky: The Liberation of Guine and Cape Verde, Aspects of an African Revolution* (London: Zed Press, 1981), pp. 65–88.

15. Gérard Chaliand, *Armed Struggle in Africa* (New York: Monthly Review Press, 1969), pp. 118–125.

16. Cabral, *Unity and Struggle*, p. 122; Patrick Chabal, "The Social and Political Thought of Amílcar Cabral: A Reassessment," *Journal of Modern African Studies* 19, no. 1 (1981): p. 37.

17. Extracts from a declaration to the Organization of the Solidarity of the People of Asia, Africa, and Latin America (OSPAAAL) General Secretariat in December 1968, in Amílcar Cabral, *Revolution in Guinea* (New York: Monthly Review Press, 1969), p. 112, emphasis added.

18. Amílcar Cabral, "The War in 'Portuguese' Guinea," *Revolution* 1, no. 2 (June 1963), emphasis added.

19. Amílcar Cabral, *Rapport General sur la Lutte de Liberation Nationale* (Conakry, July 1961). The report discusses the evolution of nationalism and the PAIGC struggle in detail.

20. Cabral, *Unity and Struggle*, pp. 125–126.

21. Ibid., pp. 201–202.

22. Ibid., pp. 201–202; Amílcar Cabral, "The Danger of Destruction from Within," message sent in March 1972 to all those holding posts of responsibility in the party, translated from French, in Aquino de Braganca and Immanuel Wallerstein, eds., *The African Liberation Reader*, vol. 3: *The Strategy of Liberation* (London: Zed Press, 1982), pp. 23–26.

23. Text of an address by Amílcar Cabral, leader of the PAIGC, group 605, box 30, folder 558, Manuscripts and Archives, Yale University Library (December 2008), p. 8.

24. Henry Bienen, "State and Revolution: The Work of Amílcar Cabral," *Journal of Modern African Studies* 15, no. 4 (1977).

25. Potholm writes along similar lines that "when the PAIGC declared the independence of Guinea-Bissau in September 1974, and when the Portuguese acknowl-

edged that independence a year later, the PAIGC was already a government in being." Potholm, *Theory and Practice*, p. 236.

26. Christopher Clapham, *African Guerrillas* (Oxford: James Currey, 1998).

27. Davidson, *No Fist Is Big Enough*, p. 125; Rudebeck, *Guinea-Bissau*, p. 133.

28. Ministry of Foreign Affairs, *Portuguese Africa: An Introduction* (Lisbon: Ministry of Foreign Affairs, 1973), p. 76.

29. Ronald Chilcote, *Emerging Nationalism in Portuguese Africa: Documents* (Stanford, Calif.: Hoover Institution Press, 1972), pp. 360–366.

30. Cabral, *Revolution in Guinea*, p. 30.

31. Bienen, "State and Revolution," p. 558.

32. Cabral found little inspiration in Marxism. He stated, "We are not a Communist or a Marxist-Leninist party.... The people now leading the peasants to the struggle in Guinea are mostly from the urban milieu and connected with the wage earning group." Cabral, *Revolution in Guinea*, p. 55; Chabal, "Social and Political Thought," p. 3.

33. Mustafah Dhada, *Warriors at Work: How Guinea Was Really Set Free* (Niwot, Colo.: University Press of Colorado, 1993), pp. 7, 10; Ronald Chilcote, *Portuguese Africa* (Englewood Cliffs, N.J.: Prentice Hall, 1967), p. 100.

34. Amílcar Cabral, "The Tactic of Division," in Aquino de Braganca and Immanuel Wallerstein, eds., *The African Liberation Reader*, vol. 2: *The National Liberation Movements* (London: Zed Press, 1982), pp. 178–179.

35. Chilcote, *Amílcar Cabral's Revolutionary Theory and Practice*, p. 67.

36. Richard Gibson, *African Liberation Movements: Contemporary Struggles Against White Minority Rule* (New York: Oxford University Press, 1972), pp. 258–259.

37. Chabal, "Social and Political Thought," p. 45; Cabral, *Revolution in Guinea*, p. 85.

38. Cabral, *Unity and Struggle*, pp. 75–78, 249–250; Davidson, *No Fist Is Big Enough*, pp. 94–95.

39. Amílcar Cabral, "Liberating Portuguese Guinea from Within," *New African* 4 (June 1965): p. 85.

40. Dhada, "Liberation War in Guinea-Bissau Reconsidered," pp. 577–578.

41. Gibson, *African Liberation Movements*, pp. 8–9.

42. "Report on the Development of the National Liberation Struggle in Guine and Cape Verde in 1964," compiled in Cabral, *Unity and Struggle*, p. 179.

43. Chaliand, *Armed Struggle in Africa*, p. 82.

44. Lars Rudebeck, "Political Mobilization for Development in Guinea-Bissau," *Journal of Modern African Studies* 10, no. 1 (May 1972): 3.

45. Rudebeck, *Guinea-Bissau*, p. 63.

46. Dhada, *Warriors at Work*, p. 248, fn 164.

47. John Cann, *Counterinsurgency in Africa: The Portuguese Way of War, 1961–1974* (Westport, Conn.: Greenwood, 1997), pp. 45–48.

48. Douglas Porch, *The Portuguese Armed Forces and the Revolution* (Stanford, Calif.: Hoover Institution Press, 1977), p. 64; Cann, *Counterinsurgency in Africa*, pp. 42–45.

49. Cann, *Counterinsurgency in Africa*, pp. 160–161.

50. Coelho, "African Troops," p. 139.

51. Davidson, *No Fist Is Big Enough*, p. 32.

52. Gibson, *African Liberation Movements*, p. 258.

53. Rudebeck, *Guinea-Bissau*, p. 132.

54. Cabral, *Revolution in Guinea*, p. 107.

55. Davidson, Slovo, and Wilkinson, *Southern Africa*, p. 60; Davidson, *No Fist Is Big Enough*, p. 124.

56. Cabral, *Revolution in Guinea*, p. 50; Gibson, *African Liberation Movements*, pp. 254–255, emphasis added.

57. Rudebeck, *Guinea-Bissau*, p. 133.

58. Davidson, *No Fist Is Big Enough*, pp. 66, 74, 76.

59. Chabal, *Amílcar Cabral*, p. 99.

60. Ibid., p. 93.

61. Ibid., pp. 100–101.

62. Porch, *Portuguese Armed Forces and the Revolution*, p. 63.

63. Cann, *Counterinsurgency in Africa*, pp. 132–134.

64. Porch, *Portuguese Armed Forces and the Revolution*, pp. 62–63.

65. Carlos Lopes, *Guinea Bissau: From Liberation Struggle to Independent Statehood* (Boulder, Colo.: Westview, 1987), p. 35.

66. Dhada argues that "Cabral's assassination had a most positive impact on PAIGC military strategy" because it increased the PAIGC's determination to fight Portugal and the PAIGC restored the leadership quickly, as seen in the launching of Operation Amílcar Cabral soon after the incident. Dhada, "Liberation War in Guinea Bissau Reconsidered," p. 590.

67. John Keegan, *World Armies*, 2nd ed. (Detroit, Mich.: Gale Research, 1979), p. 239.

68. Maxwell, *Making of Portuguese Democracy*, p. 31.

69. Hugo Gil Ferreira and Michael Marshall, *Portugal's Revolution: Ten Years On* (Cambridge: Cambridge University Press, 1986).

70. Porch, *Portuguese Armed Forces and the Revolution*, p. 31.

71. Ibid., pp. 31–32.

72. Ferreira and Marshall, *Portugal's Revolution*, p. 15.

73. Aquino de Braganca and Basil Davidson, "Independence Without Decolonization: Mozambique, 1974–1975," in Gifford and Louis, *Decolonization and African Independence*, pp. 428–431.

74. Gibson, *African Liberation Movements*, p. 3.

75. Ferreira and Marshall, *Portugal's Revolution*, p. 173.

76. See, for instance, Dhada, "Liberation War in Guinea-Bissau Reconsidered," p. 576, where he writes, "The offensive had exposed the Portuguese to possible defeat in battle. This realization helped strengthened the PAIGC's resolve to open a new front in the east."

77. See Dhada, *Warriors at Work*, p. 236n88.

78. Arreguin-Toft, *How the Weak Win Wars*.

79. Record, *Beating Goliath*.

80. Staniland, "Organizing Insurgency."

81. Dhada, "Liberation War in Guinea-Bissau Reconsidered," p. 574n583.

82. Kenneth Maxwell, "Portugal and Africa: The Last Empire," in Gifford and Louis, *Decolonization and African Independence*, p. 363.

Chapter 9

1. Fanon, *Wretched of the Earth*, p. 70.

2. Vietminh is an abbreviated name of the Viet Nam Doc Lap Dong Minh Hoi, meaning in Vietnamese the League for the Independence of Vietnam.

3. *Ho Chi Minh Selected Writings, Part II (1945–1954), Political Report at the Second National Congress of the Viet Nam Workers' Party* (February 1951).

4. Charles Kupchan, *The Vulnerability of Empire* (Ithaca, N.Y.: Cornell University Press, 1994).

5. John Prados, "Assessing Dien Bien Phu," in Lawrence and Logevall, *First Vietnam War*, p. 217.

6. David Marr, "Creating Defense Capacity in Vietnam, 1945–1947," in Lawrence and Logevall, *First Vietnam War*, p. 101; Prados, "Assessing Dien Bien Phu," p. 221.

7. There was persistent confusion over whether political officers had the same power as military commanders. Marr, "Creating Defense Capacity," pp. 79–80.

8. Vietnamese Ministry of National Defense, *Ho Chi Minh Thought on the Military* (Hanoi: Institute of Military History, 2008), p. 30.

9. Ibid., p. 28.

10. William Duiker, "Ho Chi Minh and the Strategy of People's War," in Lawrence and Logevall, *First Vietnam War*, p. 153.

11. Vo Nguyen Giap, *People's War People's Army: The Viet Cong Insurrection Manual for Underdeveloped Countries* (New York: Praeger, 1962), pp. 42–43; Hoang Van Thai, "Dien Bien Phu: Why and How?," in George Katsiaficas, ed., *Vietnam Documents: American and Vietnamese Views of the War* (Armonk, N.Y.: M. E. Sharpe, 1992), pp. 16–23.

12. Duiker, "Ho Chi Minh and the Strategy of People's War," p. 168.

13. Ibid., p. 161; Truong Chinh, *Selected Writings* (Hanoi: Foreign Language Press, 1977), p. 188; Mao Zedong, *Selected Writings of Mao Zedong* (Beijing: Foreign Languages Press, 1972), p. 186n.

14. Duiker, "Ho Chi Minh and the Strategy of People's War," pp. 161–162.

15. Ibid.
16. *Ho Chi Minh Selected Writings, Part II.*
17. Giap, *People's War People's Army.*
18. George Tanham, *Communist Revolutionary Warfare: From the Vietminh to the Viet Cong* (Westport, Conn.: Praeger, 2006), p. 3.
19. Ibid., p. 15.
20. Vietnamese Ministry of National Defense, *Ho Chi Minh Thought on the Military*, pp. 245–246.
21. Marr, "Creating Defense Capacity," p. 97.
22. Tanham, *Communist Revolutionary Warfare*, pp. 53–55.
23. Marr, "Creating Defense Capacity," p. 104.
24. Greg Lockhart, *Nation in Arms: The Origin of the People's Army of Vietnam* (Wellington: Allen and Unwin, 1990), pp. 188–221.
25. Marr, "Creating Defense Capacity," p. 84.
26. Duiker, "Ho Chi Minh and the Strategy of People's War," p. 164.
27. Douglas Pike, *War, Peace, and the Viet Cong* (Cambridge, Mass.: MIT Press, 1969), p. 119.
28. Giap, *People's War People's Army*, p. 27.
29. Lien-Hang Nguyen, "Vietnamese Historians and the First Indochina War," in Lawrence and Logevall, *First Vietnam War*, p. 48.
30. Ellen Hammer, *The Struggle for Indochina 1940–1955* (Stanford, Calif.: Stanford University Press, 1955), pp. 223–224.
31. Duiker, "Ho Chi Minh and the Strategy of People's War," p. 173.
32. Hammer, *Struggle for Indochina*, p. 223.
33. Qiang Zhai, "Transplanting the Chinese Model: Chinese Military Advisers and the First Vietnam War, 1950–1954," *Journal of Military History* 57 (October 1993): 700–703.
34. Marr, "Creating Defense Capability," pp. 83–87.
35. *Ho Chi Minh Selected Writings, Part IV* (Hanoi: Foreign Languages Publishing House, 1962), pp. 119–120.
36. Ibid., p. 338.
37. Mark Bradley, "Making Sense of the French War," in Lawrence and Logevall, *The First Vietnam War*, p. 29.
38. Hammer, *Struggle for Indochina*, p. 223, emphasis added.
39. Pentagon Papers, Gravel Edition, "US Involvement in the France-Viet Minh War, 1950–1954" (Boston: Beacon, 1971).
40. Hammer, *Struggle for Indochina*, p. 373.
41. Ibid., pp. 283–284.
42. Duiker, "Ho Chi Minh and the Strategy of People's War," pp. 169, 338n28.
43. Tanham, *Communist Revolutionary Warfare*, pp. 15–34.
44. Ibid., p. 10.
45. Duiker, "Ho Chi Minh and the Strategy of People's War," pp. 171–172.

46. See Phillip Davidson, *Vietnam at War: The History 1946–1975* (New York: Oxford University Press, 1991).

47. Tanham, *Communist Revolutionary Warfare*, pp. 44–45.

48. Ibid., pp. 46–48.

49. Bernard Fall, *Hell in a Very Small Place: The Siege of Dien Bien Phu* (Boston: Da Capo Press, 2002), p. 457n11.

50. Ibid., pp. 457–458.

51. Ibid., pp. 451–453, 455.

52. Tanham, *Communist Revolutionary Warfare*, p. 5.

53. Ibid., pp. 22–27.

54. Ibid., pp. 5, 13n1.

55. Hammer, *Struggle for Indochina*, pp. 292–293.

56. Duiker, "Ho Chi Minh and the Strategy of People's War," p. 337n.

57. Tanham, *Communist Revolutionary Warfare*, p. 7.

58. Vietnamese Ministry of National Defense, *Ho Chi Minh Thought on the Military*, p. 151.

59. The school of "colonial consensus" based on a resurgent popular imperialism in postwar France is reflected in the works of, among others, D. Bruce Marshall, *The French Colonial Myth and Constitution-Making in the Fourth Republic* (New Haven, Conn.: Yale University Press, 1973), and Tony Smith, "The French Colonial Consensus and People's War, 1946–58," *Journal of Contemporary History* 9 (1974): 217–247.

60. Kupchan, *Vulnerability of Empire*, pp. 267–296.

61. Prados, "Assessing Dien Bien Phu," p. 218.

62. Kupchan, *Vulnerability of Empire*, pp. 278–279.

63. Record, *Beating Goliath*, pp. 44–46.

64. On the role of Chinese aid, see Qiang Zhai, *China and the Vietnam Wars, 1950–1975* (Chapel Hill: University of North Carolina Press, 2000).

65. Mark Atwood Lawrence and Fredrik Logevall, "Introduction," in Lawrence and Logevall, *First Vietnam War*, p. 10.

Conclusion

1. For instance, see Hy Rothstein and John Arquilla, eds., *Afghan Endgames: Strategy and Policy Choices for America's Longest War* (Washington, D.C.: Georgetown University Press, 2012).

2. T. V. Paul, "Complex Deterrence: An Introduction," Janice Gross Stein, "Rational Deterrence Against 'Irrational' Adversaries? No Common Knowledge," and Emanuel Adler, "Complex Deterrence in the Asymmetric-Warfare Era," all in T. V. Paul, Patrick Morgan, and James Wirtz, eds., *Complex Deterrence: Strategy in the Global Age* (Chicago: University of Chicago Press, 2009).

3. Ivan Arreguin-Toft, "Unconventional Deterrence: How the Weak Deter the Strong," in Paul, Morgan, and Wirtz, *Complex Deterrence*, p. 215.

4. Johnson and Madin, "Population Models and Counterinsurgency Strategies."

5. Hy Rothstein and John Arquilla, "Understanding the Afghan Challenge," in Rothstein and Arquilla, *Afghan Endgames*.

6. Audrey Kurth Cronin, *How Terrorism Ends: Understanding the Decline and Demise of Terrorist Campaigns* (Princeton, N.J.: Princeton University Press, 2009); Seth Jones and Martin Libicki, *How Terrorist Groups End: Lessons for Countering al Qa'ida* (Santa Monica, Calif.: RAND, 2008).

7. Paul Pillar, *Terrorism and U.S. Foreign Policy* (Washington, D.C.: Brookings Institution Press, 2001).

8. Martel, *Victory in War*, p. 310.

9. Seth Jones, *In the Graveyard of Empires: America's War in Afghanistan* (New York: Norton, 2010).

10. Record, *Beating Goliath*.

11. Abdulkader Sinno, *Organizations at War in Afghanistan and Beyond* (Ithaca, N.Y.: Cornell University Press, 2008), p. 255.

12. See Stephen Biddle, "Allies, Airpower, and Modern Warfare: The Afghan Model in Afghanistan and Iraq," *International Security* 30, no. 3 (Winter 2005–2006).

13. Sinno, *Organizations at War in Afghanistan and Beyond*, pp. 256–257.

14. Kilcullen, *Accidental Guerrilla*.

15. Sinno, *Organizations at War in Afghanistan and Beyond*, p. 270.

16. Martel, *Victory in War*, p. 296.

17. Bruce Riedel, *The Search for Al Qaeda: Its Leadership, Ideology, and Future* (Washington, D.C.: Brookings Institution Press, 2008), chap. 7.

18. Jones, *In the Graveyard of Empires*, p. 332.

19. Fawaz Gerges, *The Rise and Fall of Al-Qaeda* (Oxford: Oxford University Press, 2011), p. 127.

20. John Mueller, *Overblown: How Politicians and the Terrorism Industry Inflate National Security Threats, and Why We Believe Them* (New York: Free Press, 2006).

21. Martel, *Victory in War*, p. 304.

22. John Mearsheimer, "Hollow Victory," *Foreign Policy*, November 2, 2009.

23. Jones, *In the Graveyard of Empires*, p. 329.

24. Ibid.

25. See, for instance, Dan Reiter, *How Wars End* (Princeton, N.J.: Princeton University Press, 2009); Fred Ikle, *Every War Must End* (New York: Columbia University Press, 2005).

26. Gleditsch, Salehyan, and Schultz, "Fighting at Home, Fighting Abroad."

27. Monica Duffy Toft, "Ending Civil Wars: A Case for Rebel Victory?," *International Security* 34, no. 4 (Spring 2010).

28. Charles Call, *Why Peace Fails: The Causes and Prevention of Civil War Recurrence* (Washington, D.C.: Georgetown University Press, 2012).

29. David Cunningham, *Barriers to Peace in Civil War* (Cambridge: Cambridge University Press, 2011).

30. Hashim, *Insurgency and Counterinsurgency in Iraq*, pp. 47–48.

31. Peter Mansoor, *Baghdad at Sunrise: A Brigade Commander's War in Iraq* (New Haven, Conn.: Yale University Press, 2007), p. 82.

BIBLIOGRAPHY

Abrahms, Max. "Why Democracies Make Superior Counterterrorists." *Security Studies* 16, no. 2 (2007).
———. "Why Terrorism Does Not Work." *International Security* 31, no. 2 (Fall 2006).
Adas, Michael. *Machines as a Measure of Men: Science, Technology, and Ideologies of Western Dominance.* Ithaca, N.Y.: Cornell University Press, 1989.
Adeleye, R. A. *Power and Diplomacy in Northern Nigeria, 1804–1906.* New York: Humanities Press, 1971.
Adler, Emanuel. "Complex Deterrence in the Asymmetric-Warfare Era." In Paul, Morgan, and Wirtz, *Complex Deterrence*.
Ageron, Charles-Robert. *Modern Algeria: A History from 1830 to the Present.* Trenton, N.J.: Africa World Press, 1990.
Ahmida, Ali Abdullatif. *The Making of Modern Libya: State Formation, Colonization, and Resistance, 1830–1932.* Albany: State University of New York Press, 1994.
Ajayi, J. F. A. and Michael Crowder, eds. *The History of West Africa.* New York: Columbia University Press, 1973.
Aldrich, Robert. *Greater France: A History of French Overseas Expansion.* London: Macmillan, 1996.
Allen, Charles. *Soldier Sahibs: The Men Who Made the North-West Frontier.* London: John Murray, 2000.
Alpern, Stanley. *Amazons of Black Sparta: The Women Warriors of Dahomey.* New York: New York University Press, 1998.
Anderson, Benedict. *Java in a Time of Revolution: Occupation and Resistance, 1944–1946.* Ithaca: Cornell University Press, 1972.
Anderson, George and M. Subedar. *The Expansion of British India (1818–1858).* London: G. Bell and Sons, 1918.
Argyle, W. J. *The Fon of Dahomey: A History and Ethnography of the Old Kingdom.* Oxford: Oxford University Press, 1966.
Arreguin-Toft, Ivan. *How the Weak Win Wars: A Theory of Asymmetric Conflict.* Cambridge: Cambridge University Press, 2005.
———. "Unconventional Deterrence: How the Weak Deter the Strong." In Paul, Morgan, and Wirtz, *Complex Deterrence*.
Avant, Deborah. *Political Institutions and Military Change: Lessons from Peripheral Wars.* Ithaca, N.Y.: Cornell University Press, 1994.

Aylwin-Foster, Nigel. "Changing the Army for Counterinsurgency Operations." *Military Review*, November–December 2005.
Bakker, J. I. "The Ache War and the Creation of the Netherlands East Indies State." In Hamish Ion and Elizabeth Jane Errington, eds., *Great Powers and Little Wars: The Limits of Power*. Westport, Conn.: Greenwood, 1993.
Balesi, Charles John. *From Adversaries to Comrades-in-Arms: West Africans and the French Military, 1885–1918*. Waltham, Mass.: Crossroads Press, 1979.
Bates, Robert. "The Centralization of African Societies." In Robert Bates, *Essays on the Political Economy of Africa*. Berkeley: University of California Press, 1987.
Bean, Richard. "War and the Birth of the Nation State." *Journal of Economic History* 33, no. 1 (1973).
Becker, Seymour. *Russia's Protectorates in Central Asia: Bukhara and Khiva, 1865–1924*. London: Routledge, 2004.
Belich, James. *The New Zealand Wars and the Victorian Interpretation of Racial Conflict*. Auckland: Penguin, 1986.
Bell, Bowyer. *The Myth of the Guerrilla: Revolutionary Theory and Malpractice*. New York: Knopf, 1971.
Bell, Charles. *Tibet: Past and Present*. New Delhi: Munshiram Manoharlal, 1990.
Bennett, Huw. "'A Very Salutary Effect': The Counter-Terror Strategy in the Early Malayan Emergency, June 1948 to December 1949." *Journal of Strategic Studies* 32, no. 3 (2009).
Bennett, Scott and Allan Stam. "The Declining Advantages of Democracy: A Combined Model of War Outcomes and Duration." *Journal of Conflict Resolution* 42, no. 3 (June 1998).
Bennoune, Mahfoud. *The Making of Contemporary Algeria, 1830–1987: Colonial Upheavals and Post-Independence Development*. Cambridge: Cambridge University Press, 1988.
Bensahel, Nora, et al. *After Saddam: Prewar Planning and the Occupation of Iraq*. Santa Monica, Calif.: RAND, 2008.
Betts, Raymond. *Assimilation and Association in French Colonial Theory 1890–1914*. Lincoln: University of Nebraska Press, 2005.
Biddle, Stephen. "Allies, Airpower, and Modern Warfare: The Afghan Model in Afghanistan and Iraq." *International Security* 30, no. 3 (Winter 2005–2006).
———. *Military Power: Explaining Victory and Defeat in Modern Battle*. Princeton, N.J.: Princeton University Press, 2006.
———. "Seeing Baghdad, Thinking Saigon." *Foreign Affairs* 85, no. 2 (March–April 2006).
———. "Speed Kills? Reassessing the Role of Speed, Precision, and Situation Awareness in the Fall of Saddam." *Journal of Strategic Studies* 30, no. 1 (February 2007).
Biddle, Stephen, Jeffrey Friedman, and Jacob Shapiro. "Testing the Surge: Why Did Violence Decline in Iraq in 2007?" *International Security* 37, no. 1 (Summer 2012).

Bienen, Henry. "State and Revolution: The Work of Amílcar Cabral." *Journal of Modern African Studies* 15, no. 4 (1977).
Birmingham, David. *Frontline Nationalism in Angola and Mozambique*. Trenton, N.J.: Africa World Press, 1993.
Bjorkelo, Anders. *Prelude to the Mahdiyya: Peasants and Traders in the Shendi Region, 1821–1885*. Cambridge: Cambridge University Press, 1989.
Blake, Robert. *A History of Rhodesia*. New York: Knopf, 1978.
Bloom, Mia. *Dying to Kill: The Allure of Suicide Terror*. New York: Columbia University Press, 2005.
Boahen, Bdu. *General History of Africa: Africa Under Colonial Domination 1880–1935*. Vol. 2. Berkeley: University of California Press, 1985.
Bob, Clifford. *The Marketing of Rebellion: Insurgents, Media, and International Activism*. Cambridge: Cambridge University Press, 2005.
Bowler, Peter. *Evolution: The History of an Idea*. Berkeley: University of California Press, 2003.
Bradley, Mark, "Making Sense of the French War." In Lawrence and Logevall, eds., *The First Vietnam War*.
Bridgman, Jon. *The Revolt of the Hereros*. Berkeley: University of California Press, 1981.
Briggs, Harold. *AIR20/777, Secret, Early History of Emergency. Report on the Emergency in Malaya from April, 1950 to November 1951*.
British Operation Research Section. "Far East." Memo no. 8/53.
Burton, Richard. *A Mission to Gelele, King of Dahome*. Vol. 1. London: Tylston and Edwards, 1893.
Byman, Daniel. *Deadly Connections: States That Sponsor Terrorism*. New York: Cambridge University Press, 2005.
Byman, Daniel and Kenneth Pollack. *Things Fall Apart: Containing the Spillover from an Iraqi Civil War*. Washington, D.C.: Brookings Institution Press, 2007.
CAB 129/48, C(51)26. "'The Situation in Malaya': Cabinet Memorandum by Mr. Lyttelton." In A.J. Stockwell, ed., *British Documents on the End of Empire: Malaya: Part II*. London: HMSO, 1995.
CAB 129/48, C(51)59 [December 21, 1951]. "Malaya: Cabinet Memorandum by Mr. Lyttelton." In Stockwell, ed., *British Documents on the End of Empire: Malaya: Part II*.
CAB 21/1681, MAL C(50)21 [June 17, 1950]. "The Military Situation in Malaya." In Stockwell, ed., *British Documents on the End of Empire: Malaya: Part II.*.
Cabral, Amílcar. "Liberating Portuguese Guinea from Within." *New African* 4 (June 1965).
———. *Rapport General sur la Lutte de Liberation Nationale*. Conakry, July 1961.
———. *Return to the Source*. New York: Monthly Review Press, 1974.
———. *Revolution in Guinea*. New York: Monthly Review Press, 1969.
———. "The Tactic of Division." In Aquino de Braganca and Immanuel Wallerstein, eds., *The African Liberation Reader, vol. 2: The National Liberation Movements*. London: Zed Press, 1982.

———. *Unity and Struggle: Speeches and Writings*. New York: Monthly Review Press, 1979.

———. "The War in 'Portuguese' Guinea." *Revolution* 1, no. 2 (June 1963).

Caldwell, Charles. *Small Wars: Their Principles and Practice*. London: Harrison and Sons, 1896.

Call, Charles. *Why Peace Fails: The Causes and Prevention of Civil War Recurrence*. Washington, D.C.: Georgetown University Press, 2012.

Callahan, Mary. *Making Enemies: War and State Building in Burma*. Ithaca, N.Y.: Cornell University Press, 2003.

Cann, John. *Counterinsurgency in Africa: The Portuguese Way of War, 1961–1974*. Westport, Conn.: Greenwood, 1997.

Cannadine, David. *The Rise and Fall of Class in Britain*. New York: Columbia University Press, 1993.

Carey, Peter and G. Carter Bentley, eds. *East Timor at the Crossroads: The Forging of a Nation*. Honolulu: University of Hawaii Press, 1995.

Carter, David. "A Blessing or a Curse? State Support for Terrorist Groups." *International Organization* 66, no. 1 (January 2012).

Chabal, Patrick. *Amílcar Cabral: Revolutionary Leadership and People's War*. Cambridge: Cambridge University Press, 1983.

———. "People's War, State Formation and Revolution in Africa: A Comparative Analysis of Mozambique, Guinea-Bissau, and Angola." *Journal of Commonwealth and Comparative Politics* 21, no. 3 (1983).

———. "The Social and Political Thought of Amílcar Cabral: A Reassessment." *Journal of Modern African Studies* 19, no. 1 (1981).

Chaliand, Gérard. *Armed Struggle in Africa*. New York: Monthly Review Press, 1969.

Chenoweth, Erica and Maria Stephan. *Why Civil Resistance Works: The Strategic Logic of Nonviolent Conflict*. New York: Columbia University Press, 2011.

Chilcote, Ronald. *Amílcar Cabral's Revolutionary Theory and Practice: A Critical Guide*. Boulder, Colo.: Lynne Rienner, 1991.

———. *Emerging Nationalism in Portuguese Africa: Documents*. Stanford, Calif.: Hoover Institution Press, 1972.

———. *Portuguese Africa*. Englewood Cliffs, N.J.: Prentice Hall, 1967.

Chin, Peng. *My Side of Story*. Singapore: Media Masters, 2003.

Chinh, Truong. *Selected Writings*. Hanoi: Foreign Language Press, 1977.

Chisholm, Hugh, ed. *The Encyclopædia Britannica*. Cambridge: Cambridge University Press, 1911.

Churchill, Winston. *The River War: An Account of the Reconquest of the Sudan*. Mineola, N.Y.: Dover, 2006.

Clammer, David. *The Zulu War*. Newton Abbot, UK: David & Charles, 1973.

Clapham, Christopher. *African Guerrillas*. Oxford: James Currey, 1998.

Clayton, Anthony. *France, Soldiers, and Africa*. London: Brassey's, 1988.

Cloake, John. *Templer, Tiger of Malaya: The Life of Field Marshal Sir General Templer*. London: Harrap, 1985.

Clutterbuck, Richard. *The Long, Long War: Counterinsurgency in Malaya and Vietnam*. New York: Praeger, 1966.

Coates, John. *Suppressing Insurgency: An Analysis of the Malayan Emergency, 1948–1954*. Boulder, Colo.: Westview, 1992.

Coelho, Joao Paulo Borges. "African Troops in the Portuguese Colonial Army, 1961–1974: Angola, Guinea-Bissau and Mozambique." *Portuguese Studies Review* 10, no. 1 (2002).

Coleman, James and Carl Rosberg, Jr., eds. *Political Parties and National Integration in Tropical Africa*. Berkeley: University of California Press, 1964.

Colson, Elizabeth. "African Society at the Time of the Scramble." In Gann and Duignan, *Colonialism in Africa*.

Cooke, James. *New French Imperialism, 1880–1910: The Third Republic and Colonial Expansion*. Hamden, Conn.: Archon Books, 1973.

Covarrubias, Miguel. *Island of Bali*. United Kingdom: Read Books, 2006.

Crawford, Neta. *Argument and Change in World Politics: Ethics, Decolonization, and Humanitarian Intervention*. Cambridge: Cambridge University Press, 2002.

Crawford, Timothy and Alan Kuperman, eds. *Gambling on Humanitarian Intervention: Moral Hazard, Rebellion, and Civil War*. London: Routledge, 2006.

Cronin, Audrey Kurth. "Cyber-Mobilization: The New Levee en Mass." *Parameters* (Summer 2006).

———. *How Terrorism Ends: Understanding the Decline and Demise of Terrorist Campaigns*. Princeton, N.J.: Princeton University Press, 2009.

Crowder, Michael. *West Africa Under Colonial Rule*. Evanston, Ill.: Northwestern University Press, 1968.

———, ed. *West African Resistance: The Military Response to Colonial Occupation*. New York: Africana, 1971.

Cunningham, David. *Barriers to Peace in Civil War*. Cambridge: Cambridge University Press, 2011.

Curtin, Philip. *Disease and Empire: The Health of European Troops in the Conquest of Africa*. Cambridge: Cambridge University Press, 1998.

Dahm, Bernhard. *Sukarno and the Struggle for Indonesian Independence*. Ithaca, N.Y.: Cornell University Press, 1969.

Dalzel, Archibald. *The History of Dahomy: An Inland Kingdom of Africa*. London: Frank Cass, 1967.

Davidson, Basil. *No Fist Is Big Enough to Hide the Sky: The Liberation of Guine and Cape Verde, Aspects of an African Revolution*. London: Zed Press, 1981.

———. *The People's Cause: A History of Guerrillas in Africa*. Essex, UK: Longman, 1981.

———. *A Política da Luta Armada: Libertação Nacional nas Colónias Africanas de Portugal*. Lisbon: Editorial Caminho, 1979.

Davidson, James. *The Island of Formosa: Past and Present*. Oxford: Oxford University Press, 1988.
Davidson, Janine. *Lifting the Fog of Peace: How Americans Learned to Fight Modern War*. Ann Arbor: University of Michigan Press, 2010.
Davidson, Phillip. *Vietnam at War: The History 1946–1975*. New York: Oxford University Press, 1991.
Davidson, Basil, Joe Slovo, and Anthony Wilkinson. *Southern Africa: The New Politics of Revolution*. Harmondsworth: Penguin, 1976.
de Braganca, Aquino and Basil Davidson. "Independence Without Decolonization: Mozambique, 1974–1975." In Gifford and Louis, *Decolonization and African Independence*.
Debray, Regis. *Revolution in the Revolution? Armed Struggle and Political Struggle in Latin America*. New York: Monthly Review Press, 1967.
DeNardo, James. *Power in Numbers: The Political Strategy of Protest and Rebellion*. Princeton, N.J.: Princeton University Press, 1985.
Dhada, Mustafah. "The Liberation War in Guinea-Bissau Reconsidered." *Journal of Military History* 62, no. 3 (July 1998).
―――. *Warriors at Work: How Guinea Was Really Set Free*. Niwot: University Press of Colorado, 1993.
Dittmer, Lowell. "The Legacy of Mao Zedong." *Asian Survey* 20, no. 2 (May 1980).
Dodwell, H. H., ed. *The Cambridge History of the British Empire, vol. 4: British India 1497–1858*. Cambridge: Cambridge University Press, 1929.
Donia, Robert and John Fine. *Bosnia and Hercegovina: A Tradition Betrayed*. New York: Columbia University Press, 1995.
Downes, Alexander. "How Smart and Tough Are Democracies? Reassessing Theories of Democratic Victory in War." *International Security* 33, no. 4 (2009)
Doyle, Michael. *Empires*. Ithaca, N.Y.: Cornell University Press, 1986.
Drexler, Elizabeth. *Aceh, Indonesia: Securing the Insecure State*. Philadelphia: University of Pennsylvania Press, 2008.
Duelfer, Charles and Stephen Benedict Dyson. "Chronic Misperception and International Conflict: The U.S.-Iraq Experience." *International Security* 36, no. 1 (Summer 2011).
Duffy, James. *Portugal in Africa*. Cambridge, Mass.: Harvard University Press, 1962.
Duignan, Peter and Lewis Gann. "The Pre-Colonial Economies of Sub-Saharan Africa." In Peter Duignan and Lewis Gann, eds., *Colonialism in Africa 1870–1960, vol. 4: The Economics of Colonialism*. Cambridge: Cambridge University Press, 1975.
Duiker, William. "Ho Chi Minh and the Strategy of People's War." In Lawrence and Logevall, *First Vietnam War*.
Dupuy, Ernest and Trevor Dupuy. *The Encyclopedia of Military History*. New York: HarperCollins, 1993.
Dupuy, Trevor. *The Evolution of Weapons and Warfare*. Cambridge: Da Capo Press, 1990.

Duyker, Edward. *Tribal Guerrillas: The Santals of West Bengal and the Naxalite Movement*. Delhi: Oxford University Press, 1987.

Edelstein, David. "Foreign Militaries, Sustainable Institutions, and Postwar Statebuilding." In Roland Paris and Timothy Sisk, eds., *The Dilemmas of Statebuilding: Confronting the Contradictions of Postwar Peace Operations*. London: Routledge, 2009.

———. *Occupational Hazards: Success and Failure in Military Occupation*. Ithaca, N.Y.: Cornell University Press, 2008.

Edgerton, Robert. *Africa's Armies: From Honor to Infamy*. Boulder, Colo.: Westview, 2002.

Ellul, Jacques. *Autopsy of Revolution*. New York: Knopf, 1971.

Ergang, Robert. *Europe Since Waterloo*. London: Heath, 1966.

Erickson, Sharon Nepstad. *Nonviolent Revolutions: Civil Resistance in the Late 20th Century*. Oxford: Oxford University Press, 2011.

Erlich, Haggai. *Ras Alula and the Scramble for Africa*. Lawrenceville, N.J.: Red Sea Press, 1996.

Fall, Bernard. *Hell in a Very Small Place: The Siege of Dien Bien Phu*. Boston: Da Capo Press, 2002.

Fanon, Franz. *The Wretched of the Earth*. New York: Grove Press, 1963.

Farwell, Byron. *Eminent Victorian Soldiers: Seekers of Glory*. New York: Norton, 1988.

———. *The Encyclopedia of Nineteenth-Century Land Warfare*. New York: Norton, 2001.

———. *Queen Victoria's Little Wars*. New York: Harper & Row, 1972.

Feaver, Peter and Christopher Gelpi. *Choosing Your Battles: American Civil-Military Relations and the Use of Force*. Princeton, N.J.: Princeton University Press, 2004.

Federation of Malaya. "Weekly News Summary." July 4, 1953.

Feldman, Noah. *What We Owe Iraq: War and the Ethics of Nation Building*. Princeton, N.J.: Princeton University Press, 2006.

Ferreira, Hugo Gil and Michael Marshall. *Portugal's Revolution: Ten Years On*. Cambridge: Cambridge University Press, 1986.

Fontenot, Gregory, E. J. Degen, and David Tohn. *On Point: The United States Army in Operation Iraqi Freedom*. Annapolis, Md.: Naval Institute Press, 2005.

Forbes, Archibald. *The Afghan Wars, 1839–42 and 1878–80*. London: Seeley and Co., 1982.

French, Aquino de Braganca and Immanuel Wallerstein, eds. *The African Liberation Reader, vol. 3: The Strategy of Liberation*. London: Zed Press, 1982.

Friedman, Jeffrey. "Manpower and Counterinsurgency: Empirical Foundations for Theory and Doctrine." *Security Studies* 20, no. 4 (2011).

Fukuyama, Francis. *State-Building: Governance and World Order in the 21st Century*. Ithaca, N.Y.: Cornell University Press, 2004.

Fynn, J. K. "Ghana-Asante (Ashanti)." In Crowder, *West African Resistance*.

Galula, David. *Counterinsurgency Warfare: Theory and Practice*. Westport, Conn.: Greenwood, 2006.

Gann, Lewis and Peter Duignan, eds. *Colonialism in Africa, 1870–1960*. Vol. 1. Cambridge: Cambridge University Press, 1975.
———. *The Rulers of Belgian Africa 1884–1914*. Princeton, N.J.: Princeton University Press, 1979.
Gantzel, Klaus Jürgen and Torsten Schwinghammer. *Warfare Since the Second World War*. New Brunswick, N.J.: Transaction, 2000.
Gartner, Scott Sigmund. *Strategic Assessment in War*. New Haven, Conn.: Yale University Press, 1999.
Geary, William. *Nigeria Under British Rule*. London: Routledge, 1965.
Gelpi, Christopher, Peter Feaver, and Jason Reifler. *Paying the Human Costs of War: American Public Opinion and Casualties in Military Conflicts*. Princeton, N.J.: Princeton University Press, 2009.
———. "Success Matters: Casualty Sensitivity and the War in Iraq." *International Security* 30, no. 3 (Winter 2005–2006).
George, Alexander and Andrew Bennett. *Case Studies and Theory Development in the Social Sciences*. Cambridge, Mass.: MIT Press, 2005.
Gerges, Fawaz. *The Rise and Fall of Al-Qaeda*. Oxford: Oxford University Press, 2011.
Gershovich, Moshe. *French Military Rule in Morocco: Colonialism and Its Consequences*. London: Frank Cass, 2000.
Giap, Vo Nguyen. *People's War People's Army: The Viet Cong Insurrection Manual for Underdeveloped Countries*. New York: Praeger, 1962.
Gibson, Richard. *African Liberation Movements: Contemporary Struggles Against White Minority Rule*. New York: Oxford University Press, 1972.
Gifford, Prosser and Roger Louis, eds. *Decolonization and African Independence: The Transfers of Power, 1960–1980*. New Haven, Conn.: Yale University Press, 1988.
Glass, Stafford. *The Matabele War*. London: Longmans, 1968.
Gleditsch, Kristian Skrede. "A Revised List of Wars Between and Within Independent States, 1816–2002." *International Interactions* 30 (2004).
Gleditsch, Kristian Skrede, Idean Salehyan, and Kenneth Schultz. "Fighting at Home, Fighting Abroad: How Civil Wars Lead to International Disputes." *Journal of Conflict Resolution* 52, no. 4 (2008).
Goldstein, Avery. *Deterrence and Security in the 21st Century: China, Britain, France, and the Enduring Legacy of the Nuclear Revolution*. Stanford, Calif.: Stanford University Press, 2000.
Goldstein, Joshua. *Long Cycles: Prosperity and War in the Modern Age*. New Haven, Conn.: Yale University Press, 1988.
Golob, Eugene. *The Meline Tariff: French Agriculture and Nationalist Economic Policy*. New York: Columbia University Press, 1944.
Gompert, David and John Gordon. *War by Other Means: Building Complete and Balanced Capabilities for Counterinsurgency*. Santa Monica, Calif.: RAND, 2008.
Gooch, John. *The Boer War: Direction, Experience, and Image*. London: Routledge, 2000.

———. *The Plans of War: The General Staff and Military Strategy c. 1900–1916*. New York: John Wiley, 1974.
Goodwin, Jeff. *No Other Way Out: States and Revolutionary Movements, 1945–1991*. Cambridge: Cambridge University Press, 2001.
Goody, Jack. *Technology, Tradition, and the State in Africa*. Cambridge: Cambridge University Press, 1980.
Gordon, Micheal and Bernard Trainor. *Cobra II: The Inside Story of the Invasion and Occupation of Iraq*. New York: Vintage, 2006.
Great Britain, War Office, General Staff. *Official History of the Operations in Somaliland, 1901–1904*. 2 vols. London: Harrison and Sons, 1907.
Gregorian, Raffi. "Jungle Bashing in Malaya: Towards a Formal Tactical Doctrine." *Small Wars and Insurgencies* 5, no. 3 (Winter 1994).
Guevara, Che, Rolando Bonachea, and Nelson Valdes, eds. *Che: Selected Works of Ernesto Guevara*. Cambridge, Mass.: MIT Press, 1969.
Hack, Karl. "British Intelligence and Counter-Insurgency in the Era of Decolonisation. The Example of Malaya." *Intelligence and National Security* 14, no. 2 (Summer 1999).
———. *Defense and Decolonization in Southeast Asia: Britain, Malaya, and Singapore 1941–1968*. London: Curzon, 2001.
Hafez, Mohammed. *Suicide Bombers in Iraq: The Strategy and Ideology of Martyrdom*. Washington, D.C.: U.S. Institute for Peace, 2007.
Halpern, Jack. *South Africa's Hostages: Basutoland, Bechuanaland and Swaziland*. Harmondsworth: Penguin, 1965.
Hammer, Ellen. *The Struggle for Indochina 1940–1955*. Stanford, Calif.: Stanford University Press, 1955.
Hargreaves, John. "West African States and the European Conquest." In Gann and Duignan, eds., *Colonialism in Africa*.
Harper, T. N. *The End of Empire and the Making of Malaya*. Cambridge: Cambridge University Press, 2001.
Harvey, Frank. *Explaining the Iraq War: Counterfactual Theory, Logic and Evidence*. New York: Cambridge University Press, 2011.
Hashim, Ahmed. *Insurgency and Counterinsurgency in Iraq*. Ithaca, N.Y.: Cornell University Press, 2006.
Hasrat, Bikrama Jit. *History of Bhutan: Land of the Peaceful Dragon*. Thimphu: Education Department, Royal Government of Bhutan, 1980.
Hatemi, Peter and Rose McDermott. "A Neurobiological Approach to Foreign Policy Analysis: Identifying Individual Differences in Political Violence." *Foreign Policy Analysis* 8 (2011).
Headrick, Daniel. *Power over Peoples: Technology, Environments, and Western Imperialism, 1400 to the Present*. Princeton, N.J.: Princeton University Press, 2010.
———. *The Tools of Imperialism: Technology and European Imperialism in the Nineteenth Century*. Oxford: Oxford University Press, 1981.

Hegghammer, Thomas. "The Rise of Muslim Foreign Fighters: Islam and the Globalization of Jihad." *International Security* 35, no. 3 (Winter 2010–2011).
Heinl, Robert and Nancy Heinl. *Written in Blood: The Story of the Haitian People 1492–1995*. Lanham, Md.: University Press of America, 2005.
Henty, George Alfred. *At the Point of the Bayonet: A Tale of the Mahratta War*. London: Blackie and Son, 1902.
Henze, Paul. *Layers of Time: A History of Ethiopia*. New York: St. Martin's, 2000.
Herbst, Jeffrey. *States and Power in Africa: Comparative Lessons in Authority and Control*. Princeton, N.J.: Princeton University Press, 2000.
Herring, Eric and Glen Rangwala. *Iraq in Fragments: The Occupation and Its Legacy*. Ithaca, N.Y.: Cornell University Press, 2006.
Hess, Robert. *Italian Colonialism in Somalia*. Chicago: University of Chicago Press, 1966.
Ho Chi Minh Selected Writings, Part II (1945–1954), Political Report at the Second National Congress of the Viet Nam Workers' Party. Hanoi: February 1951.
Ho Chi Minh Selected Writings, Part IV. Hanoi: Foreign Languages Publishing House, 1962.
Hoffman, Bruce. "Terrorism Trends and Prospects." In Ian Lesser et al., eds. *Countering the New Terrorism*. Santa Monica, Calif.: RAND, 1999.
Holsti, Ole. *American Public Opinion on the Iraq War*. Ann Arbor: University of Michigan Press, 2011.
Horne, Alistair. *A Savage War of Peace: Algeria, 1954–1962*. New York: New York Review of Books, 2006.
Horowitz, Michael. *The Diffusion of Military Power: Causes and Consequences for International Politics*. Princeton, N.J.: Princeton University Press, 2010.
Hosmer, Stephen. *Why the Iraqi Resistance to the Coalition Invasion Was So Weak*. Santa Monica, Calif.: RAND, 2007.
Howard, Michael. *War in European History*. Oxford: Oxford University Press, 1976.
Huntington, Samuel. "Guerrilla Warfare in Theory and Practice." In Franklin Mark Osanka, ed., *Modern Guerrilla Warfare: Fighting Communist Guerrilla Movements, 1941–1961*. New York: Free Press, 1962.
———. *Political Order in Changing Societies*. New Haven, Conn.: Yale University Press, 1968.
Ikle, Fred. *Every War Must End*. New York: Columbia University Press, 2005.
Iliffe, John. *Africans: The History of a Continent*. Cambridge: Cambridge University Press, 1995.
———. "The Organization of the Maji Maji Rebellion." *Journal of African History* 8, no. 3 (1967).
Indian Army. *The Second Afghan War, 1870–80, Abridged Official Account*. London: John Murray, 1908.
Information Services of the Somali Government. *The Somali Peninsula: A New Light on Imperial Motives*. Mogadishu: Information Services of the Somali Government, 1962.

International Crisis Group. *In Their Own Words: Reading the Iraqi Insurgency*. Middle East Report no. 50. February 15, 2006.

Issa-Salwe, Abdisalam. *The Collapse of Darwish State: The Impact of the Colonial Legacy*. London: HAAN Associates, 1994.

Jackson, Robert. *The Malayan Emergency: The Commonwealth's Wars 1948–1966*. London: Routledge, 1991.

———. "The Weight of Ideas in Decolonization: Normative Change in International Relations." In Judith Goldstein and Robert Keohane, eds., *Ideas and Foreign Policy: Beliefs, Institutions, and Political Change*. Ithaca, N.Y.: Cornell University Press, 1993.

Jacobson, Harold. "The United Nations and Colonialism: A Tentative Appraisal." *International Organization* 16, no. 1 (Winter 1962).

James, Lionel. *The Indian Frontier War: Being an Account of the Mohmund and Tirah Expeditions, 1897*. London: Heinemann, 1898.

Jardine, Douglas. *The Mad Mullah of Somaliland*. New York: Negro Universities Press, 1969.

Jervis, Robert. *Perception and Misperception in International Politics*. Princeton, N.J.: Princeton University Press, 1976.

———. *Why Intelligence Fails: Lessons from the Iranian Revolution and the Iraq War*. Ithaca, N.Y.: Cornell University Press, 2010.

Johnson, Chalmers. *Autopsy on People's War*. Berkeley: University of California Press, 1973.

———. *Peasant Nationalism and Communist Power: The Emergency of Revolutionary China 1937–1945*. Stanford, Calif.: Stanford University Press, 1962.

Johnson, Dominic. "Darwinian Selection in Asymmetric Warfare: The Natural Advantage of Insurgents and Terrorists." *Journal of the Washington Academy of Sciences* 95 (2009).

Johnson, Dominic and Joshua Madin. "Population Models and Counterinsurgency Strategies." In Raphael Sagarin and Terence Taylor, eds., *Natural Security: A Darwinian Approach to a Dangerous World*. Berkeley: University of California Press, 2008.

Johnson, Dominic and Dominic Tierney. *Failing to Win: Perceptions of Victory and Defeat in International Politics*. Cambridge, Mass.: Harvard University Press, 2006.

Johnson, Willis. *The History of Cuba*. Vol. 3. New York: B. F. Bucks and Company, 1920.

Jones, Seth. *In the Graveyard of Empires: America's War in Afghanistan*. New York: Norton, 2010.

Jones, Seth and Martin Libicki. *How Terrorist Groups End: Lessons for Countering al Qa'ida*. Santa Monica, Calif.: RAND, 2008.

Jones, Tim. "The British Army, and Counter-Guerrilla Warfare in Transition, 1944–1952." *Small Wars and Insurgencies* 7, no. 3 (Winter 1996): 265–308.

Judd, Denis. *The Lion and the Tiger: The Rise and Fall of the British Raj, 1600–1947*. Oxford: Oxford University Press, 2004.

Kahl, Colin. "In the Crossfire or the Crosshairs? Norms, Civilian Casualties, and U.S. Conduct in Iraq." *International Security* 32, no. 1 (Summer 2007).
Kanya-Forstner, A. S. *The Conquest of the Western Sudan: A Study in French Military Imperialism.* Cambridge: Cambridge University Press, 1969.
Katagiri, Nori. "Drawing Strategic Lessons from Dahomey's War." *Air Power and Space Journal—African and Francophonie* 3, no. 3 (October 2012).
———. "Suicidal Armies: Why Do Rebels Fight Like an Army and Keep Losing?" *Comparative Strategy* 32, no. 4 (2013): 354–377.
———. "Winning Hearts and Minds to Lose Control: Exploring Various Consequences of Popular Support in Counterinsurgency Missions." *Small Wars and Insurgencies* 22, no. 1 (March 2011).
Keegan, John. *The First World War.* New York: Vintage, 1998.
———. *World Armies.* 2nd ed.. Detroit, Mich.: Gale Research, 1979.
Keeley, Lawrence. *War Before Civilization: The Myth of the Peaceful Savage.* New York: Oxford University Press, 1997.
Kennedy, Paul. *The Rise and Fall of British Naval Mastery.* New York: Penguin, 2004.
Khong, Yuen Foong. *Korea, Munich, Dien Bien Phu, and the Vietnam Decisions of 1965.* Princeton, N.J.: Princeton University Press, 1992.
Khoury, Philip. *Syria and the French Mandate: The Politics of Arab Nationalism, 1920–1945.* Princeton, N.J.: Princeton University Press, 1987.
Kilcullen, David. *The Accidental Guerrilla: Fighting Small Wars in the Midst of a Big One.* Oxford: Oxford University Press, 2009.
Kilson, Martin. *Political Change in a West African State: A Study of the Modernization Process in Sierra Leone.* Cambridge, Mass.: Harvard University Press, 1966.
Knight, Ian. *The Anatomy of the Zulu Army: From Shaka to Cetshwayo, 1818–1879.* London: Greenhill Books, 1995.
Kolmas, Josef. *Tibet and Imperial China: A Survey of Sino-Tibetan Relations up to the End of the Manchu Dynasty in 1912.* Canberra: Australian National University, 1967.
Komer, R. W. *The Malayan Emergency in Retrospect: Organization of a Successful Counterinsurgency Effort.* Santa Monica, Calif.: RAND, 1972.
Koon, Heng Pek. *Chinese Politics in Malaysia: A History of the Malaysian Chinese Association.* Singapore: Oxford University Press, 1988.
Kopytoff, Igor, ed. *The African Frontier: The Reproduction of Traditional African Societies.* Bloomington: Indiana University Press, 1987.
Krepinevich, Andrew. *The Army and Vietnam.* Baltimore: Johns Hopkins University Press, 1981.
Kupchan, Charles. *The Vulnerability of Empire.* Ithaca, N.Y.: Cornell University Press, 1994.
Laitin, David. *Politics, Language, and Thought: The Somali Experience.* Chicago: University of Chicago Press, 1977.
Laitin, David and Said Samatar. *Somalia: Nation in Search of a State.* Boulder, Colo.: Westview, 1987.

Larson, Eric. *Casualties and Consensus: The Historical Role of Casualties in Domestic Support for U.S. Military Operations*. Santa Monica, Calif.: RAND, 1996.

Lawrence, Mark Atwood and Fredrik Logevall, eds. *The First Vietnam War: Colonial Conflict and Cold War Crisis*. Cambridge, Mass.: Harvard University Press, 2007.

———. "Introduction." In Lawrence and Logevall, *First Vietnam War*.

Legassick, M. "Firearms, Horses and Samorian Army Organization, 1870–1898." *Journal of African History* 7 (1966).

Lenin, Vladimir. *Collected Works*. Vol. 31, 4th English ed. Moscow: Progress, 1966.

———. "Guerrilla Warfare." *Proletary* 3, no. 5 (September 30, 1906). In Lenin, *Collected Works*. Moscow: Progress, 1965.

———. *Imperialism: The Highest Stage of Capitalism*. Broadway, NSW, Australia: Resistance Books, 1999.

———. *"Left-Wing" Communism, an Infantile Disorder*. Moscow: Foreign Languages Publishing, n.d.

———. *Selected Works*. Vol. 1: *One Step Forward, Two Steps Back*. New York: International, 1967.

LeVine, Victor. *The Cameroons from Mandate to Independence*. Berkeley: University of California Press, 1964.

Lewis, I. M. *A Modern History of Somalia: Nation and State in the Horn of Africa*. Boulder, Colo.: Westview, 1988.

Ling, Dwight. *Tunisia: From Protectorate to Republic*. Bloomington: Indiana University Press, 1967.

Lischer, Sarah Kenyon. "Security and Displacement in Iraq: Responding to the Forced Migration Crisis." *International Security* 33, no. 2 (Fall 2008).

Lockhart, Greg. *Nation in Arms: The Origin of the People's Army of Vietnam*. Wellington: Allen and Unwin, 1990.

Long, Austin. "The Anbar Awakening." *Survival* 50, no. 2 (March–April 2008).

Longrigg, Stephen. *Iraq, 1900 to 1950: A Political, Social, and Economic History*. London: Oxford University Press, 1953.

Lopes, Carlos. *Guinea Bissau: From Liberation Struggle to Independent Statehood*. Boulder, Colo: Westview, 1987.

Lyall, Jason. "Do Democracies Make Inferior Counterinsurgents? Reassessing Democracy's Impact on War Outcomes and Duration." *International Organization* 64, no. 1 (2010).

Lyall, Jason and Isaiah Wilson. "Rage Against the Machines: Explaining Outcomes in Counterinsurgency Wars." *International Organization* 63 (Winter 2009).

Mack, Andrew. "Why Big Nations Lose Small Wars: Politics of Asymmetric Conflict." *World Politics* 27, no. 2 (January 1975).

Mahnken, Thomas. *Technology and the American Way of War Since 1945*. New York: Columbia University Press, 2008.

Malayan Communist Party. "Past Errors." Party Directive of October 1951. pt. 1. *Times* (London), December 1, 1952.

Mansfield, Edward and Jack Snyder. *Electing to Win: Why Emerging Democracies Go to War*. Cambridge, Mass.: MIT Press, 2005.
Mansour, Peter. *Baghdad at Sunrise: A Brigade Commander's War in Iraq*. New Haven, Conn.: Yale University Press, 2007.
Mao Zedong. *On Coalition Government*. 3rd ed. Beijing: Foreign Languages Press, 1965.
——. *On Guerrilla Warfare*. Mineola, N.Y.: Dover, 2005.
——. "On Protracted War." In *Selected Works*, vol. 2. Honolulu, University of Hawaii Press, May 1958.
——. *Selected Writings of Mao Zedong*. Beijing: Foreign Languages Press, 1972.
Marr, David. "Creating Defense Capacity in Vietnam, 1945–1947." In Lawrence and Logevall, *First Vietnam War*.
Marshall, D. Bruce. *The French Colonial Myth and Constitution-Making in the Fourth Republic*. New Haven, Conn.: Yale University Press, 1973.
Martel, William. *Victory in War: Foundations of Modern Strategy*. 2nd ed. Cambridge: Cambridge University Press, 2011.
Marten, Kimberly. *Warlords: Strong-Arm Brokers in Weak States*. Ithaca, N.Y.: Cornell University Press, 2012.
Martin, B. G. *Muslim Brotherhoods in Nineteenth Century Africa*. Cambridge: Cambridge University Press, 1976.
Maxwell, Kenneth. *The Making of Portuguese Democracy*. Cambridge: Cambridge University Press, 1995.
——. "Portugal and Africa: The Last Empire." In Gifford and Louis, *Decolonization and African Independence*.
McCary, John. "The Anbar Awakening: An Alliance of Incentives." *Washington Quarterly* 32, no. 1 (January 2009).
McKibbin, Ross. *Ideologies of Class: Social Relations in Britain, 1880–1950*. Oxford: Oxford University Press, 1990.
McNeill, Malcolm. *In Pursuit of the "Mad" Mullah: Service and Sport in the Somali Protectorate*. London: C. Arthur Pearson Ltd., 1902.
Mearsheimer, John. "Hollow Victory." *Foreign Policy*, November 2, 2009.
Merom, Gil. *How Democracies Lose Small Wars: State, Society, and Failures of France in Algeria, Israel in Lebanon, and the United States in Vietnam*. Cambridge: Cambridge University Press, 2003.
Meyer, John, John Boli, George Thomas, and Francisco Ramirez. "World Society and the Nation-State." *American Journal of Sociology* 103 (1997).
Migdal, Joel. *Peasant, Politics, and Revolution: Pressures Towards Political and Social Change in the Third World*. Princeton, N.J.: Princeton University Press, 1974.
——. *Strong Societies and Weak States: State-Society Relations and State Capabilities in the Third World*. Princeton, N.J.: Princeton University Press, 1988.
Ministry of Foreign Affairs. *Portuguese Africa: An Introduction*. Lisbon: Ministry of Foreign Affairs, 1973.

Minority Rights Group International. *People Under Threat 2012*. London: Minority Rights Group International, 2012.

M'Leod, John. *A Voyage to Africa: With Some Account of the Manners and Customs of the Dahomian People*. London: John Murray, 1820.

Moon, Bruce. "Long Time Coming: Prospects for Democracy in Iraq." *International Security* 33, no. 4 (Spring 2009).

Mueller, John. *Overblown: How Politicians and the Terrorism Industry Inflate National Security Threats, and Why We Believe Them*. New York: Free Press, 2006.

———. *The Remnants of War*. Ithaca, N.Y.: Cornell University Press, 2004.

Muller, C. F. J., ed. *Five Hundred Years: A History of South Africa*. Pretoria, South Africa: Academica, 1969.

Munholland, John. *The Emergence of the Colonial Military in France, 1880–1905*. PhD diss., Princeton University, 1964.

Murray, Williamson and Major General Robert Scales. *The Iraq War: A Military History*. Cambridge, Mass.: Harvard University Press, 2003.

Nagl, John. *Learning to Eat Soup with a Knife: Counterinsurgency Lessons from Malaya and Vietnam*. Chicago: University of Chicago Press, 2005.

Napier, Sir W. *The Conquest of Scinde*. Vol. 2. London: T. & W. Boone, 1844.

Nguyen, Lien-Hang. "Vietnamese Historians and the First Indochina War." In Lawrence and Logevall, *First Vietnam War*.

Nkrumah, Kwame. *Neo-colonialism: The Last Stage of Imperialism*. New York: International, 1965.

Norris, J. A. *The First Afghan War, 1838–1842*. Cambridge: Cambridge University Press, 1967.

North, Douglass. *Institutions, Institutional Change, and Economic Performance*. New York: Cambridge University Press, 1990.

Obichere, Boniface. *West African States and European Expansion: Dahomey-Niger Hinterland, 1885–1898*. New Haven, Conn.: Yale University Press, 1971.

O'Hanlon, Michael. *Defense Strategy for the Post-Saddam Era*. Washington, D.C.: Brookings Institution Press, 2005.

O'Leary, Brendan. *How to Get Out of Iraq with Integrity*. Philadelphia: University of Pennsylvania Press, 2009.

Oloruntimehin, B. Olatunji. "Senegambia—Mahmadou Lamine." In Crowder, *West African Resistance*.

Ongkili, James. *Nation-building in Malaysia 1946–1974*. Singapore: Oxford University Press, 1985.

Ooi, Keat Gin. *Southeast Asia: A Historical Encyclopedia, from Angkor Wat to East Timor*. Oxford: ABC-CLIO, 2004.

Paine, S. C. M. *The Sino-Japanese War of 1894–1895: Perceptions, Power, and Primacy*. Cambridge: Cambridge University Press, 2003.

Palmer, J. A. B. *The Mutiny Outbreak at Meerut in 1857*. Cambridge: Cambridge University Press, 1966.

Panikkar, K. N. *Against Lord and State: Religion and Peasant Uprisings in Malabar, 1836–1921*. Delhi: Oxford University Press, 1989.

Pape, Robert. *Dying to Win: The Strategic Logic of Suicide Terrorism*. New York: Random House, 2005.

Paul, T. V. *Asymmetric Conflicts: War Initiation by Weaker Powers*. Cambridge: Cambridge University Press, 1994.

———. "Complex Deterrence: An Introduction." In Paul, Morgan, and Wirtz, *Complex Deterrence*.

Paul, T. V., Patrick Morgan, and James Wirtz, eds. *Complex Deterrence: Strategy in the Global Age*. Chicago: University of Chicago Press, 2009.

Pellegrin, A. "A Century of Tunisian History." In *Tunisia 54*. New York: Negro Universities Press, 1954.

Pentagon Papers. Gravel Edition. "US Involvement in the France-Viet Minh War, 1950–1954." Boston: Beacon, 1971.

Perez, Louis, Jr. *Cuba: Between Reform and Revolution*. 3rd ed. New York: Oxford University Press, 2006.

Perkin, Harold James. *Rise of Professional Society: England Since 1880*. London: Taylor & Francis, 1989.

Persell, Stuart Michael. *The French Colonial Lobby, 1889–1938*. Stanford, Calif.: Hoover Institution Press, 1983.

Pike, Douglas. *War, Peace, and the Viet Cong*. Cambridge, Mass.: MIT Press, 1969.

Pillar, Paul. *Terrorism and U.S. Foreign Policy*. Washington, D.C.: Brookings Institution Press, 2001.

Pirnie, Bruce and Edward O'Connell. *Counterinsurgency in Iraq (2003–2006)*. Santa Monica, Calif.: RAND, 2007.

Polanyi, Karl. *Dahomey and the Slave Trade: An Analysis of an Archaic Economy*. Seattle: University of Washington Press, 1966.

Porath, Yehoshua. *The Palestinian Arab National Movement: From Riots to Rebellion*. Vol. 2. London: Frank Cass, 1977.

Porch, Douglas. "Bugeaud, Gallieni, Lyautey: The Development of French Colonial Warfare." In Peter Paret, Gordon Craig, and Felix Gilbert, eds., *Makers of Modern Strategy: From Machiavelli to the Nuclear Age*. Oxford: Oxford University Press, 1986.

———. *The Portuguese Armed Forces and the Revolution*. Stanford, Calif.: Hoover Institution Press, 1977.

Potholm, Christian. *The Theory and Practice of African Politics*. Englewood Cliffs, N.J.: Prentice Hall, 1979.

Poullada, Leon. *Reform and Rebellion in Afghanistan, 1919–1929*. Ithaca, N.Y.: Cornell University Press, 1973.

Powers, John. *History as Propaganda: Tibetan Exiles Versus the People's Republic of China*. Oxford: Oxford University Press, 2004.

Prados, John. "Assessing Dien Bien Phu." In Lawrence and Logevall, *First Vietnam War*.

Prasad, S. N. *Operation Polo: The Police Action in Hyderabad, 1948*. New Delhi: Ministry of Defence, 1972.

PREM 8/1406/2, MAL C(50)12 [May 12, 1950]. In Stockwell, *Documents*, vol. 2.

Provence, Michael. *The Great Syrian Revolt and the Rise of Arab Nationalism*. Austin: University of Texas Press, 2005.

Pye, Lucien. *Guerrilla Communism in Malaya*. Princeton, N.J.: Princeton University Press, 1956.

Ralvo, Hilde, Nils Gleditsch, and Han Dorussen. "Colonial War and the Democratic Peace." *Journal of Conflict Resolution* 47, no. 4 (August 2003).

Record, Jeffrey. *Beating Goliath: Why Insurgencies Win*. Washington, D.C.: Potomac Books, 2007.

———. *Dark Victory: America's Second War Against Iraq*. Annapolis, Md.: Naval Institute Press, 2004.

Redding, Sean. *Sorcery and Sovereignty: Taxation, Power, and Rebellion in South Africa, 1880–1963*. Athens: Ohio University Press, 2006.

Reiter, Dan. *Crucible of Beliefs: Learning, Alliances, and World Wars*. Ithaca, N.Y.: Cornell University Press, 1996.

———. *How Wars End*. Princeton, N.J.: Princeton University Press, 2009.

Reiter, Dan and Curtis Meek. "Determinants of Military Strategy, 1903–1994: A Quantitative Empirical Test." *International Studies Quarterly* 43, no. 2 (June 1999).

Reiter, Dan and Allan Stam. *Democracies at War*. Princeton, N.J.: Princeton University Press, 2002.

Renda, Mary. *Taking Haiti: Military Occupation and the Culture of U.S. Imperialism, 1915–1940*. Chapel Hill: University of North Carolina Press, 2001.

Rennie, Surgeon. *Bhotan and the Story of the Dooar War*. London: John Murray, 1866.

Reno, William. *Warfare in Independent Africa*. Cambridge: Cambridge University Press, 2011.

Riedel, Bruce. *The Search for Al Qaeda: Its Leadership, Ideology, and Future*. Washington, D.C.: Brookings Institution Press, 2008.

Roberts, Robert. *The Classic Slum: Salford Life in the First Quarter of the Century*. Harmondsworth: Penguin, 1990.

Ropp, Theodore. *War in the Modern World*. Baltimore: Johns Hopkins University Press, 2000.

Ross, David. "Dahomey." In Michael Crowder, ed., *West African Resistance: The Military Response to Colonial Occupation*. London: Hutchinson University Press.

Rotberg, Robert. *Haiti: The Politics of Squalor*. Boston: Houghton Mifflin, 1971.

Rothstein, Hy and John Arquilla, eds. *Afghan Endgames: Strategy and Policy Choices for America's Longest War*. Washington, D.C.: Georgetown University Press, 2012.

———. "Understanding the Afghan Challenge." In Rothstein and Arquilla, *Afghan Endgames*.

Rudebeck, Lars. *Guinea-Bissau: A Study of Political Mobilization*. Uppsala: Scandinavian Institute of African Studies, 1974.

———. "Political Mobilization for Development in Guinea-Bissau." *Journal of Modern African Studies* 10, no. 1 (May 1972).

Russell, James. *Innovation, Transformation, and War: Counterinsurgency Operations in Anbar and Ninewa Provinces, Iraq, 2005–2007*. Stanford, Calif.: Stanford University Press, 2011.

Sagarin, Rafael. "Adapt or Die: What Charles Darwin Can Teach Tom Ridge About Homeland Security." *Foreign Policy*, September–October 2003.

Sagarin, Raphael and Terence Taylor, eds. *Natural Security: A Darwinian Approach to a Dangerous World*. Berkeley: University of California Press, 2008.

Sageman, Marc. *Leaderless Jihad: Terror Networks in the Twenty-First Century*. Philadelphia: University of Pennsylvania Press, 2008.

———. "A Strategy for Fighting International Islamist Terrorists." *Annals of the American Academy of Political and Social Science* 618 (July 2008).

Salehyan, Idean. *Rebels Without Borders: Transnational Insurgencies in World Politics*. Ithaca, N.Y.: Cornell University Press, 2009.

Samatar, Ahmed. *Socialist Somalia: Rhetoric and Reality*. London: Zed Books, 1988.

Samatar, Said. *Oral Poetry and Somali Nationalism: The Case of Sayyid Mahammad 'Abdille Hasan*. Cambridge: Cambridge University Press, 1982.

Sarkees, Meredith Reid. "The Correlates of War Data on War: An Update to 1997." *Conflict Management and Peace Science* 18, no. 1 (2000).

Sarkees, Meredith Reid and Frank Wayman. *Resort to War: 1816–2007*. Washington, D.C.: CQ Press, 2010.

Scheina, Robert. *Latin America's Wars: The Age of the Caudillo, 1791–1899*. Washington, D.C.: Brassey's, 2003.

Schultz, Kenneth. *Democracy and Coercive Diplomacy*. New York: Cambridge University Press, 2001.

Sechser, Todd and Elizabeth Saunders. "The Army You Have: The Determinants of Military Mechanization, 1979–2001." *International Studies Quarterly* 54, no. 2 (June 2010).

Serena, Chad. *A Revolution in Military Adaptation: The US Army in the Iraq War*. Washington, D.C.: Georgetown University Press, 2011.

Sheik-Abdi, Abdi. *Divine Madness: Mohammed Abdulle Hassan (1856–1920)*. London: Zed Books, 1993.

Shennan, Margaret. *Out in the Midday Sun: The British in Malaya 1880–1960*. London: John Murray, 2000.

Shillington, Kevin, ed. *Encyclopedia of African History*. Vol. 1. New York: Fitzroy Dearborn, 2005.

———, ed. *Encyclopedia of African History*. Vol. 3. New York: Fitzroy Dearborn, 2005.

Shimko, Keith. *The Iraq Wars and America's Military Revolution*. New York: Cambridge University Press, 2010.

Shinar, Pessah. "Abd al Qadir and Abd al Krim: Religious Influences on Their Thought and Action." *Asian and African Studies* 1 (1965).

Short, Anthony. *The Communist Insurgency in Malaya 1948–1960*. London, Frederick Muller, 1975.

Singer, David and Melvin Small. *Resort to Arms: International and Civil War, 1816–1980*. Beverly Hills, Calif.: Sage, 1982.

Sinno, Abdulkader. *Organizations at War in Afghanistan and Beyond*. Ithaca, N.Y.: Cornell University Press, 2008.

Skocpol, Theda. *States and Social Revolutions: A Comparative Analysis of France, Russia, and China*. Cambridge: Cambridge University Press, 1979.

Skrine, Francis and Edward Ross. *The Heart of Asia: A History of Russian Turkestan and the Central Asian Khanates from the Earliest Times*. London: Methuen, 1899.

Smith, Alastair. "Fighting Battles, Winning Wars." *Journal of Conflict Resolution* 42, no. 3 (June 1998).

Smith, Rupert. *The Utility of Force: The Art of War in the Modern World*. London: Knopf, 2007.

Smith, Tony. "The French Colonial Consensus and People's War, 1946–58." *Journal of Contemporary History* 9 (1974).

Smithers, A. J. *The Kaffir Wars 1779–1877*. London: Leo Cooper, 1973.

Staniland, Paul. "Between a Rock and a Hard Place: Insurgent Fratricide, Ethnic Defection, and the Rise of Pro-State Paramilitaries." *Journal of Conflict Resolution* 56, no. 1 (2012).

——. "Organizing Insurgency: Networks, Resources, and Rebellion in South Asia." *International Security* 37, no. 1 (Summer 2012).

Stanley, Elizabeth and John Sawyer. "The Equifinality of War Termination: Multiple Paths to Ending War." *Journal of Conflict Resolution* 53, no. 5 (October 2009).

Stein, Janice Gross. "Rational Deterrence Against 'Irrational' Adversaries? No Common Knowledge." In Paul, Morgan, and Wirtz, *Complex Deterrence*.

Stephenson, Robert. *Population and Political Systems in Tropical Africa*. New York: Columbia University Press, 1968.

Stiglitz, Joseph and Linda Bilmes. *Three Trillion Dollar War: The True Conflict of the Iraq Conflict*. Rochester, N.Y.: Boydell & Brewer, 2008.

Stockwell, A. J. "'A Widespread and Long-Concocted Plot to Overthrow Government in Malaya?' The Origins of the Malayan Emergency." *Journal of Imperial and Commonwealth History* 21, no. 3 (1993).

Stubbs, Richard. "Guerrilla Strategies and British Counterinsurgency Strategies of the 1950s and 1960s: Why Was the War Lost?" In C. C. Chin and Karl Hack, eds., *Dialogues with Chin Peng: New Light on the Malayan Communist Party*. Singapore: Singapore University Press, 2004.

———. *Hearts and Minds in Guerrilla Warfare: The Malayan Emergency, 1948–60.* New York: Oxford University Press, 1990.

Stumm, Hugo. *Russia in Central Asia: Historical Sketch of Russia's Progress in the East up to 1873, and of the Incidents Which Led to the Campaign Against Khiva.* London: Harrison and Sons, 1865.

Sugar, Peter. *Industrialization of Bosnia-Herzegovina, 1875–1918.* Seattle: University of Washington Press, 1963.

Sullivan, Patricia. *Who Wins? Predicting Strategic Success and Failure in Armed Conflict.* Oxford: Oxford University Press, 2012.

Sunderland, Riley. *Antiguerrilla Intelligence in Malaya.* Santa Monica, Calif.: RAND, 1964.

———. *Army Operations in Malaya, 1947–1960.* Santa Monica, Calif.: RAND, 1964.

———. *Organizing Counterinsurgency in Malaya, 1947–1960.* Santa Monica, Calif.: RAND Corporation, 1964.

———. *Winning the Hearts and Minds of the People, Malaya: 1948–1960.* Santa Monica, Calif.: RAND, 1964.

Taber, Robert. *War of the Flea: The Classic Study of Guerrilla Warfare.* Washington, D.C.: Brassey's, 2002.

Tanham, George. *Communist Revolutionary Warfare: From the Vietminh to the Viet Cong.* Westport, Conn.: Praeger, 2006.

Taw, Jennifer Morrison. *Mission Revolution: The U.S. Military and Stability Operations.* New York: Columbia University Press, 2012.

Taylor, Jean Gelman. *Indonesia: Peoples and Histories.* New Haven, Conn.: Yale University Press, 2003.

Thai, Hoang Van. "Dien Bien Phu: Why and How?" In George Katsiaficas, ed., *Vietnam Documents: American and Vietnamese Views of the War.* Armonk, N.Y.: M. E. Sharpe, 1992.

Thompson, Leonard. "Cooperation and Conflict: The Zulu Kingdom and Natal." In Monica Wilson and Leonard Thompson, eds., *The Oxford History of South Africa*, vol. 1. Oxford: Oxford University Press, 1969.

———. *A History of South Africa.* New Haven, Conn.: Yale University Press, 1996.

Thompson, Robert. *Defeating Communist Insurgency: Experiences from Malaya and Vietnam.* New York: Praeger, 1966.

Thomson, Janice. *Mercenaries, Pirates, and Sovereigns: State-Building and Extraterritorial Violence in Early Modern Europe.* Princeton, N.J.: Princeton University Press, 1996.

Tilly, Charles. *From Mobilization to Revolution.* New York: McGraw-Hill, 1978.

Tilman, Robert. "The Non-Lessons of the Malayan Emergency." *Military Review* 46 (December 1966).

Times, London, December 1, 1952.

Toft, Monica Duffy. "Ending Civil Wars: A Case for Rebel Victory?" *International Security* 34, no. 4 (Spring 2010).

Touval, Saadia. *Somali Nationalism: International Politics and the Drive for Unity in the Horn of Africa*. Cambridge, Mass.: Harvard University Press, 1963.

Trinquier, Roger. *Modern Warfare: A French View of Counterinsurgency*. New York: Praeger, 1961.

Turney-High, Harry. *Primitive War: Its Practice and Concepts*. Columbia: University of South Carolina Press, 1971.

Turney-High, Harry. *The Military: The Theory of Land Warfare as Behavioral Science*. West Hanover, Mass.: Christopher Publishing House, 1981.

Ucko, David. "Countering Insurgents Through Distributed Operations: Insights from Malaya 1948–1960." *Journal of Strategic Studies* 30, no. 1 (February 2007).

———. *The New Counterinsurgency Era: Transforming the U.S. Military for Modern Wars*. Washington, D.C.: Georgetown University Press, 2009.

U.S. Army and Marine Corps. *Field Manual 3-24*. Washington, D.C.: U.S. Department of the Army, 2006.

U.S. Department of Defense. "Measuring Stability and Security in Iraq" (July 2009).

Van Creveld, Martin. *The Changing Face of War: Lessons of Combat from the Marne to Iraq*. New York: Ballantine Books, 2006.

———. *Supplying War: Logistics from Wallenstein to Patton*. Cambridge: Cambridge University Press, 1977.

Vedantam, Shankar. "Don't Send a Lion to Catch a Mouse." *Washington Post*, March 5, 2007.

Vietnamese Ministry of National Defense. *Ho Chi Minh Thought on the Military*. Hanoi: Institute of Military History, 2008.

Vimalananda, Tennakoon. *The Great Rebellion of 1818: The Story of the First War of Independence and Betrayal of the Nation*. Colombo: M. D. Gunasena, 1970.

Waltz, Kenneth. "The Politics of Peace." *International Studies Quarterly* 11, no. 3 (September 1967).

———. *Theory of International Politics*. New York: McGraw-Hill, 1979.

Weeks, Jessica. "Strongmen and Straw Men: Authoritarian Regimes and the Initiation of International Conflict." *American Political Science Review* 106, no. 2 (May 2012).

Weigley, Russell. *The American Way of War: A History of United States Military Strategy and Policy*. New York: Macmillan, 1973.

Weinstein, Jeremy. *Inside Rebellion: The Politics of Insurgent Violence*. Cambridge: Cambridge University Press, 2007.

Wendt, Alexander and Michael Barnett. "Dependent State Formation and Third World Militarization." *Review of International Studies* 19, no. 4 (1993).

Weyand, Fred and Harry Summers. "Vietnam Myths and American Realities." In *Commanders Call*. Fort Leavenworth, Kans.: U.S. Army Command and General Staff College, July–August 1976.

Wilks, Ivor. *Asante in the Nineteenth Century: The Structure and Evolution of a Political Order*. London: Cambridge University Press, 1975.

Williams, Phil. *Criminals, Militias, and Insurgents: Organized Crime in Iraq.* Carlisle, Pa.: Strategies Studies Institute, 2009.

Woddis, Jack. *New Theories of Revolution.* New York: International, 1972.

Woods, Kevin, James Lacy, and Williamson Murray. "Saddam's Delusions: The View from the Inside." *Foreign Affairs* 85, no. 3 (May–June 2006).

Woolman, David. *Rebels in the Rif, Abd el Krim and the Rif Rebellion.* Stanford, Calif.: Stanford University Press, 1968.

Wright, Donald and Timothy Reese. *On Point II: Transition to the New Campaign.* Fort Leavenworth, Kans.: U.S. Army Combined Arms Center, 2008.

Wright, Lawrence. "The Rebellion Within: An Al Qaeda Mastermind Questions Terrorism." *New Yorker,* June 2, 2008.

Wright, Quincy. *A Study of War.* Vol. 1. Chicago: University of Chicago Press, 1942.

Wulf, Herbert. "Dependent Militarism in the Periphery and Possible Alternative Concepts." In Stephanie Neuman and Robert Harkavy, eds., *Arms Transfers in the Modern World.* New York, St. Martin's, 1979.

Zhai, Qiang. *China and the Vietnam Wars, 1950–1975.* Chapel Hill: University of North Carolina Press, 2000.

———. "Transplanting the Chinese Model: Chinese Military Advisers and the First Vietnam War, 1950–1954." *Journal of Military History* 57 (October 1993).

INDEX

Afghanistan, 23, 178, 267n1, 268n5, 268n12; Britain and, 36, 61, 193, 195, 197, 206, 218, 232, 246n46; insurgents in, 176, 184; Karzai and, 23, 180, 184; Mohmand War and, 226; Pakistan and, 5; people and, 183; primitive model and, 180–81; Soviet Union and, 33, 171, 195, 199, 220, 240; Taliban in, 177. *See also* Pakistan; Al-Qaeda; Taliban
Afghan model, 180
Africanization, 132, 142–43
African Party for the Independence of Guinea and Cape Verde (PAIGC), 22, 131, 138–39, 147; Cabral and, 134; conventional war and, 143–45; evolution of, 132–33, 135–40, 145; hearts and minds and, 22, 132, 135, 140–43; Portugal and, 136–37, 141; theories of asymmetric war and, 147–49; victory of, 132–33, 136–37, 144, 146
Algeria, 8, 9, 148, 216, 243n4; Fanon and, 39; French wars in, 30, 33, 34, 39, 76, 166, 171, 193, 195, 198, 206–7, 211, 227, 238; al-Qaeda in, 173; sequencing theory and, 61
Anbar Awakening. *See* Awakening movement
Arab Spring, 186
Armed Forces of the People (FARP, of Guinea-Bissau), 137, 144–45, 148
Arreguín-Toft, Ivan, 10, 14–15, 172, 174, 178, 232, 246. *See also* Strategic interaction
Asymmetric war: conditions of, 53; definition of, 14; extra-systemic war as, 6; fighting in, 4, 174; history of, 174; Indochina War and, 152; interstate, civil, and, 2; knowledge about, 17; significance of, 178; theories of, 2, 13–14, 21, 64, 75, 77, 80, 91–92, 112, 116, 127, 129, 147, 151, 166, 170, 172, 178, 185; winning in, 2, 7

Awakening movement, 95–96, 102, 105, 107–9, 176

Balance of resolve, 14, 91, 128, 147, 151, 166. *See also* Mack, Andrew
Bao Dai, 160–61, 163, 176. *See also* Indochina War
Béhanzin, 66–68, 74–76. *See also* Dahomey
Biddle, Stephen, 102, 104, 257n, 258n53, 268n12
Bin Laden, Osama, 23, 181, 182
Boer War, 38, 122, 206, 209, 261n52; First Boer War, 36, 195, 219; Second Boer War, 36, 120, 129, 196, 227
Briggs Plan, 85–86, 91, 92, 142. *See also* Resettlement
Britain, 32, 64–65, 161, 244n14, 246n46, 253nn2–4, 253n14, 254n24, 254n36, 254n40, 255n50, 259n4, 259n9, 260n12, 260n23, 260n25, 260n28, 260n31, 260n39, 261n54, 261nn56–57; colonialism of, 8; extrasystemic wars and, 33, 36, 67, 68, 193–98, 202–37 passim; in Malayan Emergency, 21, 33, 37, 79–93, 171, 175; in Somalia, 22, 45–46, 115–30, 171, 175
Burmese Wars: First Burmese War, 36, 193, 203; Second Burmese War, 36, 194, 210; Third Burmese War, 36, 195, 220
Bush administration, 96, 100, 103, 107, 114

Cabral, Amílcar: assassination of, 145; conventional war and, 144; guerrilla war and, 141, 144–45; Portugal and, 136; sequencing ideas of, 38–39; 47, 134–37; state-building by, 139–40, 148–49
Cape Verde, 22, 57, 131–32, 134–35, 138, 140
China, 5, 24, 30, 35, 36, 173, 197, 204, 211, 215, 225–33 passim, 239, 244n14, 248n2, 250n42, 255n49, 266nn29–30, 266nn32–33, 266n38, 266n40, 267n55,

China (continued)
  267n64; Anglo-Somali War, 122; Indochina War and, 37, 151–55, 159, 161, 163, 165, 168; Malayan Emergency and, 37, 80–89, 171, 198; Mao and, 37, 39; Portuguese-Guinean War and, 140, 148
Chin Peng, 81, 83, 86–87
Civil war, 1, 4, 132, 170, 185, 186, 243n5, 244n11, 248n10, 255n1, 268nn27–28, 269n29; in China, 37; in Colombia, 38; extrasystemic war and, 3, 5; in Iraq, 94, 97, 104
Coalition of the willing, 95
Cold War, 6, 10; end of, 4, 173, 186; Indochina War and, 168; United States and, 10
Colonialism, 31, 39, 216, 219–29 passim, 247n20, 248n36, 252n33, 259n3; of Britain, 81, 116; of France, 64, 151; of Portugal, 135–38, 141
Conventional model, 12, 18, 48, 49–52, 54, 55; in Dahomean War, 21, 63–66, 72, 77, 171; examples of, 201–37; failure of 49, 62; in Iraq War, 101
Conventional strategy, 15, 49–50, 52, 54–55; in Dahomean War, 67, 77; in Iraqi War, 112; in Malayan Emergency, 92; in Portuguese-Guinean War, 148; preference for, 35
Correlates of War (COW), 2, 207, 223, 224
Counterinsurgency (COIN): of Britain, 32, 80–88 passim, 175; examples of, 233; literature of, 13, 16, 30, 79, 243n4, 244n20, 245n28, 246n47, 247nn17–18, 247nn24–25, 248n5, 250n40, 253nn1–2, 253nn4–5, 254n22, 255n48, 255n53, 256n4, 257n33, 257nn37–38, 258n40, 258n49, 258n52, 258n59, 259n67, 259n72, 263n47, 264nn48–49, 264n63, 268n4, 269n30; of Portugal, 136, 142–43; strategy of, 42, 54, 105, 107, 113; of United States, 6, 21, 34, 95, 100, 102, 104, 107–9, 112–14, 178–82
Counterterrorism, 6, 177, 245n38, 246n3, 253n12
Covert operation, 121

Dahomey, 18, 171, 196, 222, 251n6, 251n12, 251n15, 251nn17–19, 252n29, 252n34, 252n42; conventional model and, 21, 50, 61, 63–78
Darwinism, 25–27
Degenerative model, 12, 20, 48, 54–55, 187, 201; Aceh War as, 195, 216; Anglo-Somali War as, 197; Bosnia as, 195, 217; failure of, 49, 55, 60–62; Iraq War as, 21, 94–95, 99, 109, 112, 114, 171, 199, 241; Maji-Maji Revolt, 197, 229; Maori War as, 194, 212; Moplah Rebellion as, 197, 234; Philippine War as, 196, 227; Saya San's Rebellion as, 197, 235; Second Boer War as, 36, 196, 227; Second Mahdist War as, 196; Second Senegalese War as, 36, 196, 221; South West African Revolt as, 197, 229
De Lattre line, 162–63
Democracy, 10, 78, 245n32, 262n8, 264n68; of Britain, 93, 129–30; in Iraq, 97, 111, 256n9; revolutionary democracy in Guinea, 140; in small wars, 14; weakness of 11, 15–16, 243n6, 245n30, 245n38, 246n47
Dervish, 22, 117, 128–29, 225–26; conventional war and, 124–27; guerrilla war and, 119, 121–23
Dien Bien Phu, 23, 150, 154, 163–65, 168, 248n28, 265nn5–6, 265n11, 267n49

Evolution, 63–64, 169, 175–76, 181, 187, 189, 204, 239, 246n, 251n, 262n; in Anglo-Somali War, 115, 118, 123, 127; in Indochina War, 172; in Iraqi War, 100; in Malayan Emergency, 87; in Portuguese-Guinean War, 132–33, 136, 148; in sequencing theory, 12–13, 15, 17, 22, 28–29, 43, 45, 47, 55–56, 60; theories of, 20, 25–27, 32, 172, 188
External support, 14, 16, 41–42, 79, 113, 148, 168, 178–79. *See also* Record, Jeffrey

Fanon, Frantz, 38–39, 150. *See also* Algeria
Field Manual, 3–24, 105, 258n47
France, 30, 31, 34, 37, 39, 138, 148, 171, 243n4, 244n14, 250n42, 251n5, 251n7, 266n39, 267n59; colonialism of, 8; in Dahomey, 21, 63–78; and extrasystemic wars and, 33, 36, 193–99, 206, 208, 211, 214, 215, 219, 220, 222, 232, 236, 238; in Indochina, 23, 33, 150–54, 156–68 passim, 176; in Somalia, 116

Giap, Vo Nguyen, 153–54, 157–58; Ho Chi Minh and, 155, 157, 168; influence of, 37; Mao and, 37; military operations of, 162–63, 165–66; sequencing theory and, 161

Guerrilla strategy, 10, 15, 49–51, 62, 178; in Anglo-Somali War, 22, 123; in Dahomean War, 63, 75, 77; definition of, 7; in degenerative model, 54–56, 112–13; examples of, 227, 233, 235, 239; in Indochina War, 155, 168; in Malayan Emergency, 37, 79, 83, 92, 171; in Portuguese-Guinean War, 144, 148; in primitive model, 52–54

Guevara, Che, 37–39

Gulf war, 100–101, 113

Haqqani network, 23, 176–77, 181

Hearts and minds: in Afghanistan, 182; in Dahomey, 74; in Malaya, 37, 80, 82–83, 86–87; PAIGC and, 22, 132, 137; United States and, 176; Vietminh and, 46, 176. *See also* Local support

Ho Chi Minh: in Indochina War, 150–60 passim, 166, 265n3, 265n8, 265n10, 265n12, 265n14, 266n16, 266n20, 266n26, 266n31, 266n35, 266n42, 266n45, 267n56, 267n58; learning of, 47; Mao's influence on, 37

Huntington, Samuel, 15, 41

Hussein, Saddam, 22, 99, 100–102, 104

Hybrid war, 3, 5–6

Imperialism, 3, 226, 249n12, 249n15; of Britain, 116; of France, 159, 250n2, 251n4, 267n59; Lenin and, 28, 39, 247n10; of Portugal, 133; of United States, 231; war against, 57, 248n36

Improvised explosive device (IED), 100, 103

Independence, 141, 158, 193, 202, 203, 120, 229, 239

Indochina War, 18, 33, 35, 37, 46, 266n, 285n; progressive model in, 23, 150–52, 172, 176. *See also under* Giap, Vo Nguyen; Ho Chi Minh; Maoist model

Indochinese Communist Party (ICP), 23, 37; guerrilla war and, 155; state-building and, 158

Indonesia, 9, 31, 151; Aceh in, 216; al-Qaeda in, 173; war with East Timor, 53, 199, 239–40; war with Netherlands, 33, 61, 193, 196, 198, 204, 223, 236

Innovation, 26–27, 103, 133, 258n; in strategy, 28–29, 83, 85, 89, 95, 107; in tactics, 4, 19, 32, 188; in weapons, 43. *See also* Learning

Intelligence: counterinsurgency and, 10, 16, 41–42, 51, 179; by France, 71; in Guinea-Bissau, 133, 143; in Indochina War, 152, 157, 164; in Malaya, 82, 84–85, 89–90, 253n14; of United States, 96, 103, 113–14, 182, 257n30

International Security Assistance Force, 5, 23, 170, 171, 177, 180

Inter-Services Intelligence (of Pakistan), 179, 184

Interstate war, 2–4, 186; characteristics of, 6; extra-systemic war and, 5, 14

Iran, 102, 176, 179, 184, 232

Iraq, 1, 26, 79, 94–98, 105, 109–10, 112–14, 199, 241; armed forces in, 95, 99–103, 107, 109; Britain and, 197, 232–33; conventional war in, 21; extrasystemic war in, 3, 10, 18–19, 21, 24, 32, 33, 61, 95; government of, 22, 97–99, 105, 109–11; insurgents in, 94, 96, 106, 111, 171, 174, 188; Iran and, 97; people in, 96, 104, 107–9, 176; United States and, 34, 44, 95–100, 108, 181

Japan, 29, 37, 81, 151, 196, 197, 225, 230

Kaffir Wars: Eighth Kaffir War, 36, 209; Ninth Kaffir War, 36, 217; Seventh Kaffir War, 36, 208

Karzai, Hamid, 23, 180, 184

Kissinger, Henry, 7, 23, 183

Lamarckism, 26–27

Learning, 4, 26, 34, 247n24, 247n28, 253n1, 253n17, 254n29, 255n41, 259n72; in Indochina War, 152; in Malayan Emergency, 80, 87; of sequencing theory, 34–35, 38. *See also* Innovation; Nagl, John

Lenin, Vladimir, 20, 138, 247n, 250n, 263n, 283n; sequencing ideas of, 28–29, 35, 37–39, 139, 152, 172, 237; state-building and, 43, 53, 157–58

Liberation Front of Mozambique (FRELIMO), 132, 239

Libya, 1, 174, 187, 197, 233
Local support, 45, 57, 81, 87, 90, 107, 142, 208. *See also* Hearts and minds
Lyall, Jason, 2, 10, 16–17, 172. *See also* Mechanization; Wilson, Isaiah

Mack, Andrew, 14. *See also* Balance of resolve
Mahdist War, 18, 196, 221; First Mahdist War, 36, 195, 219–20; Second Mahdist War, 36, 120, 196, 225
Malayan Communist Party (MCP), 37, 79–93 passim
Malayan Emergency, 18, 33, 35, 142, 237, 253n8, 253nn11–13, 254n18, 254n32, 254n34, 254n38, 255n44, 255n47, 255n57; primitive model in, 21, 61, 79, 87, 91, 171, 198
Malayanization, 90
Maliki, Nuri al, 111
Mao Zedong, 244n13, 246n1, 247n14, 247n16, 248n4, 250n39, 265n13; Malayan Emergency and, 37, 41; in Indochina War, 153–54, 158, 166; people's war and, 29, 35, 40, 51, 54, 58; sequencing ideas and, 20, 25, 29, 37–39, 53, 83, 151–53, 172, 237. *See also* Maoist model
Maoist model, 12, 13, 20, 48, 56–58, 60, 201; Indochina War as, 23, 153; Indonesian War as, 198, 236; Mau Mau Rebellion as, 198, 237; Namibian War as, 199, 240; people's war and, 35, 38; Portuguese-Guinean War as, 22, 131–33, 137, 146, 171, 199, 239; Spanish-Cuban War as, 196, 224; success of, 21, 49, 61–62, 170; Tunisian War as, 198, 237
Martel, William, 7, 98, 177, 182–83, 244n21, 256n2, 256n5, 256n17, 257n22, 268n8, 268n16, 268n21. *See also* Victory
Marxism, 28, 35, 37–38, 138–39, 146, 263n32
Mechanization, 2, 10, 14, 16–17, 113, 172, 179. *See also* Lyall, Jason; Wilson, Isaiah
Merom, Gil, 15–16, 243n, 245n. *See also* Democracy
Min Yuen, 81, 86, 90

Nagl, John, 32, 87, 112. *See also* Innovation; Learning.
National Armed Forces (FAN, of Guinea-Bissau), 137, 145, 148

Nationalism, 174, 232, 235, 239, 248n, 259n, 260n, 262n, 263n
Nation-building. *See* State-building
Navarre, Henri, 163, 167
Nonviolence, 19, 246n
Norm, 32, 178, 258n; against war, 16; of decolonization, 31–32; of self-determination, 20, 31

Obama administration, 182
Omar, Mullah, 170, 183. *See also* Afghanistan; Pakistan; Taliban
Operation Enduring Freedom, 180
Operation Iraqi Freedom, 95, 98, 100–101

PAIGC. *See* African Party for the Independence of Guinea and Cape Verde
Pakistan, 5, 176; Afghanistan and, 23, 179, 184; Balucchi and, 207; al-Qaeda and, 173
Paul, T. V., 2, 174, 175. *See also* Asymmetric war
Political will. *See* Balance of resolve
Popular Movement for the Liberation of Angola (MPLA), 132, 239
Population transfer. *See* Briggs Plan; Resettlement
Portugal, 8, 32, 196, 197, 239, 261n1, 261n3, 264n66, 264n69, 264n72, 264n75, 265n82; in Angola, 33, 199, 239; in Guinea-Bissau, 33, 39, 132–49, 199; in Mozambique, 33, 199, 239. *See also* PAIGC
Premature model, 12, 20, 48–49, 55–56, 60–62; Anglo-Somali War and, 22, 115–16, 118–19, 125–27, 171; Italo-Libyan War and, 197, 233; Syrian War and, 197, 232
Primitive model, 12–13, 52–55, 99; in Afghanistan, 23, 169–71, 176–87 passim; in Anglo-Somali War, 118; examples of, 204–40 passim; failure of, 49, 62; in Malayan Emergency, 79–80, 83, 87
Progressive model, 12, 48, 58–60, 187, 201; Angolan War as, 199, 239; Indochina War as, 23, 150–55, 172, 198, 236; Mozambique as, 199, 239; success of, 21, 49, 61–62, 170

Record, Jeffrey, 16, 113, 148, 168. *See also* External support
Republican Guard (of Iraq), 99–101, 171

Resettlement, 83, 85–86, 233. *See also* Briggs Plan

Revolution in military affairs, 100, 103–4, 114

Qaeda, Al-: in Afghanistan, 5, 23, 171, 178, 182; evolution of, 32; in Iraq, 108; in Pakistan, 173; Taliban and, 5, 169–70, 176–78, 181, 183–84. *See also* Afghanistan; Pakistan; Taliban

Sayyid (in Somalia, also known as the Mullah), 117–28, 259n7, 260n21, 260n24, 260n35, 261n43

Scramble for Africa, 64, 221, 224

September 11 tragedy, 11, 176, 178, 181

Shia, 97, 102, 107

Small war: definition of, 5; democracy and, 2, 15, 16; fighting in, 4–5, 53, 92, 173; insurgency and, 1; in Somalia, 120; United States and, 34; victory of, 10

Somalia, 3, 5, 61, 186, 259n, 260n; in Anglo-Somali War, 22, 115–30, 175; United States and, 10, 33, 199, 240–41, 246

Somaliland Field Force, 121

Sons of Iraq. *See* Awakening movement

Soviet Union, 1, 140, 159; in Afghanistan, 9, 33, 199, 240

Staniland, Paul, 16, 148, 245n40, 258n58, 265n80

State-building, 244n24, 249n22; in Afghanistan, 170; in Anglo-Somali War, 126; in Dahomean War, 64, 72; examples of, 235; in Indochina War, 23, 158, 159, 172; in Iraqi War, 99, 110–11; in Malayan Emergency, 88; in sequencing theory, 43, 46–47, 57, 62, 169, 172, 187

Strategic assessment, 27, 247n9

Strategic interaction, 2, 14–15, 77, 91, 92, 112, 128, 148, 151, 167, 172, 178. *See also* Arreguin-Toft, Ivan

Sunni, 95–97, 102–9 passim, 112–13, 171, 188, 199

Surge, the: in Afghanistan, 179, 182; in Iraq, 34, 95, 107–9, 112, 114

Syria, 1, 61, 173–74, 187, 197–98, 232–33, 234–35

Taliban: al-Qaeda and, 5, 169–70, 176–78, 181, 183–84; in Afghanistan, 23, 184; fall of, 181–83; in Pakistan, 184; resurgence of, 181–82, 184; United States and, 177–78, 185, 188

Templer, Gerald, 83, 85, 87–89

Terrorism, 244n25, 245n41, 246n3, 258n42, 258n61, 268nn6–7, 268n20; in Afghanistan, 171, 177, 183, 202; definition of, 5; examples of, 6; insurgency and, 6, 8; in Iraq, 94, 105, 111; in Malayan Emergency, 82; in Portuguese-Guinean War, 142; and sequencing theory, 25, 30, 185; as strategic innovation, 4

Thompson, Robert, 82

Transformation. *See* Revolution in military affairs

Tunku, Abdul Rahman, 88, 90

Underdog: in asymmetric war, 2, 10, 14, 29; Dahomey as 67, 77; Iraqi insurgents and, 113; PAIGC as, 148; victory of, 16–17; Vietminh as, 150, 172

United Nations, 31, 96, 110, 135, 140, 240, 247n

United States: Afghanistan and, 5, 23, 169–70, 173–83, 185; Britain and, 82; Caco Revolt and, 197, 231; extrasystemic war and, 1, 5, 10, 24, 34; France and, 168; Iraq and, 3, 33, 95–98, 100–113 passim, 199, 257n, 277n; military strategy of, 244, 291n; the Philippines and, 196, 197; Somalia and, 33, 199, 240; Spanish-Cuban War and, 224; Vietnam War and, 9, 243n, 284n

United States Department of Defense, 258n, 266n

Victory, 3, 43, 46, 49–51, 53, 55, 60–61, 115, 186, 244nn21–22, 245n38, 249n25, 255n43, 255n48, 256n2, 256n5, 256n11, 256n17, 257n22, 268n8, 268n16, 268nn21–22, 268n27; of Britain, 87, 92, 116, 129; of China, 79, 84; definition of, 7–8, 14, 96, 170, 176–77, 184; examples of, 201–15, 220–26, 230–33, 237–38; of France, 66–67, 70, 72, 76, 78; by insurgents, 9, 10, 12–13, 16–17, 20, 22, 29, 34, 47, 62, 109, 113, 187; of MCP, 79, 82; of PAIGC, 132–33, 136–38, 143–44, 146–49; of al-Qaeda, 24, 170; by states, 9, 15; of United States, 10, 97–98, 111, 176–77,

Victory (*continued*)
   182–85; of Vietminh, 23, 150–51, 153–54, 157, 159, 161, 163. *See also* Martel, William

Vietminh, 37, 172, 265n2, 266n18; in Indochina War, 23, 150–68; strategy of, 37, 46. *See also* Giap, Vo Nguyen; Ho Chi Minh; Indochina War; Progressive model

Vietnam War, 9, 265nn5–6, 265n10, 266n29, 266n33, 266n37, 267n64, 267n65

Waltz, Kenneth, 1–2, 25

War termination, 14, 91, 185–86, 245n31

Weapon of the weak, 67. *See also* Guerrilla strategy

Wilson, Isaiah. *See* Lyall, Jason; Mechanization

Zulu, 38, 61; First Zulu War, 36, 193, 206; Second Zulu War, 36, 195, 218; Third Zulu War, 36, 197, 222

ACKNOWLEDGMENTS

I would like to thank a number of individuals and institutions for supporting this book's completion. I am most grateful to Avery Goldstein, at the University of Pennsylvania, who provided me with a great deal of support, guidance, and commitment at various stages of this project. He also did so much to pave the way for me to become a teacher and scholar of political science and international security. Daryl Press asked a series of sharp questions to eliminate weaknesses in my argument and made good suggestions to improve the book's quality. Edward Mansfield helped me sharpen my argument and recommended several ways to make this book an attractive project. Jennifer Amyx read all the chapters, challenged many of my ideas and assumptions, and was always there for me when I needed advice and encouragement.

I received a lot of support from my colleagues at the United States Air War College, Maxwell Air Force Base, where I completed this book. I am grateful to Christopher Hemmer, my department chair, and Jeffrey Record, who took time out of their busy schedules to read several chapters of the manuscript and provided constructive ideas on each. Chris helped not only by reading the chapters with great care for logic, structure, and balance but also by making sure to keep a wonderful workplace in which to complete the project. Jeff's prior works—followed by my numerous conversations with him—helped me tremendously to appreciate the role of external support in asymmetric conflict and understand serious issues associated with the war in Iraq. Dean Mark Conversino, Colonel Daniel Baltrusaitis (retired), Colonel Al Hunt, Colonel Pete McCabe (retired), and Colonel Jeffrey J. Smith all gave me the necessary institutional and collegial support for this project. David Sorenson and Zach Zwald always responded positively to my request to comment on chapters and kept me focused on important issues. Of course, I benefited from many stimulating intellectual exchanges with my colleagues in the Department of International Security Studies.

Outside Maxwell, I learned so much from the seminal work on asymmetric war from T. V. Paul, who kindly read chapters of this book and provided constructive comments. My additional thanks go to Kanji Akagi, Ian Beckett, Con Crane, Peter Feaver, Ryan Grauer, Tim Junio, Stuart Kaufman, Masashi Nishihara, Colonel David O'Meara (retired), Paul Staniland, Dominic Tierney, Major John P. Williams (USMC, retired), and Andrew Yeo, who read early chapters of the book, shared their ideas and experiences with me, or offered support at various stages of research and writing. Richard Betts deserves special mention as someone who so deeply influenced my career on security studies. I also want to thank many of my former students at the Air War College who stimulated our seminars with a number of great contributions and advanced my thoughts about national security affairs the past few years.

My research was supported by several organizations, including the Air War College, the USAF Institute for National Security Studies, the Smith Richardson Foundation, the Matsushita International Foundation, the Christopher H. Browne Center for International Politics, and the Department of Political Science of the University of Pennsylvania. Dr. Nishihara allowed me to write this book as a visiting fellow at the Research Institute for Peace and Security in Tokyo, and I was also fortunate to be able to explore nontraditional security issues as the Straus Research Fellow at the Center for Defense Information in Washington, D.C. The RAND Corporation gave me a great opportunity to interact with specialists in this field as a summer research associate. In addition, I benefited greatly from scholarly comments on this and related projects during my presentations at the New Faces conference held at Duke University, USAF Air Command and Staff College, Texas Tech University, University of the Philippines–Diliman, National Defense Academy of Japan, and Keio University in Tokyo. Parts of this book appeared earlier in *African Security Review*, *Small Wars and Insurgencies*, *Air and Space Power Journal*, and *Comparative Strategy*.

At the University of Pennsylvania Press, William Finan has been a strong supporter of this project as my editor from the beginning. I am deeply indebted to him for his continuous encouragement and interest in this book. I also thank Joseph Dahm, Erica Ginsburg, Caroline Hayes, and Rachel Taube for their assistance in preparing the final manuscript.

Finally but not least, my family in both Japan and the United States has been a tremendous source of support and made a many sacrifices for this project. Among many family members to whom I am grateful, I thank most my wife, Mariko, who understands the value of this book to my career and to our life and who has always been there for me to support it. This book is dedicated to her.

Disclaimer

Views expressed in this book belong solely to the author and do not necessarily represent the official policy of the United States government, Department of Defense, Department of the Air Force, Air University, or Air War College.